John Timbs

Stories of Inventors and Discoverers in Science and the Useful Arts

A Book for Old and Young

John Timbs

Stories of Inventors and Discoverers in Science and the Useful Arts
A Book for Old and Young

ISBN/EAN: 9783744750943

Printed in Europe, USA, Canada, Australia, Japan

Cover: Foto ©berggeist007 / pixelio.de

More available books at **www.hansebooks.com**

MR. BABBAGE'S DIFFERENCE ENGINE.

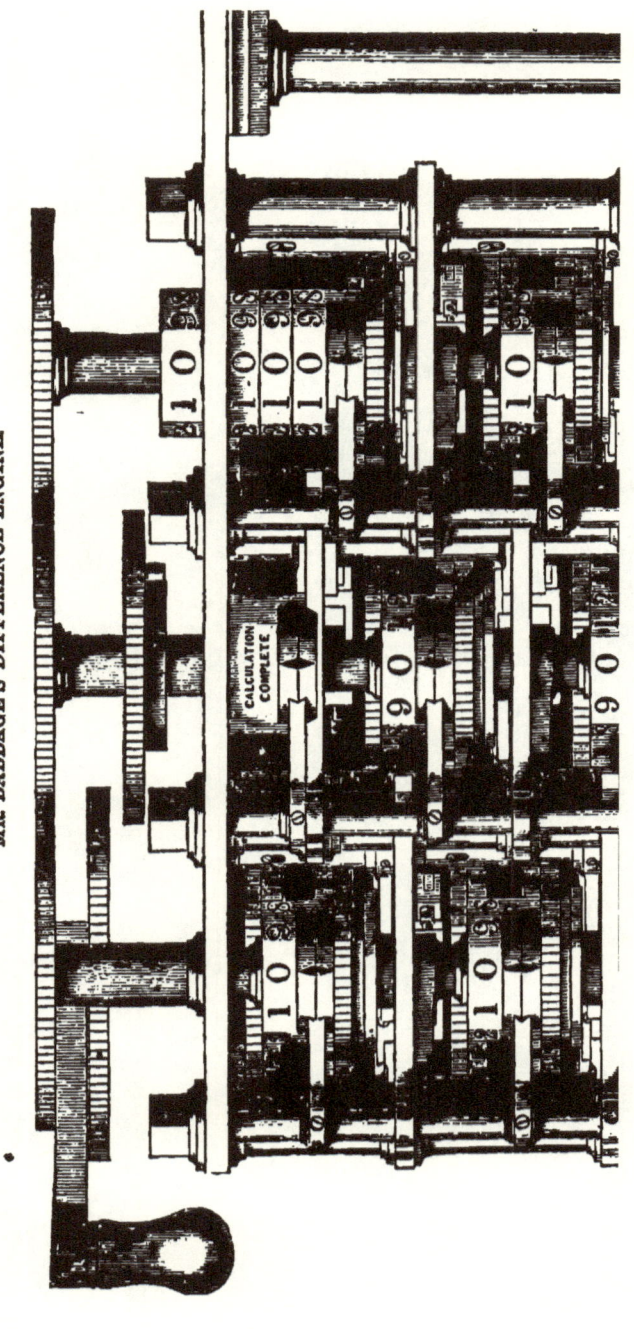

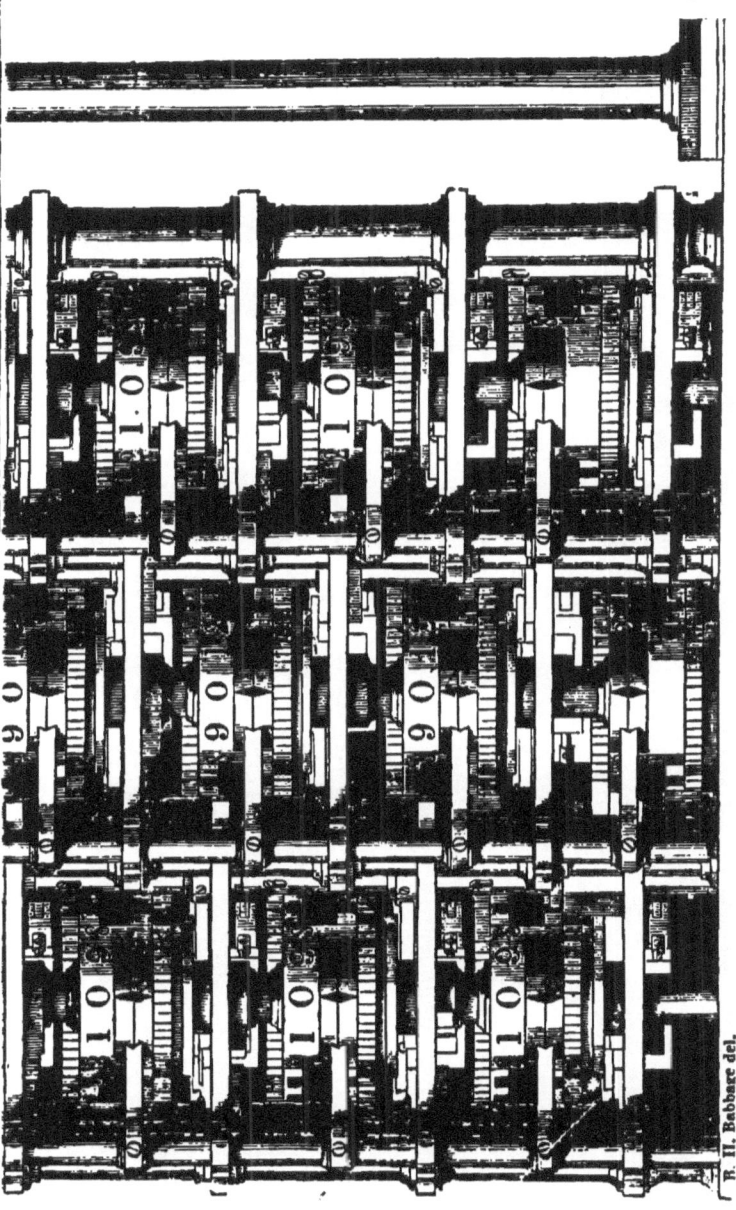

R. H. Babbage del.

Small portion of Mr. BABBAGE's DIFFERENCE ENGINE, No. 1 (CALCULATING MACHINE), the property of Government; in the Museum of King's College, Somerset House.

STORIES OF

INVENTORS AND DISCOVERERS

IN

SCIENCE AND THE USEFUL ARTS.

A BOOK FOR OLD AND YOUNG.

BY JOHN TIMBS, F.S.A.

With Illustrations.

The First Practical Steam-boat.

"Justice exacts that those by whom we are most benefited should be most honored."—Dr. Johnson's *Rambler*.

NEW YORK:

HARPER & BROTHERS, PUBLISHERS,

FRANKLIN SQUARE.

1860.

TO THE READER.

Sir Humphrey Davy, in his last work of charming philosophy, remarks: "The beginning of civilization is the discovery of some useful arts, by which men acquire property, comforts, or luxuries. The necessity or desire of preserving them leads to laws and social institutions. The discovery of peculiar arts gives superiority to particular nations; and the love of power induces them to employ this superiority to subjugate other nations, who learn their arts, and ultimately adopt their manners; so that, in reality, the origin as well as the progress and improvement of civil society is founded in *mechanical and chemical inventions.*"* This remark was made thirty years ago; and the foresight of the author is proved by his words having since become still stronger evidence of his position than at the time they were written. You will not, therefore, be surprised to find the majority of these "Stories of Inventors and Discoverers" selected from the recorded triumphs of Mechanics and Chemistry.

Although the Sixty Narratives which are the staple of the present volume range through ages—from Archimedes to Isambard Kingdom Brunel—they, for the most part, consist of modern instances. The earlier records have, however, proved rich in what may be termed the *Curiosities* of Invention, among which it is not difficult to find many a germ of later success. In many cases, too, the moderns have repaid what they owed to

* *Consolations in Travel; or, the Last Days of a Philosopher.* By Sir Humphrey Davy, Bart.

their predecessors by throwing new light upon some of the boasted wonders of ancient ingenuity; and this mode of illustration has been specially attended to in the present work. In each instance also it has been sought, as far as practicable, to bring the narrative down to the science of our own time.

The *antiquities* of such subjects are curious, and interesting to a large class of readers: as in the cases of Printing and Gunpowder; the Art of Navigating the Air and Living under Water; the marvels of Automata; and a host of "Secret Inventions" besides those of John Napier.

Occasionally it has been but justice to set in their proper light the merits of old workers—as in "The True History of Friar Bacon," who was a reformer of science centuries before his more illustrious namesake, Francis Lord Bacon. In the "Story of Paracelsus," too, a proper estimate is attempted of his discoveries, which have been, in some instances, obscured by his quackery.

To the next group of Inventors—of the times of the Civil War and the Restoration—a sort of romantic interest attaches; whether in the philosophical pursuits of Prince Rupert beside his forge in the keep of Windsor Castle, or in importing "Rupert's Drops;" in the recreations of Sir Samuel Morland, "Master of Mechanics" to Charles II., or in the *Century of Inventions* by the Marquis of Worcester, who by this rational means beguiled the captivity in the Tower of London to which his loyalty had consigned him. His "Water-commanding Engine" is believed to have been one of the results of that period.

In "the separate, simultaneous, and yet mutually dependent progress of industry" in the latter half of last century, several instances have been gathered, at the head of which is that of "Watt, who, poor in worldly wealth, but possessed of mental riches vouchsafed to few, was then wishing to realize an idea destined to effect more surprising results in the history of Britain than the wars, alliances, and legislation of centuries."* Then, what a series of sufferings and conflicts with jealousy and ignorance can be traced in the progress of the Cotton Manufacture, consummated by Watt's great invention!

To a somewhat earlier period belong the perils of John Lombe in his furtive journey to Piedmont, to bring over Silk-throwing machinery; and the story of Lee's invention of the Stocking-frame, traceable to the tenderest feeling of man—his sympathy for "the sole part of all his joys."

In another group of narratives we see how brilliant was the success of Davy's Safety-Lamp, and how miserable the fate of poor Carcel; and how hard was the battle which the projectors of Gas-lighting had to fight with Parliament-men and men of science ere the new light broke forth upon the world.

Next we have the Era of Engineering, in which our country was improved by Canals, Light-houses, and Harbors, Bridges, Breakwaters, and Docks — by Brindley, Smeaton, Telford, and Rennie, whose fortunes, as here narrated, are so many cheering lessons to striving genius.

The Steam-boat yields a long and interesting chapter —from the records of nearly four centuries since to the fate of Symington, whose invention led to the earlier accomplishment of Steam Navigation in another country.

The Railway proved, however, a more secure success through the genius of George Stephenson, "once a locomotive stoker in the north of England, and afterward one of the most distinguished engineers of modern times,"

* James Sime, M.A.

succeeded by his not less distinguished son, Robert Stephenson, whose genius matured the system which his father had originated. To this group also belong the Brunels, father and son, the latter famed for his Railway Works and Iron Ship-building.

The arch-chemic art of Photography, aided by the science of the Stereoscope, forms the next chapter; and the work concludes with an account of the Electric Telegraph, its anticipation and consummation, which is crowded with incident.

Throughout the following pages acknowledgment is made of the respective authorities for the facts and statements in the several narratives, the choice of which has been dictated by impartiality and anxiety to be just.

In tracing the fortunes of Inventors and Discoverers, it is painful to note how many have become "Martyrs of Science;" a phrase sometimes misapplied, and which, there is reason to hope, will at no very distant time be inapplicable. A brighter era is at hand. "Thirty years ago there was not a single literary or scientific man who enjoyed a pension from the crown, or (with one exception) was distinguished by any mark of honor from the sovereign. This is happily no longer the case; for since 1830 there have been conferred for intellectual services thirty titles of honor, and we now find on the Civil List the names of nearly fifty distinguished persons. These liberal reforms naturally led to others; institutions as well as individuals now share in the generosity of the state:"* and that scientific men may long continue to receive such honors from a country which so largely owes its pre-eminence to the applied sciences, is the fervent hope of   THE AUTHOR.

* Address of Sir David Brewster, Principal of Edinburgh University, 1859.

CONTENTS.

LIST OF ENGRAVINGS.

MR. BABBAGE'S DIFFERENCE ENGINE.

The Frontispiece represents the face of that small portion of Mr. Babbage's Difference Engine which is now standing in the Museum of King's College.

In correction of the closing sentence of the last paragraph in page 207, it should be stated that the portion of the engine in King's College is in order, and is capable of calculating to five figures, and two orders of differences, at the rate of 12 or 14 arguments and corresponding tabular numbers per minute; and neither the number of orders of differences, nor the number of digits, would make any difference in its rate of work.

Without numerous carefully lettered and figured mechanical drawings, it would be impossible properly to describe the elaborate mechanism of this engine; it has, indeed, been found impossible for one competent mechanic, who has fully mastered every portion, to explain the machine itself to another equally competent mechanic without the devotion of considerable time.

There is a very commonly entertained, and certainly a very natural notion that Mr. Babbage's "Analytical Engine" (see page 207) is an improvement (we were going to say a mere improvement) on his "Difference Engine."

This is altogether a mistake, there being scarcely less connection between a clock and a steam-engine: the two entirely different engines of Mr. Babbage merely follow one another in order of time, though, of course, the mechanical experience he acquired during the progress of the one must have been of the greatest assistance while contriving the separate portions of the other.

STORIES

OF

INVENTORS AND DISCOVERERS.

THE INVENTIONS OF ARCHIMEDES.

IT is scarcely possible to view the vast steam-ships of our day without reflecting that to a great master of Mechanics, upward of 2000 years since, we in part owe the invention of the machine by which these mighty vessels are propelled upon the wide world of waters. This power is an application of "the Screw of Archimedes," the most celebrated of the Greek geometricians. He was born in Sicily, in the Corinthian colony of Syracuse, in the year 287 B.C., and, when a very young man, was fortunate enough to enjoy the patronage of his relative, Hiero, the reigning Prince of Syracuse.

The ancients attribute to Archimedes more than forty mechanical inventions, among which are the endless screw; the combination of pulleys; an hydraulic organ, according to Tertullian; a machine called the *helix*, or screw, for launching ships; and a machine called *loculus*, which appears to have consisted of forty pieces, by the putting together of which various objects could be framed, and which was used by boys as a sort of artificial memory.

Archimedes is said to have obtained the friendship and confidence of Hiero by the following incident. The king had delivered a certain weight of gold to a workman to be made into a crown. When the crown was made and sent to the king, a suspicion arose in the royal mind that the gold had been adulterated by the alloy of a baser metal, and he applied to Archimedes for his assistance in detecting the imposture: the difficulty was

to measure the bulk of the crown without melting it into
a regular figure; for silver being, weight for weight, of
greater bulk than gold, any alloy of the former in place
of an equal weight of the latter would necessarily increase the bulk of the crown; and at that time there was
no known means of testing the purity of metal. Archimedes, after many unsuccessful attempts, was about to
abandon the object altogether, when the following circumstance suggested to his discerning and prepared mind
a train of thought which led to the solution of the difficulty. Stepping into his bath one day, as was his custom, his mind doubtless fixed on the object of his research, he chanced to observe that, the bath being full,
a quantity of water of the same bulk as his body must
flow over before he could immerse himself. He probably perceived that any other body of the same bulk
would have raised the water equally; but that another
body of the same weight, but less bulky, would not have
produced so great an effect. In the words of Vitruvius,
"as soon as he had hit upon this method of detection, he
did not wait a moment, but jumped joyfully out of the
bath, and running forthwith toward his own house, called out with a loud voice that he had found what he
sought. For as he ran, he called out, in Greek, Eureka!
Eureka! 'I have found it out! I have found it out!'"
When his emotion had sobered down, he proceeded to
investigate the subject calmly. He procured two masses of metal, each of equal weight with the crown—one
of gold, and the other of silver; and having filled a vessel very accurately with water, he plunged into it the
silver, and marked the exact quantity of water that overflowed. He then treated the gold in the same manner,
and observed that a less quantity of water overflowed
than before. He next plunged the crown into the same
vessel full of water, and observed that it displaced more
of the fluid than the gold had done, and less than the
silver; by which he inferred that the crown was neither
pure gold nor pure silver, but a mixture of both. Hiero
was so gratified with this result as to declare that from
that moment he could never refuse to believe any thing
Archimedes told him.*

* Galileo, while studying the hydrostatical treatise of Archimedes,

Traveling into Egypt, and observing the necessity of raising the water of the Nile to points which the river did not reach, as well as the difficulty of clearing the land from the periodical overflowings of the Nile, Archimedes invented for this purpose the Screw which bears his name. It was likewise used as a pump to clear water from the holds of vessels; and the name of Archimedes was held in great veneration by seamen on this account. The screw may be briefly described as a long spiral with its lower extremity immersed in the water, which, rising along the channels by the revolution of the machine on its axis, is discharged at the upper extremity. When applied to the propulsion of steam vessels, the screw is horizontal; and, being put in motion by a steam-engine, drives the water backward, when its reaction, or return, propels the vessel.

The mechanical ingenuity of Archimedes was next displayed in the various machines which he constructed for the defense of Syracuse during a three years' siege by the Romans. Among these inventions were catapults for throwing arrows, and balistæ for throwing masses of stone; and iron hands or hooks attached to chains, thrown to catch the prows of the enemy's vessels, and then overturn them. He is likewise stated to have set their vessels on fire by burning-glasses: this, however, rests upon modern authority, and Archimedes is rather believed to have set the ships on fire by machines for throwing lighted materials.

After the storming of Syracuse, Archimedes was killed by a Roman soldier, who did not know who he was. The soldier inquired; but the philosopher, being intent upon a problem, begged that his diagram might not be disturbed; upon which the soldier put him to death. At his own request, expressed during his life, a sphere in-

wrote his Essay on the Hydrostatic Balance, in which he describes the construction of the instrument, and the method by which Archimedes detected the fraud committed by the jeweler in the composition of Hiero's crown. This work gained for its author the esteem of Guido Ubaldi, who had distinguished himself by his mechanical and mathematical acquirements, and who engaged his young friend to investigate the subject of the centre of gravity in solid bodies. The treatise on this subject, which Galileo presented to his patron, proved the source of his future success in life.

scribed in a cylinder was sculptured on his tomb, in memory of his discovery that the solid contents of a sphere is exactly two thirds of that of the circumscribing cylinder; and by this means the memorial was afterward' identified. One hundred and fifty years after the death of Archimedes, when Cicero was residing in Sicily, he paid homage to his forgotten tomb. "During my quæstorship," says this illustrious Roman, "I diligently sought to discover the sepulchre of Archimedes, which the Syracusans had totally neglected, and suffered to be grown over with thorns and briers. Recollecting some verses, said to be inscribed on the tomb, which mentioned that on the top was placed a sphere with a cylinder, I looked round me upon every object at the Agragentine Gate, the common receptacle of the dead. At last I observed a little column which just rose above the thorns, upon which was placed the figure of a sphere and cylinder. This, said I to the Syracusan nobles who were with me, this must, I think, be what I am seeking. Several persons were immediately employed to clear away the weeds, and lay open the spot. As soon as a passage was opened, we drew near, and found on the opposite base the inscription, with nearly half the latter part of the verses worn away. Thus would this most famous, and formerly most learned city of Greece have remained a stranger to the tomb of one of its most ingenious citizens, had it not been discovered by a man of Arpinum."

To Archimedes is attributed the apophthegm, "Give me a lever long enough, and a prop strong enough, and with my own weight I will move the world." This arose from his knowledge of the possible effects of machinery; but, however it might astonish a Greek of his day, it would now be admitted to be as theoretically possible as it is practically impossible. Archimedes would have required to move with the velocity of a cannon ball for millions of ages to alter the position of the earth by the smallest part of an inch. In mathematical truth, however, the feat is performed by every man who leaps from the ground; for he kicks the world away when he rises, and attracts it again when he falls back.*

---

* Ozanam has taken the trouble to calculate the time which would be required to move the earth one inch; he makes it 3,653,745,176,803 centuries.

Under the superintendence of Archimedes was also built the renowned Galley for Hiero. It was constructed to half its height by 300 master workmen and their servants in six months. Hiero then directed that the vessel should be perfected afloat; but how to get the vast pile into the water the builders knew not, till Archimedes invented his engine called the Helix, by which, with the assistance of very few hands, he drew the ship into the sea, where it was completed in six months. The ship consumed wood enough to build sixty large galleys; it had twenty tiers of bars, and three decks; the middle deck had on each side fifteen dining apartments, besides other chambers, luxuriously furnished, and floors paved with mosaics of the story of the *Iliad*. On the upper deck were gardens, with arbors of ivy and vines; and here was a temple of Venus, paved with agates, and roofed with Cyprus wood: it was richly adorned with pictures and statues, and furnished with couches and drinking vessels. Adjoining was an apartment of box-wood, with a clock in the ceiling, in imitation of the great dial of Syracuse; and here was a huge bath set with gems called Tauromenites. There were also, on each side of this deck, cabins for the marine soldiers, and twenty stables for horses; in the forecastle was a fresh-water cistern, which held 253 hogsheads; and near it was a large tank of sea-water, in which fish were kept. From the ship's sides projected ovens, kitchens, mills, and other offices, built upon beams, each supported by a carved image nine feet high. Around the deck were eight wooden towers, from each of which was raised a breastwork full of loop-holes, whence an enemy might be annoyed with stones; each tower being guarded by four armed soldiers and two archers. On this upper deck was also placed the machine invented by Archimedes to fling stones of 300 pounds weight, and darts eighteen feet long, to the distance of 120 paces; while each of the three masts had two engines for throwing stones. The ship was furnished with four anchors of wood and eight of iron; and "the Water-Screw" of Archimedes, already mentioned, was used instead of a pump for the vast ship, "by the help of which one man might easily and speedily drain out the water, though it were very deep." The whole ship's company consisted of an immense multitude, there being in the forecastle alone 600 seamen. There were placed on board her 60,000 bushels of corn, 10,000 barrels of salt fish, and 20,000 barrels of flesh, besides the provisions for her company. She was first called the Syracuse, but afterward the Alexandria. The builder was Archias, the Corinthian shipwright. The vessel appears to have been armed for war, and sumptuously fitted for a pleasure yacht, yet was ultimately used to carry corn. The timber for the mainmast, after being in vain sought for in Italy, was brought from England. The dimensions are not recorded, but they must have exceeded those of any ship of the present day: indeed, Hiero, finding that none of the surrounding harbors sufficed to receive his vast ship, loaded it with corn, and presented the vessel, with its cargo, to Ptolemy, King of Egypt; and on arriving at Alexandria it was hauled ashore, and nothing more is recorded respecting it. A most elaborate description of this vast ship has been preserved to us by Athenæus, and translated into English by Burchett, in his *Naval Transactions.*

Archimedes has been styled the Homer of Geometry; yet it must not be concealed that he fell into the prevailing error of the ancient philosophers, that geometry was degraded by being employed to produce any thing useful. "It was with difficulty," says Lord Macaulay, "that he was induced to stoop from speculation to practice. He was half ashamed of those inventions which were the wonder of hostile nations, and always spoke of them slightingly, as mere amusements, as trifles in which a mathematician might be suffered to relax his mind after intense application to the higher parts of his science."

THE MAGNET AND THE MARINER'S COMPASS.

THE vast service of which Magnetism is to man may be said to have commenced by supplying him with that invaluable instrument, *the Mariner's Compass*. Mr. Hallam characterizes it as "a property of a natural substance, which, long overlooked, even though it attracted observation by a different peculiarity, has influenced by its accidental discovery the fortunes of mankind more than all the deductions of philosophy."

Before we describe the discovery of the Compass, we shall briefly explain the source from which its power and usefulness are derived. The Magnet is a metallic body, possessing the remarkable property of attracting iron and some other metals. It is said to have been found abundantly at Magnesia, in Lydia, from which circumstance its name may have been derived. The term *native magnet* is applied to the *loadstone*, which appears to be derived from an Icelandic term, *leider-stein*, signifying the *leading-stone*, so designated from the stony particles found connected with it. India and Ethiopia formerly furnished great quantities of this native magnet. Tiger Island, at the mouth of the Canton River, in China, is in great measure made up of this ore, as mariners infer from the circumstance of the needles of their compasses being much affected by their proximity to the island. In the earliest times there were reputed to be five distinct kinds of loadstone—the Ethiopian, the Magnesian, the Bœotic, the Alexandrian, and the Natolian. It is also found abundantly in the iron mines of Sweden, in America, and sometimes, though rarely, among the iron ores of England. The ancients also believed the loadstone to be of two species, male and female. "We read," says Tomlinson, "of its being used in the Middle Ages medicinally —to cure sore eyes and to procure purgation. Even in modern times plasters have been made from this ore, and much other quackery has been perpetrated by its means."*

The attracting power of the Magnet was known at a very early period, as references are made to it by Aristotle, and more particularly by Pliny, who states that ignorant persons call it *ferrum vivum*, or quick iron, a name somewhat analogous to our loadstone. The same author appears to have been acquainted with the power of the magnet to communicate properties similar to its own to other bodies. The polarity of the Magnetic Needle, that is, the power of taking a particular direction when freely suspended, escaped the notice of the Greeks and Romans of antiquity, but the Chinese appear to have been acquainted with it from a very early date.

We are not surprised to find so mysterious an agency as the Magnet exercises to have been referred to accidental origin. The ancient Greeks represent one Magnes, a shepherd, leading his flocks to Mount Ida: he stretched himself upon the green-sward to take repose, and left his crook, the upper part of which was made of iron, leaning against a large stone. When he awoke and arose to depart, he found, on attempting to take up his crook, that the iron adhered to the stone. He communicated this fact to some philosophers of the time, and they called the stone, after the name of the shepherd, Magnes, *the Magnet*, which it retains to the present day. It is, however, denominated among many nations the *love-stone*, from its apparent affection for iron.

A tradition of very ancient date still exists among the Chinese respecting a mountain of magnetic ore† rising in the midst of the sea, whose intensity of attraction is so great as to draw the nails and iron bands, with which the planks of the ship are fastened together, from their places with great force, and cause the ship to fall to pieces.

---

* It has been observed that the smallest natural magnets generally possess the greatest proportion of attractive power. The magnet worn by Sir Isaac Newton, in his ring, weighed only three grains, yet it was able to take up 746 grains, or nearly 250 times its own weight; whereas magnets weighing above two pounds seldom lift more than five or six times their own weight.

† European writers in general attribute the history of magnetic mountains to the Moors; and reference to the supposition may be found even in writers of the seventeenth century.

This tradition is very general throughout Asia; and the
Chinese historians place the mountain in Tchang-hai, the
southern sea, between Tunkin and Cochin China. Ptolemy also, in a remarkable passage in his Geography,
places this mountain in the Chinese seas. In a work attributed to St. Ambrose, there is an account of one of the
islands of the Persian Gulf, called Mammoles, in which
the magnet is found; and the precaution necessary to be
taken (of building ships without iron) to navigate in that
vicinity is distinctly specified. It should also be added
that the Chinese writers place this magnetic mountain
in precisely the same geographical region that the author of the voyages of Sinbad the Sailor does, which is
to be regarded as a confirmation of the Oriental origin
of a great number of tales, half fiction, half fact, which
are so universally diffused among the legendary literature
of every language as to seem indigenous in each of them.

It is extremely probable (says Humboldt) that Europe
owes the knowledge of the northern and southern directing powers of the Magnetic Needle—the use of the
Mariner's Compass—to the Arabs, and that these people
were, in turn, indebted for it to the Chinese. In the
Chinese historical Szuki of Szumathsian, who lived in the
earlier half of the second century before our era, we meet
with an allusion to the "magnetic cars," which the emperor had given more than 900 years earlier to the em-
bassadors from Tunkin and Cochin China, that they might
not miss their way on their return home.* In the fourth
century of our era, Chinese ships employed the magnet
to guide their course safely across the open sea; and it
was by means of these vessels that the knowledge of the
compass was carried to India, and from thence to the
eastern coasts of Africa. The Arabic designations *Zoron*
and *Aphron* (south and north), which Vincenzius of Beau-
vais gives, in his *Mirror of Nature*, to the two ends of
the Magnetic Needle, indicate, like many Arabic names
of stars which we still employ, the channel and the people from whom Western countries received the elements
of their knowledge. In Christian Europe, the first men-

* Maurice, in his *Indian Antiquities*, describes this instrument as a
sort of magnetic index, which the Chinese called *Chimans;* a name by
which they at this day denominate the Mariner's Compass.

tion of the use of the Magnetic Needle occurs in the politico-satirical poem called *La Bible*, by Guyot of Provence, in 1190; and in the description of Palestine, by Jacobus of Vitry, Bishop of Ptolemais, between 1204 and 1215. Dante (in his *Par.*, xii., 29) refers, in a simile, to the needle (*ago*) "which points to the star." Navarrete quotes a remarkable passage in the Spanish *Leyes de las Partidas* of the middle of the thirteenth century: "The needle which guides the seaman in the dark night, and shows him, both in good and in bad weather, how to direct his course, is the intermediate agent (*medianera*) between the loadstone (*la piedra*) and the north star."

Humboldt considers it striking that the use of the south direction of the Needle should have been first applied in eastern Asia, not to navigation, but to land traveling. In the anterior part of the Magnetic Wagon, a freely-floating needle moved the arm and hand of a small figure, which pointed toward the south. Klaproth, whose researches upon this curious subject have been confirmed by Biot and Stanislas Julien, adduces an old tradition, according to which the Magnetic Wagon was already in use in the reign of the Emperor Honngti, presumed to have lived 2600 years before our era; but no allusion to this tradition can be found in any writers prior to the early Christian ages.

The Magnetic Wagon was used as late as the fifteenth century. Several of these carriages were carefully preserved in the Chinese imperial palace, and were employed in the building of Buddhist monasteries in fixing the points toward which the main sides of the edifice should be directed.

As the excessive mobility of the Chinese Needles floating upon water rendered it difficult to note down the indications which they afforded, another arrangement was adopted in their place as early as the twelfth century of our era, in which the Needle, which was freely suspended in the air, was attached to a fine cotton or silken thread, and by means of this more perfect apparatus, the Chinese, as early as the beginning of the twelfth century, determined the amount of the western variation of the needle. From its use on land, the Compass was finally adapted to maritime purposes. When it had become general throughout the Indian Ocean, along the shores of Persia and Arabia, it was introduced into the West, in the twelfth century, either directly through the influence of the Arabs, or through the agency of the Crusaders, who, since 1096, had been brought into contact with Egypt and the true Oriental regions. The most essential share in its use seems to have belonged to the Moorish pilots, the Genoese, Venetians, Majorcans, and Catalans. The old story, that Marco Polo first brought the Compass into Europe, has long been disproved: as he traveled from 1271 to 1295, it is evident, from the testimony we have quoted, that the Compass was, at all events, used in European seas from sixty to seventy years before Marco Polo set forth on his journeyings.

Dr. Gilbert, who was physician in ordinary to Queen Elizabeth, states that P. Venutus brought a Compass from China in 1260. Gilbert bestowed much attention upon magnetism, and to some extent inculcated the doctrine of gravitation, by comparing the earth to a great magnet. The term "poles of a magnet" arose from his theory, which is remarkably consonant with the notions of the present day.

The discovery of the Compass was long ascribed to Flavio Gioja, of Positano, in 1302, not far from the lovely town of Amalfi, on the coast of Calabria, and which town was rendered so celebrated by its widely-extended maritime laws. The Compass was then a rude and simple instrument, being only an iron needle magnetized, and stuck in a bit of wood, floating in a vessel of water; in which artificial and inconvenient form it seems to have remained till about the beginning of the fourteenth century, when Flavio Gioja made the great improvement of suspending the needle on a centre, and inclosing it in a box. The advantages of this were so great that it was universally adopted, and the instrument in its old and simple form laid aside and forgotten; hence Gioja in after times came to be considered as the inventor of the Mariner's Compass, of which he was only the improver. He lived in the reign of Charles of Anjou, who died King of Naples in 1509. It was in compliment to this sovereign (for Amalfi is in the dominion of Naples) that Gioja distinguished the north point by a fleur-de-lis; and this was one of the circumstances by which, in France, in later days, it was endeavored to prove that the Mariner's Compass was a French discovery.

Guyot of Provence, the French poet, who lived a century earlier than Flavio Gioja, or, at the latest, under St. Louis, describes the polarity of the Magnet in the most unequivocal language. Evidence of the earlier use of the Compass in European seas than at the beginning of the fourteenth century is also furnished by a nautical treatise of Raymond Lully, of Majorca, who was at once a philosophical systematizer and an analytic chemist, a skillful mariner and a successful propagator of Christianity; in 1286 he remarked that the seamen of his time employed "instruments of measurement, sea-charts, and the magnetic needle."

The application of the Compass to the purposes of navigation, doubtless, speedily led to the discovery of the Variation of the Needle. It must have been known to the Chinese as far back as the beginning of the twelfth century, as it is mentioned in a work published by a Chinese philosopher, named Keon-tsoung-chy, who wrote about the year 1111 (Sir Snow Harris's *Rudimentary Magnetism*). In the *Life of Columbus*, written by his son, it is distinctly assigned to that celebrated man; and though its amount at this period must have been small in France, Spain, etc., yet it was doubtless a very observable quantity in many of the regions visited by Columbus.

It is remarkable that Columbus noticed the Variation of the Needle for the first time when sailing across the Atlantic Ocean in his attempt to find a new world. It was on the 14th of September, 1492; he was perhaps 200 leagues from land, and the variation was a little to the west at London. It appears that Columbus perceived, about nightfall, that the needle, instead of pointing to the north star, varied about half a point, or between five and six degrees, to the northwest, and still more on the following morning. Struck with this circumstance he observed it attentively for three days, and found that the variation increased as he advanced. He at first made no mention of this phenomenon, knowing how ready his people were to take alarm; but it soon filled with consternation his pilots and mariners, who had leisure on the wide ocean for anxiety and curious wonder. It seemed as if the very laws of nature were changing as they advanced, and that they were entering another world, subject to unknown influences. They apprehended that the compass was about to lose its mysterious virtues: and without this guide, what was to become of them in a vast and trackless ocean? But Columbus was prepared with a theory to account for this deviation of the laws of nature, as the terrified sailors deemed it to be. The needle was not at fault, he said; for it did not tend to the polar star, but to some fixed and unseen point. The Variation, therefore, was not caused by any fallacy in the Compass, but by the movement of the polar star itself, which, like the other heavenly bodies, had its changes and revolutions, and every day described a circle round the pole. The high opinion that the pilots entertained of Columbus as a profound astronomer gave weight to his theory, and their alarm subsided. As yet the solar system of Copernicus was unknown; the explanation of Columbus was therefore highly plausible and ingenious, and it shows, and we admire, the perspicacity of the man who, with so little means, could trace up so fearful an effect to a cause founded partly in truth, and thus meet the emergency of the moment. The theory may at first have been advanced merely to satisfy the minds of others, but Columbus appears subsequently to have been satisfied with it himself.

The discovery of a magnetic line without variation is due to Columbus. In a letter written in 1498, he says, "Each time that I sail from Spain to the Indies I find, as soon as I arrive a hundred miles to the west of the Azores, an extraordinary alteration in the movements of the heavenly bodies, in the temperature of the air, and in the character of the ocean; I have observed these alterations with particular care, and have recognized that the needle of the Mariner's Compass, the deviation of which had been *northeast, now turned to the northwest.*"

An eloquent writer thus picturesquely illustrates the benefits of this great discovery: "In the development of the commercial spirit of the Crusades, Providence is seen in its most manifest footsteps. Sitting upon the

floods, it opens to new enterprises.   The Compass
twinkling on its card was a beam from heaven; that
tiny magnet was given as a seniory of earth and sky.
Like a new revelation, the mysteries of an unknown
world were unveiled; like a new illapse, the bold and
noble were inspired to lead the way.   Dias doubles the
Cape of Storms; De'Gama finds his course to the East
Indies; Columbus treads the Bahamas; and twelve
years do not separate these discoveries."

Franklin at his Case.

WHO INVENTED PRINTING, AND WHERE?

THE inquirers into the origin and history of this almost ubiquitous "noble craft and mystery" would seem to have arrived at this conclusion—that it is difficult to say at what period of time the art of Printing did not exist. The simplest and most natural mode of conveying an idea is by the reproduction of similar appearances from an impression of the same surface; and whether this be by a hand or foot upon snow, or by the pressure of wood or metal upon paper or vellum, it is alike *printing*. Accordingly, we find evidence that nearly four thousand years since a rude and imperfect method of printing was certainly practiced. First, seals were impressed upon a plastic material; next, symbols or characters were stamped upon clay in forming bricks (as practiced in Babylon), cylinders, and the walls of edifices. Of this art, Wilkinson and others have brought examples from Egypt; and Rawlinson and Layard from the ruins of the buried cities of Asia. Not only have the inscribed bricks been found, but the wooden stamps with which they were impressed; of these numerous specimens are in the British Museum. Here also may be seen several instruments presenting a singular instance how very nearly we may approach to an important dis-

covery, and yet miss it.   These are brass or bronze
stamps, having on their faces inscriptions in raised characters reversed.   To the back has been fastened a handle,
a loop, a boss, or a ring.   One use of these stamps has
evidently been to *print* the inscription on surfaces, by
aid of color, upon papyrus, linen, or parchment; and, as
the inscriptions show these stamps to have been of the
period when literature had become one of the pursuits
of the great, and the copying of books was a slow and
expensive process, it is strange that the Romans, by
whom these signets were used, should not have improved
upon them by engraving whole sentences and compositions upon blocks, and thence transferring them to paper.
The Chinese printing from blocks at this day closely resembles the old Roman ; and they assert that it was used
by them several centuries before it was known in Europe
—in fact, fifty years before the Christian era.

A vast interval elapses between the above attempts
and the next advance—engraving pictures upon wooden
blocks, invented toward the end of the thirteenth century
by a twin brother and sister of the illustrious family of
Cunio, lords of Italy: these consisted of nine engravings
of the " Heroic Actions" of Alexander the Great, and,
as stated in the title-page, "first reduced, imagined, and
attempted to be executed in relief, with a small knife, on
blocks of wood;" "all this was done and finished by us
when only sixteen years of age."   This title, if genuine,
presents us at once with the origin, execution, and design of the first attempts at block-printing.   The next
earliest evidence is a decree found among the archives
of the Company of Printers at Venice, dated 1441, relating to playing-cards, printed from wood blocks, the impressions being taken by means of a burnisher.   Then,
instead of a single block, a series of blocks was employed, in engravings of the *Biblia Pauperum*, the text being
printed from movable types.

We have now reached the practice of *printing* in the
present sense of the term.   The invention of the *movable
types* is disputed by many cities, but only three have the
slightest claim—Harlem, Strasburg, and Mentz: Harlem
for Lawrence Koster, who, when " walking in a suburban
grove, began first to fashion beech-bark into letters, which

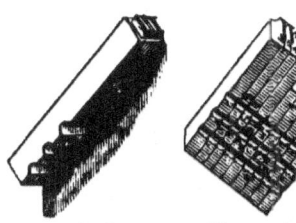

Type of a letter.   Types set up.

being impressed upon paper, reversed in the manner of a seal, produced one verse, then another, as his fancy pleased, to be for copies for the children of his son-in-law." Next, he, with his son-in-law, devised " a more glutinous and tenacious species of writing-ink, which he had commonly used to draw letters; thence he expressed entire figured pictures, with characters added," only on opposite pages, not printed on both sides. Afterward he changed beech-blocks for lead, and then for tin. The tradition adds that an unfaithful servant, having fled with the secret, set up for himself at Strasburg or Mentz; but the whole story, which claims the substitution of movable for fixed letters as early as 1430, can not be traced beyond the middle of the sixteenth century, and is generally discredited as a romantic fiction. Nevertheless, some have believed that a book called *Speculum humanæ Salvationis*, of very rude wooden characters, proceeded from the Harlem press before any other that is generally recognized. Whether movable wooden characters were ever employed in any entire work is very questionable; they appear, however, in the capital letters of some early printed books. "But," says Hallam, "no expedient of this kind could have fulfilled the great purposes of this invention, until it was perfected by founding metal types in a matrix or mould; the essential characteristic of printing, as distinguished from other arts that bear some analogy to it."

The invention is now unhesitatingly ascribed to John Gutenberg, a native of Mentz, the evidence of which does not rest upon guesses from dateless wood-cuts, but upon a legal document, dated 1439, by which it is proved that Gutenberg, being engaged " in a wonderful and unknown art," admitted certain persons into partnership, one of whom dying, his brother claimed to be admitted as his successor; and on Gutenberg's refusal, they brought an action against him as principal partner. From the evidence produced on the trial, it was proved that one of

the witnesses had been instructed by Gutenberg to " take
the *stücke* (pages) from the presses, and, by removing
two screws, thoroughly separate them from one another,
so that no man may know what it is." From this curious document (says the latest investigator of the sub-
ject*) may be learned that separate types were used ;
for if they were block, arranged so as to print four pages
(as stated in the evidence), how could they be so pulled
to pieces that no one should know what they were, or
how could the abstraction of two screws cause them to
fall to pieces ? We are here reminded that within comparatively few years screws have been substituted for
quoins, or wedges, in locking up the type in the chases,
or iron frames, which may be a revival of Gutenberg's
screw method of 400 *years since.*

It seems that some sort of presses were now used, and
the transfers no longer taken by a burnisher or roller ;
and, lastly, that the art was still a great secret at the time
when Koster was at the point of death. Hence it is manifest that the ingenuity of Gutenberg had made a vast
advance from the rude methods of the time, and had, in
fact, invented a new and hitherto unknown art.

All this took place at Strasburg, where Gutenberg resided many years ; but it did not lead to any practical result, and the *first book* was printed at Mentz, near which
the inventor was born. Thither Gutenberg returned
about the year 1450, with all his materials. His former
partnership had expired, and at Mentz he associated himself with John Fust, a wealthy goldsmith and citizen,
who, upon agreement of being taught the secrets of the
art, and admitted into the participation of the profits, advanced the necessary funds, 2020 florins. The new partnership then hired a house called Zum Jungen, and took
into their employ Peter Schœffer and others. A lawsuit
arose between the partners in 1455 ; and from a document in existence we learn that, having expended the
whole of his considerable private fortune in his experiments, Gutenberg had mortgaged his printing materials
to Fust, which is proved by the initial letters used by

---

* "Printing," by T. C. Hansard, Esq. (*Encyclopædia Britannica,*
eighth edition, 1859), in which the history and practice of the art are
lucidly traced.

Gutenberg and his partners in printing works between 1450 and 1455, being likewise used by Fust and Schœffer in the Psalter of 1457 and 1459. Gutenberg did not, however, abandon the unprofitable pursuit, but, starting anew at Mentz, carried on the business for ten years; but in 1465, on becoming one of the band of gentlemen pensioners of the Elector Adolphus of Nassau, "he finally abandoned the pursuit of an art, which, though it caused him infinite trouble and vexation, has been more effectual in preserving his name and the memory of his acts than all the warlike deeds and great achievements of his renowned master and all his house" (*Hansard*). Gutenberg died on the 24th day of February, 1468. His printing-office and materials were eventually sold to Nicholas Bechtermunze, of Elfield, whose works are greatly sought after by the curious, as they afford much proof, by collation, of the genuineness of the works attributed to his great predecessor.

Gutenberg appears to have had a troubled life. When young, he became implicated in an insurrection at Mentz, and was compelled to fly to Strasburg; there necessity compelled him to employ himself in mechanical pursuits, when he made his great discovery. On his return to Mentz, when in partnership with Fust, and Schœffer his son-in-law, he experienced the hard fate that all great inventors have to endure from the misconceptions and ingratitude of mankind. The Guild of Writers and the priests persecuted him, and even his partners joined with his enemies against him; and only his last few years were passed in peace. Posterity has endeavored, in some degree, to make amends for the ingratitude of the discoverer's contemporaries. In 1837, a statue of Gutenberg, by Thorwaldsen, was erected at Mentz, and inaugurated with great ceremony; and at high mass, in the fine old Cathedral, was displayed the first Bible printed by Gutenberg. The statue was erected by a general subscription, to which all Europe was invited to contribute. One who witnessed the ceremony writes, with honest indignation, "England literally gave nothing toward the statue of a man who has done as much as any other single cause to make England what she is."* The Guten-

* Charles Knight, in *The Old Printer and Modern Press*, 1854.

berg Society, to which all the writers of the Rhenish provinces belong, hold a yearly meeting also in Mentz, to honor the memory of the first printer, and to celebrate his discovery.

It is hard to apportion the share of honor to which each of the partners—Gutenberg, Fust, and Schœffer— is entitled in advancing their art. Gutenberg would readily suggest a new and expeditious method of manufacturing types; the practical skill of Fust as a worker

Casting the Type.

in metals, and his large pecuniary resources, would readily provide the necessary appliances; and the entire conception and execution of the *casting* of type is given to Schœffer. The only evidence shows that the partners had for some time taken casts of types in moulds of plaster; for the types of Gutenberg's earlier efforts, both at Strasburg and at Mentz, were cut out of single pieces of wood or metal with infinite labor and imperfection. Schœffer has therefore (Mr. Hansard allows) an undoubt-

ed claim to be considered as one of the three inventors
of printing; for he it was who first suggested the cutting of punches, whereby beautiful form could be stamped upon the matrix, and the highest sharpness and finish
given to the face.   Lambinet, who thinks "the essence
of the art of printing is in the engraved punch," naturally gives the chief credit to Schœffer; this is not the
generally-received opinion; but he is entitled to a place
on the right hand of Gutenberg.   It should be noted
that there is no book known which bears the conjoint
names of Gutenberg, Fust, and Schœffer, nor any which
has the imprint of Gutenberg alone; but there are several books which, from internal evidence, are unanimously attributed by the *literati* of all parties and opinions to
Gutenberg's press.

It is curious to observe that War was the means of
quickening the growth and extension of Printing.   In
1462, the storming of Mentz dispersed the workmen,
and gave the secret to the world.   In 1465 it appeared
in Italy;* in 1469, in France; in 1474, Caxton brought
it to England; and in 1477 it was introduced into Spain.

It is generally believed that William Caxton was born
in the Weald of Kent; about 1412, he was put apprentice to a mercer or merchant of London, became a traveling agent or factor in the Low Countries, and there
bought manuscripts and books, with other merchandise.
He there also learned the new art of Printing; and, securing one of Fust and Schœffer's fugitive workmen from
Mentz, he established a printing-office at Cologne, and
there printed the French original and his own translation
of the *Recuyell of the Historyes of Troy.*   He afterward
transferred his materials to England, and brought over
with him Wynkyn de Worde, who probably was the
first superintendent of Caxton's printing establishment.

---

* Near Subiaco, forty-four miles from Rome, on a hill above the
river, may be traced the ruins of Nero's villa.   It was in this villa, as
we are told by Tacitus and Philostratus, that the cup of the tyrant
was struck by lightning while he was in the act of drinking, and the
table overthrown by the shock.   In propinquity, which almost suggests a parallel, is the monastery of Santa Scolastica, the first place in
Italy in which the printing-press was set up by the German printers,
Sweynheim and Panartz: a copy of their edition of Lactantius, their
first production, dated 1465, is still preserved in the monastery.

He set up his first press at Westminster, perhaps in one of the chapels attached to the Abbey, and certainly un-der the protection of the abbot;* and he there produced *the first book printed in England, The Game of Chesse.* completed on the last day of March, 1474. His " capital work" was a *Book of the Noble Historyes of Kyng Arthur* in 1485, the most beautiful production of his press. He died in 1491, being about fourscore years of age: his industry and devotedness is recorded in the fact that he finished his translation of the *Vitæ Patrum*, from French into English, *on the last day of his life.*

Caxton was buried in the old church of St. Margaret, built in the reign of Edward I., and of which few traces remain. The parish books contain an entry of the ex-pense "for iiij torches" and "the belle" at the old print-er's " bureying ;" and the same books record the church-wardens' selling for 6s. 8d. one of the books bequeathed to the church by Caxton ! In the chancel a tablet to his memory was raised in 1820 by the Roxburghe Club.

* But a very curious placard, in Caxton's largest type, and now preserved in the library of Brazen-nose College, Oxford, shows that he printed in the Almonry; for in this placard he invites customers to "come to Westmonester in to the Almonestrye at the Reed Pale," the name by which was known a house in which Caxton is said to have lived. It stood on the north side of the Almonry, with its back against that of a house on the south side of Tothill Street. Bagford describes this house as of brick, with the sign of the King's Head: it is stated to have fallen down in November, 1845, before the removal of the other dwellings in the Almonry, to form a new line (Victoria Street) from Broad Sanctuary to Pimlico. A beam of wood was saved from the materials of the house, and from it have been made a chessboard and two sets of chessmen, as appropriate memorials of Caxton's first labor in England—*The Game and Playe of the Chesse.* According to a view of Caxton's house, engraved by G. Cooke in 1827, it was three-storied, and had a gallery or balcony to the upper floor, with a window in its bold gable.—(*Curiosities of London.*) The site of Caxton's house is now included in the Westminster Hotel Com-pany's premises.

Note.—The presses of Seth Adams & Co., of Boston, are the best now made for book-printing. For this purpose they are in general use throughout the United States, and are found, under proper man-agement, to give clear impressions of the finest wood-cuts. *Harper's Magazine,* the excellence of whose typographical execution is thought remarkable, not only here, but in Europe, is printed from Adams presses, more than forty of which are constantly in operation in the Harper Printing Establishment.

AN ADAMS POWER-PRESS.

This tablet (a chaste work by Westmacott) was originally intended to have been placed in Westminster Abbey; but the fees for its erection were so great, that application was made to the churchwardens of St. Margaret's, who, as a mark of respect to their parishioner's memory, allowed it to be placed in the church without any of the customary fees. It was proposed, several years since, to erect at Westminster a memorial statue of Caxton, but the fund raised for that purpose now enlarges the Printers' Pension Society's sphere of benevolence.

We must say a few words as to the first *Presses*. Gutenberg is thought to have felt the want of a machine of sufficient power to take the impressions of the types or blocks which he employed; nor is it supposed that, with cutting type, forming screws, making and inventing ink, he could have had time to construct a press, even had he possessed the requisite mechanical skill. His junction with Fust and Schœffer is thought to have supplied the defect.

The earliest form of printing-press very closely resembled the common screw-press, as the cheese or napkin press, with some contrivance for running the form of type, when inked, under the pressure (obtained from the screw by means of a lever inserted into the spindle), and back again when the pressure is made. The presses used in the office of Fust and Schœffer are believed to have differed in no essential form from the above, until improved in the details by Blew, a printer of Amsterdam, in 1620. Other improvements were from time to time introduced, but they were all superseded about the commencement of the present century, when the old wooden press gave way to Earl Stanhope's invention of the iron press which bears his name. Its novelty consisted in an improved application of the power to the spindle and screw, whereby it was greatly increased. Lord Stanhope also made some improvements in the process of stereotyping, and in the construction of locks for canals; he invented an ingenious machine for performing arithmetical operations; during great part of his life he studied the action of the electric fluid; and in 1779 he made public his theory of what is called "the returning stroke of lightning." Lord Stanhope bequeathed £500 to the Royal Society, of which he had been a fellow fifty-one years.

The Hand Press.

The principle of the Stanhope press has been followed out by several subsequent inventors, and improvements of mechanical detail introduced, tending to the economy of time and labor, and to precision of workmanship. The printing-press, however, proved inadequate to a rate of production equal to the demand; and as early as 1790, even before the Stanhope press was generally known, Mr. W. Nicholson patented a PRINTING-MACHINE, of which the chief points were the following: "The type, being rubbed or scraped narrower toward the bottom, was to be fixed upon a cylinder, in order, as it were, to radiate from the centre of it. This cylinder, with its type, was to revolve in gear with another cylinder covered with soft leather (the impression cylinder), and the type received its ink from another cylinder, to which the inking apparatus was applied. The paper was impressed by

The Roller.

passing between the type and impression cylinders." (*Hansard.*) Such was the first printing-machine: it was never brought into use, although most of Nicholson's plans were, when modified, adopted by after-constructors.

König, a German, conceived nearly the same idea; and meeting with the encouragement in England which he failed to receive on the Continent, constructed a printing-machine for Mr. Walter; and on the 28th of November, 1814, the readers of the *Times* were informed that they were then, for the first time, reading a newspaper printed by machinery driven by steam-power, and working at the rate of 1100 impressions per hour. In this machine the ordinary type was used, and laid upon a flat surface, the impression being given by the form passing under a cylinder of great size. This machine was, however, very complicated, and was soon superseded by that of Messrs. Applegath and Cowper, the novel features of which were, accuracy in the register (that is, one page falling precisely upon the back of the other), the method of inking the types, and the simplification of very complicated parts; and this machine, with numerous modifications by different makers, is now in general use, so that the foremost improver of the printing-machine is Augustus Applegath. The simplicity of the operation is admirable: the whole machine is put in motion by means of a strap, which passes over a wheel under the frame, and is mostly worked by steam, it requiring only two boys, one to lay on, and the other to take off the sheets.

The next great improvement was the construction of

the vertical machine by Mr. Applegath, in which he abandoned the reciprocating motion (occasioning a great waste of motive power), and instead of placing the type on a plane table, placed it on a cylinder of large dimensions, which revolves on a vertical axis, with a continuous rotatory motion. "No description," says Mr. Hansard, "can give any adequate idea of the scene presented by one of these machines in full work—the maze of wheels and rollers, the intricate lines of swift-moving tapes, the flight of sheets, and the din of machinery. The central drum moves at the rate of six feet per second, or one revolution in three seconds; the impression cylinders make five revolutions in the same time. The layer-on delivers two sheets every five seconds, consequently sixteen sheets are printed in that brief space. The diameter of an eight-feeder, including the galleries for the layers-on, is twenty-five feet. The *Times* employs two of these eight-cylinder machines, each of which averages 12,000 impressions per hour; and one nine-cylinder, which prints 16,000." Messrs. Hoe, of New York, have constructed machines differing from Applegath's Vertical chiefly in the drum and impression cylinders being horizontal: one of these machines has been constructed with ten cylinders for working the *Times* at 20,000 impressions per hour. Another American machine has been constructed to work 22,000 *double impressions* per hour.

"Could Gutenberg, if he were to rise from the dead, imagine that at the present day there would be more than 4000 presses in Europe, each house being designated by its press; and of these, 600 in the city of London alone—and 1000 printing-machines in England, supplying the printing requirements, on such a scale as this, for her populations!"—*Lecture delivered at the Royal Institution, by Mr. Henry Bradbury*, 1858.

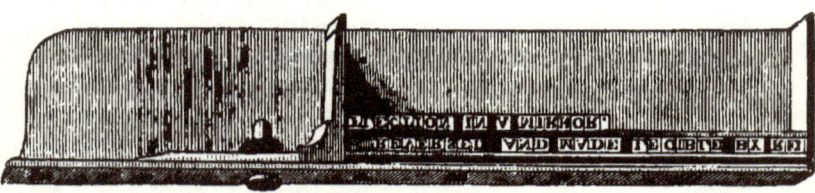

The Composing-stick.

WHO INVENTED GUNPOWDER?

"From the earliest dawnings of policy to this day," says Burke, "the invention of men has been sharpening and improving the mystery of murder, from the first rude essay of clubs and stones to the present perfection of gunnery, cannoneering, bombarding, mining." The imputed universality of the class of invention may account for the difficulty of tracing the special practice of it in the composition of Gunpowder with certainty to any period or nation. The evidence is conflicting, and it ranges from several centuries before the commencement of our era to the claim of the German monk of the fourteenth century, of whom a commemorative statue was erected so lately as the year 1853.

The earliest account extant on the subject of Gunpowder exists in a code of Gentoo laws, where it is mentioned as applied to fire-arms; this document, being of some fifteen centuries before Christ, is thought by many to have been coeval with the time of Moses! The notice occurs in the Sanscrit preface, translated by Halhed, and is as follows: "The magistrate shall not make war with any deceitful machine, nor with poisoned weapons, nor with cannon and guns, nor any kind of fire-arms." Halhed observes: "The reader, no doubt, will wonder to find a prohibition of fire-arms in records of such remote antiquity; and he will probably hence renew the suspicion which has long been deemed absurd, that Alexander the Great did absolutely meet with some weapons of this kind in India, as a passage in Quintus Curtius seems to ascertain. Gunpowder has been known in China as well as in Hindostan far beyond all periods of investigation. The word 'fire-arms' is literally translated by the Sanscrit *agnee-aster* (*agnyastra*), a weapon of fire. In their earliest form they are described to have been a kind of dart tipped with fire, and discharged by some sort of explosive compound from a bamboo. Among several extraordinary properties of this weapon, one was, that, after it had taken its flight, it divided into several separate streams of flame, each of which took effect, and which, when once kindled, could not be extinguished; but this kind of agnee-aster is now lost."

Dutens has selected many passages from Greek and Latin authors favorable to the opinion that Gunpowder was known to the ancients. He mentions the attempt of Salmoneus to imitate thunder, and of the Brahmins to do the same thing; but his most remarkable quotation is from the life of Apollonius of Tyana, written by Philostratus, showing that Alexander was prevented from extending his conquests in India because of the use of Gunpowder by a people called Oxydracæ, who repulsed the enemy "with storms of lightning and thunder-bolts, hurled upon them from above." Philostratus is not remarkable for veracity; but taking into consideration the records of Oriental history, and the fact of pyrotechny having been cultivated from time immemorial in India and China, his assertion does not seem improbable. In India and many other parts of Asia, nitre occurs in great quantity, spread over the surface of the earth. Dr. Scoffern, the experienced writer on this subject, supposes a fire lighted on such a spot: the most careless observer must have noticed the effect of the saltpetre in augmenting the flame; if then, attention having been directed to this phenomenon, charcoal and saltpetre had been mixed together purposely, Gunpowder would have been formed. The third ingredient, sulphur, is not absolutely necessary; indeed, very good Gunpowder, chemically speaking, can be made without it. Sulphur tends to increase the plasticity of the mass, and better enables it to be made into and to retain the form of grains.

It has been said that Gunpowder was used in China as early as the year A.D. 85. Sir George Staunton observes that "the knowledge of Gunpowder in China and India seemed coeval with the most distant historic events. Among the Chinese it has at all times been applied to useful purposes, as blasting rocks, etc., and in the making of fire-works; although it has not been directed through strong metallic tubes, as the Europeans did soon after they had discovered it." In short, there can be no doubt that a sort of Gunpowder was at an early period used in

China, and in other parts of Asia; and Barrow's statement that the Chinese soldiery make their Gunpowder, and every soldier prepares his own, is highly characteristic of the people. Against the claim of the Chinese to the invention, it is urged that the silence of Marco Polo respecting Gunpowder may be considered as at least a negative proof that it was unknown to the Chinese in the time of Kublai Khan.

There is nothing in the history of these people, nor in their " Dictionary of Arts and Sciences," that bears any allusion to their knowledge of cannon before the invasion of Ghengis Khan, when (in the year 1219) mention is made of *ho-pao*, or fire-tubes, the name of cannon, which are said to have killed men, and to set fire to inflammable substances; they are said, too, to have been used by the Tartars, not by the Chinese, and were probably nothing more than the enormous rockets known in India at the time of the Mohammedan invasion (*Quarterly Review*, No. 41).

Numerous documents, however, show that Gunpowder was known in the East at periods of great antiquity, whence it might have been introduced into Europe, either through the medium of the Byzantine Greeks, or by the Saracens into Spain. In a paper read about fifty-five years since before the French Institute, M. Langles maintained that the use of Gunpowder was conveyed to us by the Crusaders, who are stated to have employed it at the siege of Mecca in 690 : he contended that they had derived it from the Indians.

Mr. Hallam considers it nearly certain that Gunpowder was brought by the Saracens into Europe. Its use in engines of war, though they may seem to have been rather like our fire-works than artillery, is mentioned by an Arabic writer in the Escurial collection about the year 1249. The words which are thought to mean gunpowder are translated *pulvis nitratus*. The Moors or Arabs, in Spain, appear to have used gunpowder and cannon as early as 1312 ; and in 1331, when the King of Granada laid siege to Alicant, he battered its walls with iron bullets, discharged by fire from machines; which novel mode of warfare (says the chronicle) inspired great terror. And when Alonzo XI., King of Castile, besieged

Algesiras in 1342-3, the Moorish garrison, in defending the place, employed *truenos* (literally *thunders*), which a passage in the chronicle proves to have been a species of cannon fired with powder. And Petrarch, in a passage written before 1344, and quoted by Muratori, speaks of the art of making Gunpowder as *nuper rara, nunc communis* (recently rare, now common).

Another authority traces Gunpowder to the Arabs, but at an earlier date than hitherto mentioned, and at the same time seeks to identify it with an invention of much earlier antiquity. The celebrated Oriental scholar, M. Reinaud, has discovered an Arabic MS. of the thirteenth century, which proves that compositions identical with Gunpowder in all but the granulations were, and had been for a long time previously, in the possession of the Arabs; and that there is every probability they had obtained them from the Chinese in the ninth century. Many of these were called " Greek fire;" and comparing the account of Joinville, of the wars on the Nile in the time of St. Louis, with the Arabic recipes, there can be little doubt we are now in possession of what was then termed " Greek fire." Mr. Grove, F.R.S., who has investigated the subject experimentally as well as historically, concludes that the main element of Greek fire, as contradistinguished from other inflammable substances, was nitre, or a salt containing much oxygen; that Greek fire and Gunpowder were substantially the same thing; and that the development of the invention had been very slow and gradual, and had taken place long antecedent to the date of Schwartz, the monk of Cologne, A.D. 1320, to whom the invention of Gunpowder is generally attributed; thus adding to the innumerable, if not unexceptionable cases in which discoveries commonly attributed to accident, and to a single mind, are found, upon investigation, to have been progressive, and the result of the continually-improving knowledge of successive generations.

It was long the custom to attribute the invention of Gunpowder to our philosopher, Roger Bacon; but a passage in his *Opus Majus*, written in 1267, proves that instead of claiming the merit of the discovery, he mentions Gunpowder as a substance well known in his time, and even employed by the makers of fire-works; and he minutely describes a common cracker. In his treatise *De Secretis Operibus Artis et Naturæ*, he says, that from "saltpetre *and other* ingredients we are able to make a fire that shall burn at any distance." In another passage he indicates two ingredients, saltpetre and sulphur, and "Lura nope cum ubre," which is a transposition of the words "carbonum pulvere" (charcoal in powder). At the period when Bacon lived, Spain was the favorite seat of literature and art. Bacon is known to have traveled through Spain, and to have been conversant with Arabic, so that he might have seen the manuscript in the Escurial collection, which is at least as probable a supposition as that he saw the treatise of Marcus Græcus. Some fifty years later, 1320, is the date claimed by the Germans for the invention due to their monk, Bartholdus Schwartz, in whose honor a stone statue has been erected in the town of Freiburg, where he was born; and in reply to earlier claims to the invention, it is maintained that to Schwartz is due the merit, because he did not learn the secret from any one else.

Nearly two hundred years before this date, Humboldt states that a species of Gunpowder was used to blast the rock in the Rammelsberg, in the Hartz Mountains.

Authorized statements negative the assertion by Camden, Kennett, and other writers, that no Gunpowder was manufactured in England until the reign of Elizabeth. Its first application to the firing of artillery has been commonly ascribed to the English at the battle of Cressy, in August, 1346; but hitherto the fact has depended almost solely on the evidence of a single Italian writer, and the word "gunners" having been met with in some public accounts of the reign of Edward III. The Rev. Joseph Hunter has, however, from records of the period, shown the names of the persons employed in the manufacture of Gunpowder (out of saltpetre and "quick sulphur," without any mention of charcoal), with the quantities supplied to the king just previously to his expedition to France in June or July, 1346. In the records it is termed *pulvis pro ingeniis*; and they establish that a considerable weight had been supplied to the English army subsequently to its landing at La Hogue, and previously to the battle of Cressy; and that before Edward

III. engaged in the siege of Calais, he issued an order to the proper officers in England, requiring them to purchase as much saltpetre and sulphur as they could procure. Sharon Turner, in his *History of England*, has also shown, from an order of Richard III. in the Harleian MSS., that Gunpowder was made in England in 1483; and Mr. Eccleston (*English Antiquities*) states that the English both made and exported it as early as 1411. Nevertheless, Gunpowder long remained a costly article; and even in the reign of Charles I., on account of its dearness, "the trained bands are much discouraged in their exercising." In 1686, it appears from the *Clarendon Correspondence* that the wholesale price ranged from about £2 10s. to £3 a barrel.

John Evelyn, of Wotton, Surrey, asserts that his ancestors were the first who manufactured Gunpowder in England; but this must be regarded as the reintroduction. His grandfather transferred the patent to Sir John Evelyn's grandfather, of Godstone, in whose family it continued till the Civil Wars. As we stroll along the valley in which lies Wotton Place, we are reminded that upon the rivulet which winds through this peaceful region was once made the "warlike contrivance." Evelyn, in a letter to John Aubrey, dated February 8, 1675, says that on this stream, near his house, formerly stood many powder-mills, erected by his ancestors, who were the very first that brought that invention into England; before which we had all our powder from Flanders. He also describes the blowing-up of one of these mills, when a beam, fifteen inches in diameter, at Wotton Place, was broken; and on the blowing-up of another mill lower down, toward Sheire, there was shot through a cottage a piece of timber, "which took off a poor woman's head as she was spinning."

The Manufacture of Gunpowder may be described from a visit by Dr. Scoffern to one of her majesty's mills at Waltham, in the Essex Marshes. First, as to the ingredients. The saltpetre (principally imported from Bengal) is boiled in large pans, evaporated, and crystallized; and the charcoal is prepared from the alder and willow, which abound in the neighborhood. These processes are conducted in buildings at some distance from the Gunpowder Mills, whither the materials are carried, by water, in covered boats, to the works. There the saltpetre, brimstone, and charcoal are ground separately in mills, each consisting of a pair of heavy circular stones slowly revolving on a stone bed. Next the ingredients are conveyed to "the Mixing House," where visitors wear over-shoes.   Here, in bins, are the saltpetre, brimstone, and charcoal, weighed in the exact proportions: saltpetre 75, brimstone 10, and charcoal 15, in every 100 parts.   Of the three ingredients, 42 lbs. are placed in a hollow drum, which revolves rapidly, and contains a fly-pan, which rotates in an opposite direction; in about five minutes a complete mixture is effected, and the charge is received in a bag tied over the lower orifice of the drum.

The "composition" is next taken to "the Incorporating Mills," and is now a combustible compound, to obtain its explosive power by the ingredients being thoroughly incorporated. The mill consists of a pair of circular stones ("runners"), weighing about 3½ tons each, and slowly rolling over the powder, which is placed on the stone bed of the mill, surrounded by a huge wooden basin. The powder is previously damped, as it could not be safely ground dry; about 7 pints of water ("liquor") being added to the charge of 42 lbs. of powder during 3½ hours, the time of grinding. To insure this with precision, and to obviate the chance of any irregularity in a clock, the waterwheel which works two of these mills in one house also marks its revolutions on a dial, so that the attendant can never be mistaken in the time the charge has been "on"—a most important point, where the over-grinding of the too dry powder might cause it to explode. Sometimes a portion of the wood-work of the roof, or mill, becoming detached—such as a cog of the wheel—and falling into the pan, acts as a skid on one of the runners, and by friction produces heat enough to cause a mass of powder to explode. As a protection, over each house containing a pair of mills is suspended a flat board, which, in case of an explosion, is first blown upward, and, being connected by wires with a cistern of water over the pan of the fellow mill, upsets the same, and drowns the Gunpowder. The attendants are as little as possible in these mills, and only work by daylight.

More hazardous processes, however, follow. The powder thus incorporated is in hard, flat lumps, and has again to be reduced to dust in the "Breaking-down House," by conveying it down an inclined plane, through rollers, which crush nearly 500 lbs. in the hour. The powder is then taken to "the Press House," and there, between gun-metal plates, is pressed in thin cakes to one third its bulk by a power of 700 tons in a hydraulic press. The cakes are roughly broken up, and sent in baskets to "the Granulating Mill," where the powder is again broken down into grains, the size being regulated by sieves. The floor is covered with hides fastened down with copper nails, and the mill can be started or stopped by a rope passing through the wall, which is bomb-proof. The powder is then dried, by heat, in "the Stoving-room," which is flanked externally by "traversers" (mounds of earth 30 feet thick), to confine explosion, should it happen, as much as possible to one house. Lastly, the powder is sifted in "the Dusting House," where the sieves revolve with great velocity; the dust escapes through the meshes, and the Gunpowder is drawn off through a sort of tap, into barrels, for packing. The finest powder is "glazed" by black-lead being shaken up with it; but cannon powder has not this finish.

C

THE BAROMETER: TORRICELLI AND PASCAL.

THE invention of the Barometer is one of the most curious events in the history of philosophy. No new discovery, not even those substantiated by the telescope, ever knocked so hard at the door of a received system, or in a manner which so imperiously demanded admission. The circumstances attending it are briefly these:

The phenomena of the common Pump had been well known for more than a century at least before the Christian era. The mode of explanation was simply the well-known maxim that "Nature abhors a vacuum;" but no attempt had been made to discover why. Sir John Herschel observes, that "if any such abhorrence existed, and had the force of an *acting cause* which could urge water a single foot into a pipe, there is no reason why the same principle should not carry it up two, three, or any number of feet; none why it should suddenly stop at a certain height, and refuse to rise higher, however violent the suction might be—nay, even fall back, if purposely forced up too high."

It is related that the engineers of Cosmo de Medicis, wishing to raise water higher than thirty-two feet by means of a sucking-pump, they found it impossible to take it higher than thirty-one feet. Galileo, the Italian sage, was applied to in vain for a solution of the difficulty. It had been the belief of all ages that the water followed the piston from the horror which nature had of a vacuum; and Galileo improved the dogma* by telling the engineers that this horror was not felt, or at least not shown, beyond heights of thirty-one feet! At

* The above story is told in several different ways (it has been said, for instance, that the answer of Galileo was ironical); but, whichever may be true, it is most probable that it led him to abandon the theory of nature's horror, though without substituting any other. It has been thought that, before his death, Galileo suspected the true explanation.

his desire, however, his disciple, Torricelli, investigated the subject. He found that when the fluid raised was mercury, the horror of a vacuum did not extend beyond thirty inches, because the mercury would not rise to a greater height; and hence he concluded that a column of water thirty-one feet high, and one of mercury thirty inches, exerted the same pressure upon the same base, and that the antagonistic force which counterbalanced them must in both cases be the same; and having learned from Galileo that the air was a heavy fluid, he concluded, and published the conclusion in 1645, that the weight of the air was the cause of the rise of water to thirty-one feet, and of mercury to thirty inches. He then filled a tube, more than three feet long, and open at one end only, with mercury; and then, stopping the open end with the finger, he placed the tube in an open vessel of mercury, with the open end downward. On removing the finger, the mercury in the tube sank until it stood in the tube at about twenty-eight inches higher than the mercury in the vessel. He thus constructed what is at this time considered the best form of the barometer.

In 1646, Pascal, the young philosopher of Clermont, repeated these experiments at Rouen, before more than 500 persons, among whom were five or six Jesuits of the college, and he obtained precisely the same results as Torricelli, with whose explanation, however, he did not become acquainted until the following year, when, assuming that the mercury in the Torricellian tube was suspended by the weight or pressure of the air, he suggested that it would necessarily fall in ascending a high mountain, by the diminution of the superincumbent column of air. At his request, his relative, M. Perier, tried the barometer at the summit and the base of the mountain of Puy de Dome, in Auvergne; the result was, that the mercury, which, at the base, stood twenty-six and a quarter inches (French), was only twenty-three and a sixth inches at the summit. Pascal afterward found the same result sensibly shown in the ascent of a church tower and of a private house.

After this important experiment was made, Pascal intimated that different states of the weather would occa-

sion differences in the barometer, according as it was cold,
hot, dry, or moist; and M. Perier tested this opinion by
observations made at Clermont from 1649 to 1651. Corresponding observations were made at the same time at
Paris and at Stockholm; and from these it appeared that
the mercury rises in cold, cloudy, and damp weather, and
falls when the weather is hot and dry, and during rain
and snow; but still with such irregularities, that no general rule could be established. At Clermont, the difference between the highest and lowest state of the mercury was one inch three and a half lines; at Paris, the
same; and at Stockholm, two inches two and a quarter
lines.

The discovery was, however, at first much misconceived, and even disputed, till the question was finally
decided by an appeal to a *crucial instance;* one of the
first, if not the very first, on record in physics. "It was
then seen," says Sir John Herschel, "as by a *glaring
instance*, that the maintenance of the mercury in the
tube was the effect of a perfectly definite external cause,
while its fluctuations from day to day, with the varying
state of the atmosphere, strongly corroborated the notion
of its being due to the pressure of the external air on the
surface of the mercury in the reservoir."

The truth of the thing is just this: air, though comparatively light, is positively heavy, having a weight of
its own. The above experiments showed that a square
inch of it, carried up from the surface of the earth to the
top of the atmosphere, is no less than fifteen pounds in
weight. It is this weight of the atmosphere, fifteen
pounds on every square inch, that pushes water into the
void left by the up-drawn piston of a pump; and there
is, of course, a limit beyond which it can not push the
water, namely, the point of height at which the column
of water in the pump-tube is exactly balanced by the
weight of the atmosphere. It is just a question of balance: fifteen pounds can only support fifteen pounds—a
thing which every body now understands, thanks to Galileo, Torricelli, and Blaise Pascal, the seer, the discoverer, and verifier of the fact.

Pascal evinced such early sagacity, that, at the age of
eleven, he was ambitious of teaching as well as learning;

and he then composed a little treatise on the refractions of sounds of vibrating bodies when touched by the finger. One day he was found alone in his chamber tracing with charcoal geometrical figures on the wall; and on another occasion he was surprised by his father just when he had succeeded in obtaining a demonstration of the 32d proposition of the first book of Euclid—that the three angles of a triangle are equal to two right angles. Astonished and overjoyed, his father rushed to his friend, M. Railleur, to announce the extraordinary fact; and the young geometer was instantly permitted to study, unrestrained, the Elements of Euclid, of which he soon made himself master without any extrinsic aid. From the geometry of planes and solids he passed to the higher branches of the science; and before he was sixteen years of age he composed a treatise on the Conic Sections, which evinced the most extraordinary sagacity. When scarcely nineteen years of age, too, Pascal contrived a machine to assist his father in making the numerical calculations which his official duties in Upper Normandy required.

In later life, Pascal found researches in geometry an occupation well fitted to give serenity to a heart bleeding from the wounds of his beloved associates. He had for some time renounced the study of the sciences, when, during a violent attack of toothache, which deprived him of sleep, the subject of the cycloid forced itself upon his thoughts. Fermal, Roberval, and others, had trodden the same ground before him; but in less than eight days, and under severe suffering, he discovered a general method of solving this class of problems by the summation of certain series; and as there was only one step from this discovery to that of Fluxions, Pascal might, with more leisure and better health, have won from Newton and from Leibnitz the glory of that great invention.

Pascal's treatise on the weight of the whole mass of air forms the basis of the modern science of Pneumatics. In order to prove that the mass of air presses by its weight on all the bodies which it surrounds, and also that it is elastic and compressible, Pascal carried a balloon half filled with air to the top of the Puy de Dome. It gradually inflated itself as it ascended; and when it reached the summit it was quite full and swollen, as if

fresh air had been blown into it, or, what is the same
thing, it swelled in proportion as the weight of the column of air which pressed upon it was diminished.   When
again brought down, it became more and more flaccid ;
and when it reached the bottom, it resumed its original
condition.   In the above treatise, Pascal shows that all
the phenomena and effects hitherto ascribed to the horror of a vacuum arise from the weight of a mass of air ;
and—after explaining the variable pressure of the atmosphere in different localities and in its different states, and
the rise of water in pumps—he calculates that the whole
mass of air round our globe weighs 8,983,889,440,000,-
000,000 French pounds.

Seeing that little more than two centuries have elapsed
since the exposition of this great principle of Hydrostatics was clearly established, we are not surprised to find
that the science in the Dark Ages enabled the ancient
magicians to impose upon their dupes with unimpeach-
able certainty.   To name a few of the most celebrated
instances : the magic cup of Tantalus, which he could
never drink though the beverage rose to his lips ; the
fountain in the island of Andros, which discharged wine
for seven days, and water for the rest of the year ; the
fountain of oil, which burnt out to welcome the return
of Augustus from the Sicilian war ; the empty urns, which,
at the annual feast of Bacchus, filled themselves with wine,
to the astonishment of the assembled strangers ; the
glass tomb of Belus, which, after being emptied by
Xerxes, could never again be filled ; the weeping statues
of the ancients, and the weeping virgin of modern times,
whose tears were uncourteously stopped by Peter the
Great when he discovered the trick ; and the perpetual
lamps of the ancient temples, were all the obvious effects
of hydrostatical pressure.

THE AIR-PUMP AND THE AIR-GUN.

IMMEDIATELY after the discovery of the principle of the Barometer by Torricelli, in the pressure of the air on the general surface, followed that of Otto von Guericke, whose aim seems to have been to decide the question whether a vacuum could or could not exist, by endeavoring to make one.* The first Air-pump constructed by Guericke was exhibited by him at the Imperial Diet of Ratisbon in 1654. It was an exhausting syringe, attached underneath a spherical glass receiver, and worked somewhat like a common pump. The syringe was entirely immersed in water, to render it air-tight. The imperfection of his mechanism, however, enabled Guericke only to diminish the aerial contents of his receiver, not entirely to empty them; but the curious effects produced by even a partial exhaustion of air speedily excited attention, and induced our illustrious countryman, Robert Boyle, to construct an air-pump, in which the syringe was so far improved that the water could be dispensed with: he also first applied rack-work to the syringe. In the Journals of the Royal Society, January 2d, 1660, we find Boyle's Air-pump referred to as his Cylinder, and "that Mr. Boyle be desired to show his Experiments of the Air," which are printed in the Society's *Transac-*

* This ingenious and ardent cultivator of science, who was born at Magdeburg, in Saxony, in the beginning of the seventeenth century, in his original attempts to produce a vacuum, used first to fill his vessel with water, which he then sucked out by a common pump, taking care, of course, that no air entered to replace the liquid. It was by first filling it with water that Guericke expelled the air from the copper globe, the two closely fitting hemispheres comprising which six horses were then unable to pull asunder, although held together by nothing more than the pressure of the external atmosphere. This curious proof of the force or weight of the air, which was exhibited before the Emperor Ferdinand III. in 1634, is commonly referred to by the name of the experiment of the *Magdeburg Hemispheres*. Guericke, however, afterward adopted the method of exhausting a vessel of its contained air by the air-pump.

tions. The Air-pump constructed by Boyle was presented to the Society by him in 1662, and it is now in the museum at Burlington House: the pump consists of two barrels.

We have the testimony of a French *savant* of the nineteenth century, M. Sibes, that the Air-pump in Boyle's hands became a new machine; and Professor Baden Powell considers that "he reduced it nearly to its present construction." It is true that the second syringe and the barometer gauge were afterward added by Hawksbee, and several minor improvements were made by Hooke, Mariotte, Gravesande, and Smeaton. All the alterations which have been made since the time of the invention, however important, relate to the mechanism only, and not to the principle on which the pump acts.

Dr. Hutton has grouped these effects and phenomena of the Air-pump. In the exhausted receiver, heavy and light bodies fall equally swiftly: so a guinea and a feather fall from the top of a tall receiver to the bottom exactly together. Most animals die in a minute or two: however, vipers and frogs, although they swell much, live an hour or two, and, after being seemingly quite dead, revive in the open air. Snails survive about ten hours; efts, two or three days; leeches, five or six. Oysters live for twenty-four hours. The heart of an eel, taken out of the body, continues to beat for great part of an hour, and that more briskly than in the air. Warm blood, milk, gall, etc., undergo a considerable internescence and ebullition. Eggs of silkworms hatch *in vacuo.* Vegetation stops. Fire is extinguished; the flame of a candle usually going out in one minute, and charcoal in about five minutes. Red-hot iron seems, however, not to be affected; sulphur and gunpowder are not lighted by it, only fused. A match, after lying seemingly extinct for a long while, revives on readmitting the air. A flint and steel strike sparks of fire as copiously as in air. Magnets and magnetized needles act as in air. Heat may be produced by attrition. Camphor will not take fire; and gunpowder, though some of the grains of a heap of it be kindled by a burning-glass, will not give fire to the contiguous grains. Glowworms lose their light in proportion as the air is exhausted; but, on readmitting the air, they presently recover. A bell, on being struck, is not heard to ring, or very faintly. Water freezes. A syphon will not run; and electricity appears like the Aurora Borealis.

De la Croix relates the following instance of sagacity in a cat, who, even under the receiver of an Air-pump, discovered the means of escaping a death which appeared to all present inevitable. "I once saw," he relates, "a lecturer upon experimental philosophy place a cat under the glass receiver of an Air-pump for the purpose of demonstrating that life can not be supported without air and respiration. The lecturer had already made several strokes with the piston in order to exhaust the receiver of its air, when the cat, who began to feel herself very uncomfortable in the rarefied atmosphere, was fortunate enough to discover the source from whence her uneasiness proceeded. She placed her paw upon the hole through which the air escaped, and thus prevented any more from passing out of the receiver. All the exertions of the philosopher were now unavailing : in vain he drew the piston; the cat's. paw effectually prevented its operation. Hoping to effect his purpose, he again let air into the receiver, which as soon as the cat perceived, she withdrew her paw from the aperture; but whenever he attempted to exhaust the receiver, she applied her paw as before. The spectators clapped their hands in admiration of the cat's sagacity ; and the lecturer was compelled to remove her, and substitute another cat that possessed less penetration for the cruel experiment."

Although the Air-pump is scarcely two centuries old, yet the Air-gun, which is so nearly allied to it in the construction of its valve and condensing syringe, existed long antecedent to it; for it is recorded that an Air-gun was made for Henry IV., by Marim, of Lisseau, in Normandy, as early as 1408; and another was preserved in the armory at Schmetau, bearing the date of 1474. The Air-gun of the present day is different. Bishop Wilkins mentions "the Wind Gun" as a late ingenious invention, which discharges with force "almost equal to our powder guns."

Professor Helmholtz, one of the latest illustrators of this instrument, thus lucidly explains its theory : "Into the chamber of an Air-gun we squeeze, by means of a condensing air-pump, a great quantity of air. When we afterward open the cock of the gun, and admit the compressed air into the barrel, the ball is driven out of the latter with a force similar to that exerted by ignited powder. Now we may determine the work consumed in the pumping-in of the air, and the living force which, upon firing, is communicated to the ball, but we shall never find the latter greater than the former. The compressed air has generated no working force, but simply gives to the bullet that which has been previously communicated to it. And while we have pumped for perhaps a quarter of an hour to charge the gun, the force is expended in a few seconds when the bullet is discharged; but, because the action is compressed into so short a time, a much greater velocity is imparted to the ball than would be possible to communicate to it by the unaided effort of the arm in throwing it."

We may here relate a curious wager which Sir Robert Moray, at the request of Charles II., brought forward at a meeting of the Royal Society in 1671. It was, that the king wagered £50 to £5 "for the compression of air by water." It was accordingly resolved that Mr. Hooke should prepare the necessary apparatus for the experiment, which Sir Robert Moray said "might be done by a cane, so contrived that it should take in more and more water, according as it should be sunk deeper and deeper into it." The minutes of a subsequent meeting record the successful performance of the experiment, and that it "was acknowledged his majesty had won the wager."

LIVING UNDER WATER: THE DIVING-BELL.

WHEN we consider the vast amount of treasure which has been from time to time lost in the depths of the sea, we shall not be surprised at the variety of the means which have been devised for the recovery of the hidden wealth. The principal of these contrivances is the Diving-bell, with the operations of which the public have become familiar by the exhibition of an improved bell at our Polytechnic Institution;* but the history of the invention, as well as the primitive means by which it was preceded, present many interesting instances of ingenuity directed to humane and praiseworthy purposes.

In remote ages (says Professor Beckmann) divers were kept in ships to assist in raising anchors, and goods thrown overboard in times of danger; and, by the laws of the Rhodians, they were allowed a share of the wreck proportioned to the depth in which they had gone in search of it. In war, they were often employed to destroy the works and ships of the enemy; divers also fished for

---

* For twenty years (1839–1859) there was exhibited at the Polytechnic Institution, No. 300 Regent Street, London, a diving-bell, which was put in operation daily. This bell was manufactured by Cottam and Hallen, and cost about £400. It is of cast iron, and weighs 3 tons; 5 feet in height, and 4 feet 8 inches in diameter at the mouth. Within is affixed a knocker, under which is painted:

"More air, knock once;
Less air, knock twice;
Pull up, knock three times."

The bell is about one third open at the bottom, has a seat all round for the divers, is lit by twelve openings of thick plate glass. It is suspended by a massive chain to a large swing-crane, with a powerful crab, the chain having compensation weights, and working into a well beneath. The air was supplied from two powerful air-pumps, of eight-inch cylinder, conveyed by the leather hose to any depth; the divers being seated in the bell, it was moved over the water, and directly let down within two feet of the bottom of the tank, and then drawn up, the whole occupying only two minutes and a half. The tank and the adjoining canals held 10,000 gallons of water. Each person descending in the bell paid 1s.; and it has produced £1000 in one year.

pearls. The statements of their remaining under water unassisted by apparatus for procuring air are, however, greatly exaggerated; they speak of six hours, whereas six minutes is the longest time of submersion recorded in modern times.

Dr. Halley, in a paper in the *Philosophical Transactions* on " the Art of living under Water," describes the divers for sponges in the Archipelago taking down in their mouths a piece of sponge soaked in oil, by which they were enabled to dive for a longer period than without it. As the bulk of the sponge must diminish the quantity of air which the diver could contain in his mouth, it does not appear probable that this practice could assist respiration.

In connection with diving by the unassisted powers of the body, Professor Faraday relates this curious fact: The lungs are, in their natural state, charged with a large quantity of impure air; this being a portion of the carbonic acid gas which is formed during respiration, but which, after such expiration, remains lodged in the involved passages of the pulmonary vessels. By breathing hard for a short time, as a person does after violent exercise, this impure air is expelled, and its place is supplied by pure atmospheric air, by which a person will be enabled to hold his breath much longer than without such precaution. Dr. Faraday states that, although he could only hold his breath, after breathing in the ordinary way, for about three quarters of a minute, and that with great difficulty, he felt no inconvenience, after making eight or ten forced respirations to clear the lungs, until the mouth and nostrils had been closed more than a minute and a half; and that he continued to hold breath to the end of the second minute. A knowledge of this fact may enable a diver to remain under water at least twice as long as he otherwise could do. Possibly the exertion of swimming may have the effect of clearing the lungs, so that persons accustomed to diving may unconsciously avail themselves of this preparatory measure.

The advantage of breathing condensed air, and thereby obtaining a larger supply of oxygen in the same bulk than with air of the ordinary pressure, is shown also in the following fact: After one of the disastrous occur-

rences at the works of the Thames Tunnel, Mr. Brunel,
the engineer, descended in a diving-bell to examine the
breach made by the irruption of the river into the tunnel.   The bell was lowered to the mouth of the opening,
a depth of about thirty feet; but the breach was too narrow to allow it to go lower, in order that the shield and
other works, which lay eight or ten feet deeper, might
be examined from the bell.   Mr. Brunel therefore took
hold of the rope, and dived below the bell for the purpose.   After he had remained under water about two
minutes, his companion in the bell became alarmed, and
gave a signal which caused Brunel to rise.   On doing so,
he was surprised to find how much time had elapsed;
and, on repeating the experiment, he ascertained that he
could with ease remain fully two minutes under water, a
circumstance accounted for by the condensation of the
air in the bell, from which his lungs were supplied, by the
pressure of a column of water nearly thirty feet high,
which would condense the air into little more than one
half of its usual bulk.

Plans for enabling persons to remain for a longer period under water than is possible by the natural powers
of the body are of very old date.   Aristotle is supposed
to intimate that in his time divers used a kind of kettle
to enable them to continue longer under water; but this
interpretation is disputed.   Beckmann states that the
oldest information we have respecting the use of the Diving-bell in Europe is that of John Taisnier, quoted in
Schott's *Technica Curiosa*, Nuremberg, 1664, in which
Taisnier relates: "Were the ignorant vulgar told that
one could descend to the bottom of the Rhine, in the
midst of the water, without wetting one's clothes, or any
part of one's body, and even carry a lighted candle to
the bottom of the water, they would consider it altogether as ridiculous and impossible.   This, however, I saw
done at Toledo in Spain, in the year 1538, before the Emperor Charles V. and almost ten thousand spectators.
The experiment was made by two Greeks, who, taking a
very large kettle suspended by ropes with the mouth
downward, fixed beams and planks in the middle of its
concavity, upon which they placed themselves, together
with a candle.   The kettle was equipoised by means of

lead fixed round its mouth, so that, when let down toward the water, no part of its circumference should touch the water sooner than another, else the water might easily have overcome the air included in it, and have converted it into moist vapor; but if the vessel were gently drawn up, the men continue dry, and the candle is found burning." Schott calls the machine "an aquatic kettle;" he also describes "an aquatic armor," which would enable those who were covered with it to walk under water; and the former apparatus is represented, showing a man walking into the water with a covering like a small diving-bell over his head, descending nearly to his feet.

In England, besides the supposed contrivance of a Diving-machine by Roger Bacon, it is evident that the Diving-bell was known at a very early period. It is described more than once in the works of Lord Bacon as a machine used to assist persons laboring under water upon wrecks, by affording a reservoir of air to which they might resort whenever they required to take breath. "A hollow vessel was made of metal, which was let down equally to the surface of the water, and thus carried with it to the bottom of the sea the whole air it contained. It stood upon three feet like a tripod, which were in length somewhat less than the height of a man, so that the diver, when he was no longer able to contain his breath, could put his head into the vessel, and, having breathed, return again to his work" (*Novum Organum*, lib. ii., p. 850).

The next use of the bell occurred in America, where, in 1642, it was used by one Edward Bedall, of Boston, to weigh the *Mary Rose*, which had sunk the previous year. Bedall made use of two tubs, "upon which were hanged so many weights (600 lbs.) as would sink them to the ground." The experiment succeeded, and the guns, ballast, goods, hull, etc., were all transported into shoal water, and recovered.

Some curious information on submarine operations was published in 1688 by Professor Sinclair, of Glasgow, showing how "to buoy up a ship of any burden from the ground of the sea;" and stating that the late Marquis of Argyle, "having obtained a patent of the king on one of the Spanish Armada, which was sunk near the Isle of

Mull, anno 1588, employed James Colquhoun, of Glasgow," who, " not knowing the Diving-bell, went down several times, the air from above being communicated to his lungs by a long pipe of leather." The Armada ships sunk near Mull, according to the accounts of the Spanish prisoners, contained great riches; and this information excited from time to time the avarice of speculators, and gave rise to several attempts to procure part of the lost treasure. About 1664, an ingenious gentleman, the Laird of Melgim, " went down with a Diving-bell, and got up three guns." Sinclair also proposed to raise wrecks by the buoyancy of arks or boxes, open at the bottom, which were to be sunk full of water, and then filled with air, either by sending down casks of air, by bellows and a long tube, or otherwise. He alludes to the occasional use of casks for the purpose of raising vessels, and explains why, when at a great depth, they are liable to be crushed by the pressure of the water; showing that, by allowing the water to enter by a hole in the lower part of the cask, it would so compress the air as to produce an equilibrium of pressure, and thereby preserve it from fracture.

About twenty years after this, William Phipps, the son of a blacksmith of Pemaquid, in the United States, and who had been brought up as a ship-carpenter at Boston, formed a project for searching and unloading a rich Spanish wreck near the Bahamas, when Charles II. gave him a frigate to obtain the treasure. He sailed in 1683; but, being unsuccessful, returned in great poverty, though with a firm conviction of the practicability of his scheme. He then endeavored to procure a vessel from James II., failing in which he opened a subscription. At first he was laughed at; but at length the Duke of Albemarle, son of the celebrated General Monk, advanced Phipps a considerable sum toward the second outfit; and having collected the remainder, he set sail in 1687, in a ship of 200 tons burden, and reaching the wreck, when nearly worn out with fruitless labor he brought up, from six and seven fathoms depth, treasure of £300,000, of which Phipps received for his share £16,000, the Duke of Albemarle £90,000, and the subscribers received the remainder. Some envious persons then endeavored to persuade

the king to seize both the ship and the cargo, under a
pretense that Phipps, when he solicited his majesty's permission, had not given accurate information respecting
the business; but James nobly replied that he knew
Phipps to be an honest man, and that he and his friends
should share the treasure among them: the king afterward knighted Phipps, who had previously been made
High Sheriff of New England. In 1691 he was made
governor of his native colony. He was uneducated, and
knew not how to read or write until he had grown to
manhood; but, by strong native abilities and restless enterprise, he rose to distinction. He is erroneously said
to have been the founder of the Mulgrave family, of
which the present head is the Marquis of Normanby;
which mistake has, doubtless, arisen from one of the early members of that family, Captain Constantine John
Phipps, commander of the unsuccessful Arctic Expedition in 1773, having been raised to the British Peerage
as Baron Mulgrave, of Mulgrave, co. York, in 1790.

Among the oldest representations of Diving apparatus,
Beckmann mentions a print in editions of Vegetius on
War, published in 1511 and 1532, representing a diver
with a cap, from which rises a long leathern pipe, terminating in an opening which floats upon the surface of
the water. Beckmann also names a figure, in Lorini's
work on Fortification, 1607, which nearly resembles the
modern Diving-bell, and consists of a square box, bound
with iron, which is furnished with windows, and a seat
for the diver. Lorini, who was an Italian, does not lay
claim to the invention of this apparatus.

In 1617, Francis Kessler described his Water-armor,
intended for diving, but which Beckmann states to have
been useless. In 1671, Witsen taught, better than any of
his predecessors, the construction and use of the Diving-
bell, which, however, he erroneously says was invented
at Amsterdam. About 1679, Borelli, the celebrated physician of Naples, invented an apparatus by which persons might go a considerable depth under water, remain
there, move from place to place, and sink or rise at pleasure; and also a boat in which two or more persons might
row themselves under water; but the practicability of
these machines has been much controverted.

Dr. Halley, in the paper in the *Philosophical Trans-actions* already quoted, describes the defects of the Diving-bell as previously used, and suggests a remedy for them. This paper alone would be sufficient, although it does not enter into the early history of the machine, to contradict the erroneous statement which has been made, that Halley was the *inventor* of the Diving-bell.

In its simplest form, the Diving-bell is a strong, heavy vessel of wood or metal, made perfectly air and water-tight at the top and sides, but open at the bottom. If such a vessel be gradually lowered into the water in a perfectly horizontal position, the air which it contains can not escape, and therefore the vessel can not become full of water. This may be readily illustrated by plung-ing a glass tumbler in an inverted position into a vessel of water, and placing a bit of cork under the glass. If a bit of burning matter be laid upon the cork float, it will continue to burn, although the glass and all that it con-tains be plunged far beneath the water, thereby proving that the upper part of the cavity of the glass is occupied by air, and not by water. In this experiment, however, it will be observed that the water does fill a small part of the cavity of the glass, and that it rises more into it when it is plunged to a considerable depth than when the rim is only just immersed beneath the surface. This is occasioned by the condensation of the air contained in the glass, which, being very elastic and compressible, is condensed into a smaller space than it would occupy under the ordinary pressure of the atmosphere.

We have now illustrated the principle of the Diving-bell: let us proceed to its application. When the bell is used for descending to a very small depth, as the press-ure of the water is small, it will not rise in the bell to a sufficient height to be inconvenient; but at the depth of thirty feet the pressure is so great as to compress the air into one half its original volume, so that the bell will be-come half full of water; and at a greater depth the air will be still more compressed, and the water will rise pro-portionally higher in the bell. This condensation of the air does not materially interfere with respiration, pro-vided the descent of the bell be very gradual, as the air then insinuates itself into the cavities of the body, and balances the pressure from without. The principal effect of the increased pressure is a pain in the ears, since the Eustachian tube does not allow the condensed air immediately to find its way into the cavities of the ear, so that the pressure on the outside of the tympanum is for a time unbalanced by a corresponding pressure from within, and occasions a sensation like that of having quills thrust into the ears. This continues until the pressure of the air in the mouth, which at first has a tendency to keep the aperture of the Eustachian tube closed, forces it open; an action which is accompanied by a noise like a slight explosion. The condensed air then enters the interior cavities of the ear, and by restoring the equilibrium of pressure on each side the tympanum, removes the pain, which will return, and be remedied in the same manner, if the bell should descend to a greater depth. But, while the mere condensation of the air in the bell does not render it unfit for respiration, it would soon become so if no means were provided for renewing it from time to time, as it becomes vitiated by repeated respiration. Dr. Halley provided a remedy for the inconvenience by supplying the bell with fresh air without raising it to the surface. The air was conveyed in two thirty-gallon barrels, weighted with lead to make them sink readily. Each had an open bung-hole in the lower end, to allow water to enter during their descent, so as to condense the air. There was also a hole in the upper end of each barrel, to which was fitted an air-tight leathern hose. These air-barrels were attached to tackle, by which they were by two men let down and raised alternately, like two buckets in a well; and by lines attached to the lower edge of the bell, they were so guided in their descent that the mouth of the hose always came directly to the hand of a man who stood upon the stage suspended from the bell. As the apertures of the hose were, during the descent, always below the level of the barrels, no air could escape from them; and when they were turned up by the attendant, so as to be above the level of the water in the barrels, the air rushed out with great force into the bell, the barrels becoming at the same time full of water.

By sending down the air-barrels in rapid succession,

the air was kept in so pure a state that Halley and four other persons remained in the bell, at a depth of nine or ten fathoms, for more than an hour and a half at a time, without injurious consequences; and Halley states that he could have remained there as long as he pleased for any thing that appeared to the contrary. Halley observed that it was necessary to be set down gradually at first, and to pause at about the depth of twelve feet, to drive out, by the admission of a supply of air, the water which had entered the bell. When the Diving-bell was at the required depth, he let out, by a cock in the top of the bell, a quantity of hot impure air equal to the quantity of fresh air admitted from the barrels. This foul air rushed up from the valve with such force as to cover the surface of the sea with a white foam. So perfect was the action of this apparatus that Halley says he could, be removing the hanging stage, lay the bottom of the sea so far dry, within the circuit of the bell, that the sand or mud did not rise above his shoes. Through the strong glass window in the top, when the sea was clear, and especially when the sun shone, sufficient light was transmitted to allow a person in the bell to write or read; and when the sea was troubled or thick, which occasioned the bell to be as dark as night, a candle was burnt in it. Halley sometimes sent up orders with the empty air-barrels, writing them with an iron pen on plates of lead.

Halley, having by these ingenious contrivances removed the principal difficulties attending the use of the diving-bell, foresaw its extensive application: as fishing for pearl, diving for coral, sponges, and the like, in far greater depths than had hitherto been thought possible; also for laying the foundations of moles, bridges, etc., upon rocky bottoms; and for the cleaning and scrubbing of ships' bottoms, when foul, in calm weather at sea, to which purposes the Diving-bell has, since the date of Halley's paper (1717), been applied.

The next improver of the Diving-bell was Martin Triewald, "captain of mechanics, and military architect to his Swedish-majesty," who had the sole privilege of diving upon the coasts of the Baltic belonging to the King of Sweden. His bell was of copper, tinned inside, smaller than that of Dr. Halley, and managed by two men. A stage for the diver to stand upon was suspended at such a depth below it that the man's head would be but little above the level of the water, where the air is cooler and fitter for respiration than in the upper part of the bell; and a spiral tube was attached to the inside of the bell, with a wide aperture at the bottom, and a flexible tube and mouth-piece at the top, so that when the diver was up in the bell he might inhale cool air from the lower part, exhaling the foul air by his nostrils. In lieu of windows of flat glass, Triewald used convex lenses, such as are employed to this day,* to admit light to the bell.

In 1775, Mr. Spalding, a grocer of Edinburgh, made certain improvements upon Halley's bell, in recovering part of the cargo of a vessel lost on the Fern Islands. Spalding's bell was of wood; and to sink it he used, in addition to the weights attached to the rim, a large balance-weight suspended by a rope from the centre, and which, by pulleys, the divers employed to anchor the bell at any required level; and by hauling in the rope while the weight was at the bottom, the persons in the bell might lower themselves at pleasure. Another improvement was a horizontal partition near the top of the bell, which divided off a chamber, with valves, to be filled either with water or with air from the lower part of the bell, so as to alter the specific gravity of the whole machine, and thereby cause it to ascend or descend at pleasure. This bell also had an air apparatus like Halley's; ropes were used instead of seats in the bell, so that the divers could raise themselves to the surface unassisted from above; the bell could be removed at will from the point at which it descended, and a long-boat carried the signal-lines and the tackle for working the air barrels. Mr. John Farey, jr., has improved upon Spalding's appa-

---

* These convex glasses have been known to produce extraordinary effects. Thus, in 1828, Mr. Mackintosh, contractor for the government works at Stonehouse Point, Devon, had to descend in the Diving-bell with workmen to lay the foundation of a sea-wall. The bell was fitted with convex glasses in the upper part; and Mr. Mackintosh states that on several occasions, in clear weather, he witnessed the sun's rays so concentrated as to burn the laborers' clothes when opposed to the focal point, and this when the machine was twenty-five feet under the surface of the water.—*From the MS. Journal of the Bristol Nursery Library.*

ratus, by making the upper chamber of the bell without valves, and used it as a reservoir of condensed air, to be filled by forcing-pumps in the partition, besides other provisions.

Smeaton first employed the Diving-bell in civil engineering operations in repairing the foundations of Hexham Bridge in 1779. His bell was an oblong box of wood, and supplied with a gallon of air a minute by a forcing-pump fixed at the top, which was not covered with water, the river being shallow. In 1788 Smeaton used a cast iron bell in repairing Ramsgate Harbor, the air being supplied through a flexible tube from a forcing-pump in a boat. Rennie improved the apparatus for moving the bell in any direction; and in 1817 the wreck of the *Royal George* at Spithead was first surveyed by the aid of the Diving-bell.

Many plans have been proposed for enabling a man to walk beneath the surface of the water by means of water-proof coverings for the head and upper part of the body, or of strong vessels in which every part but the arms should be incased; a supply of air being either transmitted from above by a flexible pipe, or contained in the cavities of the protecting armor. This apparatus may be conveniently used at small depths; but at any considerable depth it is both dangerous and inconvenient, because the strength necessary to enable it to bear the pressure of the water is incompatible with the flexibility essential to the free use of the limbs. Dr. Halley invented a leaden cap for the diver's head, the front glazed for the eyes; it contained a supply of air for two minutes, and had affixed to it a pliable pipe, the other end being fastened to the bell, whence fresh air was conveyed to the diver.

At Newton-Bushel, in Devonshire, a gentleman contrived an apparatus consisting of a large strong leather water-tight case, holding half a hogshead of air, and adapted to the legs and arms, with a glass in front, so that when the case was put on the wearer could walk about easily at the bottom of the sea, examine a wrecked vessel, and deliver out the goods; the inventor of this apparatus used it forty years, and thereby acquired a large fortune.

Mr. Klingert, in 1798, constructed at Breslau tin-plate armor for the head and body, leather jacket, and water-tight drawers brass hooped; and a helmet with two pipes, one for inhaling, and the other for the escape of foul air. The body was kept down by weights. Contrivances of this kind, in which water-proof India-rubber cloth has been applied, are very numerous. In 1839 Mr. Thornthwaite made a hollow belt of India-rubber cloth, with a small strong copper vessel attached, and into which air is forced by a condensing syringe; the belt is put on collapsed, and the diver descends; but when he desires to rise, by a valve he lets out the condensed air from the copper vessel into the belt, which, as it expands, buoys up the diver to the surface.

Extraordinary substitutes have been sometimes made for the regularly-constructed Diving-bell. Thus, in the memorable recovery of treasure and stores from the wreck of the *Thetis*, which sank in a cove southeast of Cape Frio in 1830, and was not attempted to be raised until fifteen months after, by the officers and crew of H.M.S. *Lightning*, the Diving-bell consisted of a one-ton ship's water-tank, with eight inches of iron riveted to the bottom in order to give it more depth, and having attached to it eighteen pigs of ballast (17 cwt.) to sink it. Yet, with such a means of survey, often rendered unmanageable by the swell of the South Atlantic rolling into the cove of nearly perpendicular granite rocks, from 100 to 200 feet high, fifteen sixteenths of the property were recovered. A model of this enterprise may be seen in the United Service Institution Museum, Scotland Yard. For the achievement Captain Dickson received the gold medal of the Society of Arts.

One of the latest improvements upon the old Diving-bell—the *Nautilus* Submarine Machine, an American invention—has been successfully employed by engineers. It is nearly cylindrical, with a spherical top; and the working apparatus, on board a barge floating near, consists of a steam-boiler, a cylinder or reservoir, and a condensing or air pump. The workmen being stationed in the machine, water is admitted into two chambers, to serve as ballast and cause the Nautilus to descend to the bottom, meanwhile air being drawn through hose from the

reservoir in the barge.   As soon as the air thus drawn
is sufficiently condensed, a cover to the bottom is raised,
and communication obtained.   Not only do persons thus
remain under water for a considerable time, but should
the hose communicating with the reservoir become dis-
connected, no danger can ensue to those in the machine,
as they can, by means of the compressed air within the
bell itself, expel a portion of the water, and thus rise to
the surface.

AUTOMATA AND SPEAKING MACHINES.

THE amusing species of ingenuity which is requisite for the construction of these machines has been exercised to great extent. The name *Automaton* is derived from two Greek words meaning *self-moved*, and is generally applied to all machines which are so constructed as to imitate any actions of men or the lower animals, and are moved by wheels, weights, and springs.

The most ancient Automata are the Tripods which Homer mentions as having been constructed by Vulcan for the banqueting-hall of the gods, and which advanced of their own accord to the table, and again returned to their place. Self-moving Tripods are mentioned by Aristotle; and Philostratus informs us, in his Life of Apollonius, that this philosopher saw and admired similar pieces of mechanism among the sages of India. Beckmann hints that these Tripods were only small tables, or dumbwaiters, which had wheels so contrived that they could be put in motion, and driven to a distance, on the smallest impulse, like the fire-pans in the country beer-houses of Germany, at which the boors light their pipes.

That Dædalus made Statues which could not only walk, but required to be tied up that they might not move, is related by Plato and Aristotle. The latter speaks also of a wooden Venus, which moved about in consequence of quicksilver being poured into its interior; and before this method was known in Europe, Kircher proposed to put a small wagon in motion by adding to it a pipe filled with quicksilver, and heating it with a candle placed below it. Calistratus, the tutor of Dædalus,* however, states that his statues received their mo-

* Dædalus, having been banished from Athens for killing his nephew, of whose rising genius he was envious, took refuge in Crete, and here constructed the celebrated Labyrinth, in the windings of which he was subsequently confined as close prisoner by Minos, whom he had displeased. His unrivaled resource, however, did not forsake him; he manufactured for himself and his son Icarus waxen wings,

tion from the mechanical powers, which is more probable than the opinion of Beckmann, that their being in a position " as if ready to walk gave rise to the exaggeration that they possessed the power of locomotion." "This opinion," Sir David Brewster observes, " however, can not be maintained with any show of reason; for if we apply such a principle in one case, we must apply it in all, and the mind would be left in a state of utter skepticism respecting the inventions of ancient times" (*Natural Magic*, p. 265).

It is related by Aulus Gellius, on the authority of Favorinus, that Archytas of Tarentum (about 400 B.C.) constructed a Wooden Pigeon which was capable of flying. Favorinus states that when it had once alighted, it could not resume its flight; and Aulus Gellius adds, that it was suspended by balancing, and animated by a concealed *aura*, or spirit.

Of Albertus Magnus it is related that, among other prodigies, he constructed a Head of Brass, which is not only said to have moved, but to have answered questions! It is said to have occupied Albertus thirty years in its construction; and that his disciple, Thomas Aquinas, was so frightened when he saw the head, that he broke it to pieces; when Albertus exclaimed, " Periit opus triginta annorum." Of contemporary date is the legendary story of " Friar Bacon's Brazen Head." It is pretended he discovered that if he could make a head of brass which should speak, and hear it when it spoke, he might be able to surround all England with a wall of brass. Bacon, with some assistance, accomplished his object, but with this drawback—the head was warranted to speak in the course of one month, but it was quite uncertain when; and if they heard it not before it had

with which they flew over the sea. The father arrived safely in Sicily; but the son, in spite of his father's example and admonition, flew so high that his wings were melted by the sun, and he fell into the sea, which from him was called the Icarian Sea. It was the ancient custom to deify the authors of any useful inventions. Now Dædalus was especially famous for the sails of ships; and " though they did not place him in the heavens, yet they have promoted him as near as they could, feigning him to fly aloft in the air, whereas *he did but fly in a swift ship,* as Diodorus (and Eusebius) relates the historical truth on which that fiction is grounded."—*Bishop Wilkins.*

D

done speaking, all their labor would be lost. Bacon, wearied with three weeks' watching, set his man Miles to watch, with strictest injunction to awake him if the head should speak. The fellow heard the head at the end of one half hour say, "Time is;" at the end of another, "Time was;" and at the end of another half hour, "Time's past;" when down it fell with a tremendous crash; but the blockhead of a servant thought his master would be angry if he disturbed him for such trifles! Now Robert Recorde states that on the above account Bacon was considered to be a necromancer, "which never used that arte," but was an expert geometer and mathematician, as will be shown in a future page.

Among the earliest pieces of modern mechanism was the curious Water-clock presented to Charlemagne by the Calif Haroun-al-Raschid. In the dial-plate were twelve small windows corresponding with the divisions of the hours, indicated by the opening of the windows, which let out little metallic balls, which struck the hour by falling upon a brazen bell. The doors continued open till twelve o'clock, when twelve little knights, mounted on horseback, came out at the same instant, and, after parading round the dial, shut all the windows, and returned to their apartments.

The next automaton was the Artificial Eagle, which John Müller, or Regiomontanus, constructed, and which flew to meet the Emperor Maximilian when he arrived at Nuremberg, June 7th, 1470. This eagle is said to have soared aloft, and met the emperor at some distance from the city; then to have returned and perched upon the town-gate, and to have stretched out its wings and saluted the emperor when he approached! Another of Müller's prodigies was an Iron Fly, put in motion by wheel-work, and which flew about and leaped upon a table! But as none of Müller's contemporary writers speak of these pieces of mechanism, the tale of them is suspected to have been invented by Peter Ramus, who was never at Nuremberg till the year 1571.

The Emperor Charles V. is known to have amused himself in his later years with Automata, made for him by an artist of Cremona. Among the prodigies which he wrought for the emperor were figures of armed men and horses attacking with spears, while others beat drums and played flutes; besides, also, wooden sparrows which flew to and from their nests, and minute corn-mills which could be concealed in a glove.

It will hardly excite surprise to find that the artists who produced Automaton figures were in some instances suspected of practicing the black art, and thus fell victims to their own ingenuity. A melancholy incident, arising from the prevalence of this opinion, even so late as 1674, is related by Bonnet, in his *History of Music*. Alex, an ingenious Provençal mathematician and mechanician, had discovered the sympathy of sound in two instruments tuned in unison. To illustrate his discovery, he constructed an Automaton Skeleton, placed a guitar in its hand, while by a mechanical contrivance the fingers moved, as though playing it: he then set it at a window, and at a proper distance played another guitar, which produced sound in the instrument held by the figure. The inhabitants of Aix (the town in which this was exhibited), believing that the skeleton really performed on the guitar, denounced Alex as a sorcerer, and he was condemned by the Parliament to be burnt alive, together with his figure.

In the *Memoirs of the Academy of Sciences*, 1729, is described a set of Automaton Actors representing a pantomime. But previously to this, M. Camus had constructed, for the amusement of Louis XIV., a small coach drawn by two horses, etc. The coachman smacked his whip, and the horses set off, drawing the coach about a table; and when opposite the king, it stopped, the page got down and opened the door, on which a lady alighted, with a courtesy presented a petition to the king, and then re-entered the carriage. The page then shut the door, the carriage proceeded, and the servant, running after it, jumped up behind it (Hutton's *Mathematical Recreations*). This is by no means inconceivable, but is somewhat hard to believe.

Among the results of the development of the natural sciences in the seventeenth and eighteenth centuries was the attempt to build Automaton figures which should perform the functions of animals by means of wheels and pinions. Thus, General Degennes, who invented machines in navigation and gunnery, and constructed clocks without springs or weights, made a peacock, which could walk about as if alive, pick up grains of corn from the ground, eat and digest them.

This automaton is thought to have suggested to M. Vaucanson the idea of constructing his celebrated Duck, perhaps the most wonderful piece of mechanism ever made. It resembled a living duck in size and appearance, ate and drank with avidity, performed the quick motions of the head and throat peculiar to the living animal, and, like it, dabbled in the water, which it drank with its bill. It produced also the sound of quacking. In its anatomical structure every bone of the real duck had its representative, and exhibited its proper movement, as its wings were anatomically correct; and the automaton picked up corn, swallowed it, and, being digested by a chemical solution, the food was conveyed away by tubes. This famous automaton was repaired by Robert Houdin, the Parisian conjuror, who, on examining the mechanism of the Duck, found the trick to be as simple as it was interesting. "A vase," he tells us, "containing seed steeped in water, was placed before the bird. The motion of the bill in dabbling crushed the food, and facilitated its introduction into a pipe placed beneath the lower bill. The water and seed thus swallowed fell into a box placed under the bird's stomach, which was emptied every three or four days. The other part of the operation was thus effected: bread-crumb, colored green, was expelled by a forcing pump, and carefully caught on a silver salver as the result of artificial digestion" (Houdin's *Memoirs*, 1859).

Vaucanson's Automata were imitated by one Du Moulin, a silversmith, who traveled through Germany in 1752. Beckmann saw several of these automata, and among them an Artificial Duck, which was able to drink and move: its ribs were made of wire, and covered with duck's feathers, and the motion was communicated through the feet of the duck by means of a cylinder and fine chains, as in a watch.

Vaucanson also constructed a Flute-player, which really played on the flute by projecting air with its lips against the embouchure, producing the different octaves by expanding and contracting their opening; forcing more or less air in the manner of living performers, and regulating the tones by its fingers.

Of these automata, or rather *androides*, the Flute-player of Vaucanson is the only one of which a correct description has been preserved, a particular account of its mechanism having been published in the Memoirs of the French Academy. The figure was about five feet six inches high, and was placed upon an elevated square pedestal. The air entered the body by three separate pipes, into which it was conveyed by nine pairs of bellows, which expanded and contracted in regular succession by means of an axis of steel turned by the machine. The three tubes, which conveyed the air from the bellows, after passing through the lower extremities of the figure, united at the chest, and, ascending from thence to the mouth, passed through two artificial lips. Within the cavity of the mouth was a small movable tongue, which, by its motion at proper intervals, admitted or intercepted the air in its passage to the flute. The fingers, lips, and tongue derived their specific movements from a steel cylinder turned by clockwork. The cylinder was divided into fifteen equal parts, which, by means of pegs pressing upon a like number of levers, caused the other extremities to ascend. Seven of these levers directed the fingers, having rods and chains fixed to their ascending extremities; which, being attached to the fingers, made them ascend in proportion as the other extremity was pressed down by the motion of the cylinders, and *vice versâ*. Three of the levers served to regulate the ingress of the air, being so contrived as to open and shut, by means of valves, the communication between the lips and reservoir, so that more or less strength might be given, and a higher or lower note produced, as occasion required. The lips were directed by four similar levers, one of which opened to give the air a freer passage, another contracted them, a third drew them backward, and the fourth pushed them forward. The remaining lever was employed in the direction of the tongue, which, by its motion, shut or opened the mouth of the flute. The varied and successive motions performed by this ingenious androides were regulated by a contrivance no less simple than efficacious. The axis of the steel cylinder or barrel was terminated by an endless screw composed of twelve threads, above which was placed a small arm of copper, with a steel stud made to fit the threads of the worm, which, by its vertical motion, was continually pushed forward. Hence, if a lever were moved by a peg placed on the cylinder in any one revolution, it could not be moved by the same peg in the succeeding revolution, in consequence of the lateral motion communicated by the worm. By this means the size of the barrel was considerably reduced; and the statue not only poured forth a varied selection of instrumental harmony, but exhibited all the evolutions of the most graceful performer.

It is curious to find that Vaucanson's uncle reproached him by telling him that to construct the Flute-player would be a great waste of time, and he did not set about the work until he lacked employment to while away the time after a long illness. He also made a Flageolet-player, who beat a tambourine with one hand. The flageolet had only three holes, by half stopping which some notes were made: the force of wind required to produce the lowest note was one ounce; the highest, 56 lbs. French.

Jacques Vaucanson, the maker of these Automata, was a native of Grenoble, born in 1709. His mother took him one day to a fête, when, peeping through a crack in the partition of a room, he saw part of the works of a clock which hung against the wall; he was much struck, and, on his next visit, he drew with a pencil as much as he could see of the clock-springs and the escapement; and by aid of some poor tools, he soon made a clock. Then he made a sort of baby-house chapel, with figures, which he caused to move. At length he devoted all his time to studying anatomy, music, and mechanics. He grew to be so celebrated, that the King of Prussia tried to attach Vaucanson to his court: he, however, remained in France, where Cardinal Fleury made him inspector of silk manufactures, for which he greatly improved the machinery. This rendered Vaucanson unpopular, and he was nearly killed by an incensed mob. He died in 1782, having bequeathed his curious collection of machines to Louis XVI.

Next deserve to be mentioned the Writing Boy of the older, and the Piano-forte-player of the younger Droz; which latter, when performing, followed its hands with its eyes, and at the conclusion of the piece bowed courteously to the audience. Droz's Writing Boy was publicly exhibited in Germany some years ago. Its wheel-work is so complicated that no ordinary head would be sufficient to decipher its manner of action; when, however, we learn that the Boy and its constructor, being suspected of the black art, lay for a time in the Spanish Inquisition, and with difficulty obtained their freedom, we may infer that in those days even the mystery of such a toy was great enough to excite doubts as to its natural origin.

M. Maillardet next constructed an Automaton Boy, which both wrote and drew with a pencil, kneeling on one knee. When the figure began to work, an attendant dipped the pencil in ink, and adjusted the drawing-paper upon a brass tablet. Upon touching a spring, the figure proceeded to write or to execute landscape drawings. Maillardet also constructed a Magician, who answered questions inscribed in oval medallions upon a wall; one of which the spectator having selected, it was shut up in a spring drawer. The magician then rose, consulted his book, and striking a wall with his wand, two folding doors flew open, and displayed the answer to the question. The door again closed, and the drawer opened to return the medallion. The machinery being wound up, the movements in about an hour answered fifty questions; and the means by which the medallions acted upon the machinery, so as to produce the proper answers to the questions which they contained, is stated to have been very simple. Maillardet likewise constructed other automata, including a Spider, made of steel; and a Caterpillar, Lizard, Mouse, and Serpent, all with their natural movements. In London he exhibited in Spring Gardens.

Musical automata have obtained great celebrity. Maelzel, the inventor of the Metronome, exhibited in 1809 an automaton trumpeter of his construction. From a tent he led out a figure in the uniform of a trumpeter of the Austrian dragoon regiment Albert, his trumpet being at his mouth. Having pressed the figure on the left shoulder, it played the Austrian cavalry march, the signals, and a march and allegro by Weigl, accompanied by the whole orchestra. The dress of the figure was then changed into that of a French trumpeter of the Guard, when it played the French cavalry march, all the signals, a march of Dussek's, and an allegro of Pleyel, all accompanied by the orchestra. The sound of the trumpet was pure, and more agreeable than that which the ablest musician could produce from that instrument, because the breath of man gives the inside of the trumpet a moisture which is prejudicial to the purity of the tone. Maelzel publicly wound up his instrument only twice, and this was on the left hip. His most famous work was his *Panharmonica*, a band of forty-two wind-instrument players, for which Cherubini deigned to compose, and Beethoven wrote his Battle symphony. Maelzel died in

1855. Marreppe, in 1837, produced his automaton violin-player at Paris, which played airs *à la Paganini;* the interior was filled with small cranks, by which the motions were given to the several parts of the automaton at the conductor's will.

In the *speaking machines* of antiquity, the head of Orpheus in the island of Lesbos, and the tripod at Delphi, the answers were probably conveyed by the priests; and Charles II. and his court were similarly deceived by a Popish priest in an adjoining chamber answering through a pipe the question proposed to the wooden head by whispering in its ear.

The principle of a speaking-machine has, however, been developed. Bishop Wilkins, in his *Mathematical Magic,* illustrating the mode by which articulate sounds may be produced from automata, says: " Walchius thinks it possible entirely to preserve the voice, or any words spoken in a hollow trunk or pipe; and that, this pipe being rightly opened, the words will come out of it in the same order wherein they were spoken, somewhat like that cold country where the people's discourse doth freeze in the air all winter, and may be heard in the next summer or at a great thaw; but this conjecture will need no refutation."

Van Helmont, one of the first persons who wrote upon the adaptation of the organs of the voice to the articulation of the letters, considered that the letters of the alphabet constituted the order in which articulate sounds were naturally produced by the structure of the tongue and larynx; that, when one letter was uttered, the tongue was in its proper position for the pronunciation of the subsequent one. Thus, as several different sounds are formed merely by raising or depressing the tongue slightly, as in the sounds *Aw, Ah, Ae, A, E,* it was easy to produce them by means of a tube with a reed, and terminating with a bell. Mr. Willis has effected this by using a long tube with a reed, capable of being lengthened or shortened at pleasure. In the pronunciation of the vowels, *i, e, a, o, u,* it required to be shortest with the first, and in uttering the subsequent letters to be gradually lengthened. In this way it was easy to measure the length necessary for each note. When *Ae* was pronounced, the tube was 1 inch long; *Aw*, 3·8 inches; *Ah*, 2·2 inches; *A*, 0·6 inch; *E*, 0·3 inch. A speaking machine invented in Germany pronounced distinctly *mamma, papa, mother, father, summer*. The instrument consists of a pair of bellows, to which is adapted a tube terminating in a bell, the aperture of which is regulated by the hand, so as to produce the articulate sounds. This machine was exhibited at the Royal Institution in 1835 by Professor Wheatstone.

De Kempelen, the inventor of the automaton chess-player, also constructed a speaking automaton, in which he ultimately succeeded so far as to make it pronounce several sentences, among the best of which were, " Romanorum imperator semper Augustus;" " Leopoldus secundus;" " Vous êtes mon ami;" " Je vous aime de tout mon cœur." It was some years, however, before he could accomplish more than the simple utterance of the sounds *o, ou*, and *e*. Year after year, we are told, was devoted to this machine; but *i* or *u*, or any of the consonants, refused to obey his summons. At length he added at the open extremity of the vocal tube an apparatus similar in action and construction to the *human mouth with its teeth*, when he quickly succeeded in making it not only pronounce the consonants, but words, and even the sentences quoted above. He had previously imitated the tongue and its actions. The fact is interesting, not only as a rare instance of human ingenuity, but also as exhibiting in a most striking light the beautiful adaptation of parts to their respective functions; and that so perfect are the contrivances in Nature for particular ends, that, in order to arrive at any thing like an imitation of those functions, we must follow closely the method she employs.

In 1843 there was exhibited before the American Philosophical Society a speaking machine, susceptible of various movements by means of keys, and thus made to enunciate various letters and words; in enunciating the simple sounds could be seen the movements of the mouth, the parts of which were made of caoutchouc. The inventor, Mr. Reale, in a phrensy, destroyed this instrument, which it had taken him sixteen years to construct.

Three years later, in 1846, there was shown at the

D 2

Egyptian Hall, Piccadilly, the *Euphonia* of Professor Faber, of Vienna, the result of twenty-five years' labor. It consisted of a draped bust and waxen-faced figure, in which the sounds were produced by striking on sixteen keys, and thus were enunciated words. A small pair of bellows was worked with the nozzle into the back part of the head, and the mouth formations were of caoutchouc.

Now, the several attempts of Cagniard la Tour, Biot, Müller, and Steinle to produce articulate sounds, or even to imitate the human voice, have not been very successful; but M. Faber's machine—with its bellows worked by a pedal, and its caoutchouc imitation of the larynx, tongue, nostrils, and a set of keys by which the springs are brought into action—is considered the nearest approach to perfect success.

Reviewing the results of the Automata of the last century, Professor Helmholtz observes: "This inventive genius was boldly chosen, and was followed up with an expenditure of sagacity which has contributed not a little to enrich the mechanical experience which a later age knew how to take advantage of. We no longer seek to build machines which shall fulfill the thousand services required of *one* man, but strive, on the contrary, that a machine shall perform *one* service, but shall occupy, in doing it, the place of a thousand men."

Nevertheless, the above passion for automatic exhibitions introduced among the higher order of artists habits of nice and accurate execution in the formation of the most delicate pieces of machinery; and the same combination of the mechanical powers which in one century enriched only the conjuror who used them, is in another employed in extending the power and promoting the civilization of our species.

Robert Houdin is one of the latest adepts in automatic art. He was born at Blois, the son of a watchmaker, and had such early mechanical tastes that he professes to have come into the world metaphorically, "with a file or hammer in his hand." His aptitude showed itself in early efforts to train mice and canary-birds, to construct ingenious toys and model apparatus; and he perfected himself at Paris as a mechanist. In 1844 he made himself widely known by exhibiting an Automaton Writer, which attracted the notice of Louis Philippe and his family. The figure drew, as well as wrote answers to questions, and by a curious coincidence its performance on this occasion was particularly ominous. When the Comte de Paris requested it to draw a crown, the Automaton began drawing the outline demanded, but its pencil broke, and the crown could not be finished. Houdin was going to recommence the experiment, when the king, declined, with thanks. "As you have learned to draw," he said to the Comte de Paris, "you can finish this for yourself." This incident is characteristic as regards the tact of the king.

Houdin, in his *Memoirs*,* relates the following remarkable proof of his assiduity in this mechanical phase of his life. He had received an order from a merchant of St. Petersburg to construct an Automaton Nightingale, and he agreed for a large sum to make a perfect imitation of the above bird. This undertaking offered some serious difficulties; for, he tells us, though he had already made several birds, their singing was quite arbitrary, and he had only consulted his own taste in arranging it. The imitation of the nightingale's pipe was much more delicate, for he had to copy notes and sounds which were almost inimitable. Fortunately, it was the season for this skillful songster, and Houdin resolved to employ him as his teacher. He went constantly to the wood of Romainville, the skirt of which almost joined the street in which he lived; and, laying himself on a soft bed of moss in the densest foliage, he challenged his master to give him lessons. (The nightingale sings both by night and day, and the slightest whistle, in tune or not, makes him strike up directly.) Houdin wanted to imprint on his memory the musical phrases with which the bird composes its melodies. The following are the most striking among them: *Tiou-tiou-tiou, ut-ut-ut-ut-ut, tchitchou, tchitchou, tchit-tchit, rrrrrrrrrrrrrouit*, etc. Houdin had to analyze these strange sounds—these numberless chirps, these impossible "rrrrrouits," and recompose them by a musical process. To imitate this flexibility of throat, and reproduce the harmonious modulations, Houdin made

* *Memoirs of Robert Houdin, Embassador, Author, and Conjuror.* Written by himself. 1859.

a small copper tube, about the size and length of a quill, in which a steel piston, moving very freely, produced the different sounds required; this tube represented in some respects the nightingale's throat. This instrument had to work mechanically: clockwork set in motion the bellows, opened or closed a valve which produced the twittering, the modulation, and the sliding notes, while it guided the piston according to the different degrees of speed and depth wanted. Houdin had also to impart motion to the bird: it must move its beak in accordance with the sounds it produced, flap its wings, and leap from branch to branch, which, however, was purely a mechanical labor.

After repeated experiments, Houdin succeeded in creating a system half musical, half mechanical, which only required to be improved by fresh studies from nature. Provided with this instrument, Houdin hurried off to the wood of Romainville, where, seating himself under an oak, near which he had often heard a nightingale sing, he wound up the clockwork, and it began playing in the midst of profound silence; but the last notes had scarcely died away ere a concert commenced from various parts of the wood. This collective lesson did not suit his purpose, for he wished to compare and study, and could positively distinguish nothing. Fortunately for Houdin, all the musicians ceased, and one of them began a solo of dulcet sounds and accents, which Houdin most attentively followed, thus passing a portion of the night, when the conjuror returned home. His lesson had done him so much good, that the next morning he began making important corrections in his mechanism; and after five or six more visits to the wood, Houdin attained the required result—the nightingale's song was perfectly imitated.

In the Great Exhibition of 1851 was shown a mechanical curiosity—an expanding Model of a Man, the construction of which has a romantic interest. It was the invention of the Polish Count Dunin, who in early life became involved in the insurrection of his countrymen, and was banished. In his dreary exile he betook himself to mechanical pursuits, that he might expiate his offense, real or imaginary, against the Emperor of Russia,

by showing that he might be useful if he were restored
to his country.

The model represents a man 5 feet high in the proportions of the
Apollo Belvidere; from that size it can be proportionally increased to
6 feet 8 inches; and as it is intended to measure the clothing of an
army, it is capable of expansion and contraction in all its parts. The
internal mechanism is completely concealed, the figure externally
being composed of thin slips of steel and copper, by the overlapping
of which expansion or contraction is exercised, the motion being communicated by thin metal slides within the figure, these slides having
pins worked in curved grooves in circular steel plates, which are put
in revolution by a train of wheels or screws. A winding-key, turned
right or left, effects the expansion or contraction noiselessly, and in
the direction of the fibres of the muscles in the living subject. The
mechanical combinations are composed of 857 framing-pieces, 48
grooved steel plates, 163 wheels, 203 slides, 476 metal washers, 488
spiral springs, 704 sliding plates, 497 nuts, 3500 fixing and adjusting
screws, with numerous steadying pins, so that the number of pieces
is upward of 7000. For this beautiful piece of mechanism a Great-
Exhibition Council Medal was awarded to Count Dunin.

THE AUTOMATON CHESS-PLAYER.

WE have reserved for a separate chapter the origin and history of this marvelous contrivance, which, at various periods during the lapse of ninety years, has astonished and delighted the scientific world in several cities of Europe and North America. Its machinery has been variously explained. It was constructed in 1769 by M. de Kempelen, a gentleman of Presburg, in Hungary, long distinguished for his skill in mechanics. The Chess-player is a life-sized figure, clothed in a Turkish dress, sitting behind a large chest, three and a half feet long, two feet deep, and two and a half feet high. The player sits on a chair fixed to the chest, his right arm rests on the table or upper surface of the chest, and in the left he holds a pipe, which is removed during the game, as it is with this hand that he makes the moves. A chess-board, with the pieces, is placed before the figure. The exhibitor first opens the doors of the chest, and shows the interior, with its cylinders, levers, wheels, pinions, and other pieces of machinery, which *have the appearance* of occupying the whole space. This machinery being wound up, the Automaton is ready to play; and when an opponent has been found, the figure takes the first move, moves its head, and seems to look over every part of the chess-board. When it gives check to its opponent it shakes its head *thrice*, and only *twice* when it checks the queen. It likewise shakes its head when a false move is made, replaces the adversary's piece on the square from which it was taken, and takes the next move itself. In general, though not always, the Automaton wins the game. During its progress, the exhibitor often stood near the machine, and wound it up like a clock after it had made ten or twelve moves. At other times he went to a corner of the room, as if it were to consult a small square box, which stood open for this purpose.

The earliest English account of the Automaton Chess-

player that we can find is in a letter from the Rev. Mr.
Dutens to the *Gentleman's Magazine*, dated Presburg,
January 24, 1771. The writer formed an acquaintance
with the inventor, whom he terms M. de Kempett (not
Kempelen), an Aulic counselor, and director general of
the salt mines in Hungary. Mr. Dutens played a game
at chess with the Automaton at Presburg; the English
embassador, Prince Giustiniani, and several English
lords, standing round the table.

"They all," according to Mr. Dutens, "had their eyes on M. de
Kempett, who stood by the table, or sometimes removed five or six
feet from it, yet not one of them could discover the least motion in
him that could influence the Automaton. . . . . He also withdraws
to any distance you please, and lets the figure play four or five moves
successively without approaching it. The marvelous in this Automaton consists chiefly in this, that it has not (as in others, the most celebrated machines of this sort) one determined series of movements,
but that it always moves in consequence of the manner in which its
opponent moves, which produces an amazing multitude of different
combinations in its movements. M. de Kempett winds up from time
to time the springs of the arms of this automaton, in order to renew
its *motive force;* but this, you will observe, has no relation to its *guiding force* or power of direction, which makes the great merit of this
machine. In general, I am of opinion that the contriver influences
the direction of almost every stroke played by the Automaton, although, as I have said, I have sometimes seen him leave it to itself
for many moves together, which, in my opinion, is the most difficult
circumstance of all to comprehend in what regards this machine."

Mr. Staunton, the celebrated chess-player, states that
De Kempelen constructed the Automaton "merely to afford a passing amusement to the Empress Maria Teresa
and her court." Upon its completion, it was exhibited
at Presburg and Vienna; in 1783, in Paris; and in that
and the following year in London and different parts of
England, without the secret of its movements having
been discovered. "It was subsequently," says Mr. Staunton, "taken, by special invitation of the emperor, to the
court of Frederick the Great at Berlin. This prince was
devotedly attached to chess; and in a moment of liberality, he proffered an enormous sum for the purchase of the
Automaton and its secret. The offer was accepted, and
in a private interview with De Kempelen, he was furnished with a key to the mystery. In a short time, however, Frederick threw aside the novelty so dearly bought,

and for many years it lay forgotten and neglected among the lumber of his palace.

"M. Kempelen died in 1804; but in two years after, when Napoleon I. occupied Berlin, we find the Chess Automaton in the field again under a new master. On one occasion of its exhibition at this period, Napoleon himself is said to have entered the lists. After some half dozen moves, he purposely made a false move; the figure inclined its head, replaced the piece, and made a sign for Napoleon to play again. Presently he again played falsely: this time the Automaton removed the offending piece from the board, and played its own move. Napoleon was delighted; and, to put the patience of his taciturn opponent to a severer test, he once more played incorrectly, upon which the Automaton raised its arm, and, sweeping the pieces from the board, declined to continue the game."

After a second tour of the leading cities of Europe, where it was received with unabated enthusiasm, in 1819 the Automaton was again established in London, under M. Maelzel. For some years it was exhibited in Canada and the United States, and was finally understood to have returned to New York, where it was shown in the autumn of 1845.

Meanwhile there were various attempts made to discover the secret. The ingenious inventor never pretended that the Automaton itself really played the game: on the contrary, he distinctly stated that "the machine was a *bagatelle*, which was not without merit in point of mechanism, but that the effects of it appeared so marvelous only from the boldness of the conception, and the fortunate choice of the methods adopted for promoting the illusion." It was surmised that the game was played either by a person inclosed in the chest, or by the exhibitor himself; yet the chest, being nearly filled with machinery, did not appear capable of accommodating even a dwarf; nor could any mechanical communication between the exhibitor and the figure be detected. It was then thought to be influenced by a magnet, which the exhibitor disproved by placing a strong and well-armed loadstone upon the machine during the game, which did not affect the moving power. The original conjecture, that the player was concealed in the interior, was then .revived; and in 1789, Mr. J. F. Freyhere, of Dresden, published a pamphlet, in which he endeavored to explain by colored plates how the effect was produced; and he concluded "that a well-taught boy, very thin and tall of his age (sufficiently so that he could be concealed in a drawer almost immediately under the chess-board), agitated the whole."

In an earlier pamphlet, published in Paris in 1785, the writer supposed the machine was put in motion by a dwarf, a famous chess-player, his legs and thighs being concealed in two hollow cylinders, while the rest of his body was out of the box, and hidden by the robes of the figure.

Sir David Brewster, in his *Natural Magic*, describes the secret as shown in a pamphlet published anonymously, and the machine to be capable of accommodating an ordinarily-sized man; and he explains, in the clearest manner, how "the inclosed player takes all the different positions, and performs all the motions which are necessary to produce the effects actually observed." Sir David devotes eight pages of his work, with illustrative wood-cuts, to this explanation, and endeavors to show how the real player may be concealed in the chest, and partly in the figure: "as his head is above the chessboard, he will see through the waistcoat of the figure, as easily as through a veil, the whole of the pieces on the board; and he can easily take up and put down a chessman without any other mechanism than that of a string communicating with the finger of the left hand of the figure," the right hand being within the chest, to keep in motion the wheel-work for producing the noise heard during the moves, and to move the head, tap the chest, etc.

Mr. Staunton also maintains that the chess-player who directed the Automaton was really hidden in the interior; that the machinery so ostentatiously exhibited was a sham, yet so contrived that it would collapse or expand, to suit the exigencies of the hidden agent's various positions; while the chest was exhibited, he was in the figure, and when the figure, he was in the chest. While conducting a game, he sat at the bottom of the chest, with a small pegged chess-board and men on his lap, and a lighted taper affixed; within reach were a handle by which he could guide the arm of the Automaton, an elastic spring for moving its fingers, and cord in communication with bellows for producing the sound of " Check." The most ingenious part of the contrivance remains to be told. M. Mouret, the celebrated chess-player, who directed the movements of the Automaton for some years, states that the concealed player was seated immediately under the chess-board of the Automaton, and from the under side, at every one of the sixty-four squares, was suspended by the finest silk a tiny metallic ball; and as each of the chess-men had a magnet inside, when it was placed upon a square, it drew up the ball beneath, while the balls beneath the other squares remained suspended. The pieces being arranged, the Automaton opened the game; and turning the handle of the arm of the figure, and putting in motion the finger-springs, he caused it to take up the piece to be played, which was indicated by the falling ball, and when it was placed upon a square, the ball was drawn up. He then repeated the move on the small board in his lap, and thus the game proceeded.*

Thus the explanation rested until the publication of the Memoirs of Robert Houdin, who therein relates the origin and construction of the Automaton Chess-player in substance as follows:

In 1769 there fell, fighting in a revolt at Riga, an officer named Worousky, a man of great talent and energy, of short stature, but well built. He had both thighs shattered by a cannon ball, but escaped by throwing himself into a hedge behind a ditch. At nightfall Worousky dragged himself along, with great difficulty, to the adjacent house of Osloff, a physician, whose benevolence was well known; and the doctor, moved by his sufferings, attended upon and promised to conceal him. His wound was serious, gangrene set in, and his life could only be saved at the cost of half his body. The amputation was successful, and Worousky saved.

Meanwhile, M. de Kempelen, the celebrated mechanician, came to Riga to visit M. Osloff, who confided to him his secret of concealing Worousky, and begged his aid. Though startled at the request—for he knew that a reward was offered for the insurgent chief, and that the act of humanity he was about to assist in might send him to Siberia—still, M. de Kempelen, on seeing Worousky's mutilated body, felt moved with compassion, and began contriving some plan to secure his escape.

---

* Selected and abridged from the *Illustrated London News*, Dec. 23, 1845.

Dr. Osloff was a passionate lover of chess, and had played numerous games with his patient during his tardy convalescence; but Worousky was so strong at the game that the doctor was always defeated. Then Kempelen joined the doctor in trying to defeat the skillful player, but it was of no use; Worousky was always the conqueror. His superiority gave M. de Kempelen the idea of the famous Automaton Chess-player. In an instant his plan was formed, and he set to work immediately; and the most remarkable circumstance is, that this wonderful *chef-d'œuvre*, which astonished the whole world, was finished within three months.

M. de Kempelen was anxious that his host should make the first trial of his Automaton; so he invited him to play a game on the 10th of October, 1769. The Automaton represented a Turk of the natural size, wearing the national costume, and seated behind a box of the shape of a chest of drawers. In the middle of the top of the box was a chess-board, with the pieces, for play.

Prior to commencing the game, the artist opened several doors in the chest, and M. Osloff could see inside a number of wheels, pulleys, cylinders, springs, etc., occupying the larger part. At the same time he opened a long drawer, from which he produced the chess-men and a cushion, on which the Turk was to rest his arm. This examination ended, the robe of the Automaton was raised, and the interior of the body could also be inspected.

The doors being then closed, M. de Kempelen wound up one of the wheels with a key which he inserted in a hole in the chest; after which the Turk, with a gentle nod of salutation, placed his hand on one of the pieces, raised it, deposited it on another square, and laid his arm on the cushion before him. The inventor had stated that, as the Automaton could not speak, it would signify check to the king by three nods, and to the queen by two.

The doctor moved in his turn, and waited patiently till his adversary, whose movements had all the dignity of the Sultan, had moved. The game, though slow at first, soon grew animated, and the doctor found he had to deal with a tremendous opponent; for, in spite of all his efforts to defeat the figure, his game was growing quite desperate. It is true, though, that for some minutes past the doctor's attention had appeared to be distracted, and one idea seemed to occupy him. But, while hesitating whether he should impart his thoughts to his friend, the figure gave three nods. The game was over.

"By Jove!" the loser said, with a tinge of vexation, which the sight of the inventor's smiling face soon dispelled, "if I were not certain that Worousky is at this moment in bed, I should believe I had been playing with him. His head alone is capable of inventing such a checkmate. And besides," the doctor said, looking fixedly at M. de Kempelen, "can you tell me why your Automaton plays with the left hand, just like Worousky?" (The Automaton Chess-player always used the left hand—a defect falsely attributed to the carelessness of the constructor.)

The mechanician began laughing, and at length confessed to his friend that he had really been playing with Worousky.

"But where the deuce have you put him, then?" the doctor said, looking round to try and discover his opponent.

The inventor laughed heartily.

"Well, do you not recognize me?" the Turk exclaimed, holding out his left hand to the doctor in reconciliation, while Kempelen raised the robe and displayed the poor cripple stowed away in the body of the Automaton.

M. Osloff could no longer keep his countenance, and he joined the others in the laughter. But he was the first to stop, for he wanted an explanation.

"But how do you manage to render Worousky invisible?"

M. de Kempelen then explained how he concealed the living automaton before it entered the Turk's body.

"See here," he said, opening the chest; "these wheels, pulleys, and cranks, occupying a portion of the chest, are only a deception. The frames that support them are hung on hinges, and can be turned back to leave space for the player, while you were examining the body of the Automaton.

"When this inspection was ended, and as soon as the robe was allowed to fall, Worousky entered the Turk's body we have just examined, and, while I was showing you the box and the machinery, he was taking his time to pass his arms and hands into those of the figure. You can understand that, owing to the size of the neck, which is hidden by the broad and enormous collar, he can easily pass his head into this mask, and see the chess-board. I must add, that when I pretend to wind up the machine, it is only to drown the sound of Worousky's movements."

M. Houdin relates that the mutilated Pole once had the audacity, in his clockwork case, to visit St. Petersburg, and play a game of chess with the Empress Catharine, against whom he had revolted.

It is hard to reconcile these conflicting statements, unless, having allowed Houdin's account of the origin of the Automaton to be correct, we consider the other narratives to explain the modes by which the Automaton was worked after Worousky had ceased to be the prime mover of this extraordinary deception.

Substitutes for the natural limbs have been constructed with great success. In 1845, Magendie described to the French Academy a pair of artificial arms, the invention of M. Van Petersen, with one of which a mutilated soldier raised a full glass to his mouth, and drank its contents without spilling a drop of the liquor; he also picked up a pin, a sheet of paper, etc. Each arm and hand, with its articulations, weighs less than a pound; and a sort of stays is fixed round the person, and from these are cords made of catgut, which act upon the articulations, according to the motion given to the natural stump.

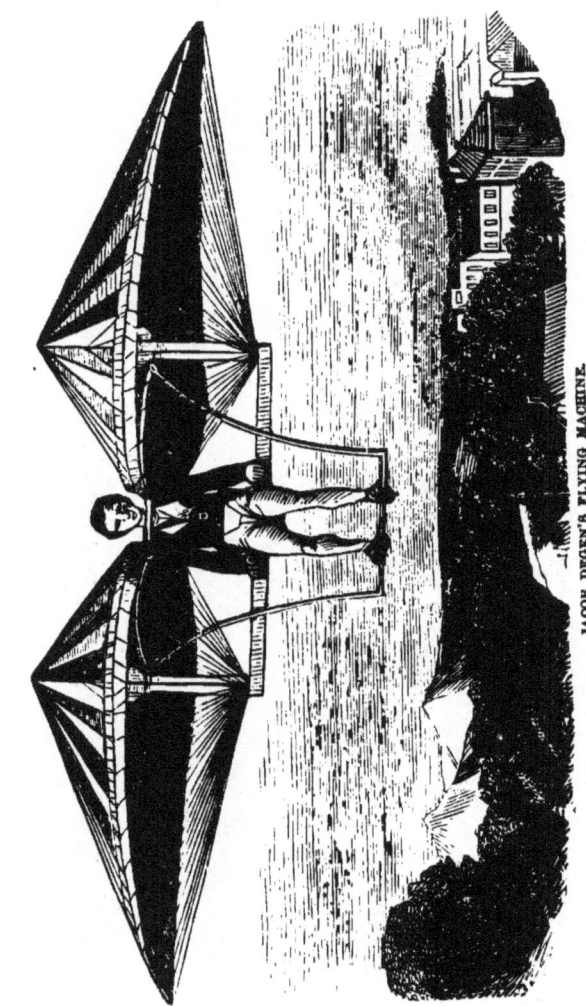

JACOB DROEN'S FLYING MACHINE.

NAVIGATION OF THE AIR: ADVENTURES WITH THE BALLOON.

The idea of constructing a machine which should enable us to rise into, and sail through the air (hence the term *aeronaut*), would seem to have occupied the human mind even in ancient times; but it was never realized until the beginning of the present, or the close of the last century.

The notion of imitating the flying of birds is very ancient. Passing over the winged gods, the stories of Abaris, Dædalus, and the like, which, with many others, might have been purely imaginative, and not traditions of any previous reality, we come to Strabo's account of the Capriobalæ, a Scythian people, who (so the word has been foolishly interpreted) raised themselves by smoke, or, more properly, heated air. The Carolinians are also mentioned by the Jesuit Cantova as having a fable about a female deity who raised herself to heaven by the smoke of a great fire. We may likewise mention the wooden pigeon of Archytas, which had air inclosed in it, and which Lucian professes to have seen raise itself in the air; the fable in British mythology of Bladud, the father of the well-known Lear, which resembles that of Dædalus; and many others, all of which serve to show that the notion of the possibility of raising a man or a machine was very widely extended in the ancient world. Roger Bacon says that there certainly is *a flying machine*, of which he knows the name of the inventor, but which he has neither seen himself, nor does he know any one who has. Van Helmont and others proved the possibility of flying by very eloquent discourses, which convinced all hearers— but not their posterity. Sometimes, however, the evidence of these ancient wonders is strangely shaken by historical fact; as in the case of Regiomontanus's wooden eagle, which flew out of Nuremberg to meet Charles V.; for, although this statement is testified by Sextus of

Ratisbon, Kircher, Porta, Schott, Gassendi, Lana, Ramus, and Bishop Wilkins, they have overlooked the fact that Regiomontanus died twenty-five years before Charles V. was born!

The Jesuit Francis Lana (A.D. 1670), among many other projects, has given, perhaps, the earliest idea of a real Balloon, as we have defined it, and his first step was purely theoretic. He proposes to raise a vessel by means of metal balls, strong enough, when expanded, to resist the pressure of the external air, but at the same time so thin as, in the same circumstances, to be lighter than their bulk of air. Had the good father made the experiment, he would have found that strength to resist the external air is incompatible with the necessary degree of thinness in the material. Still, there was one avenue to the object of pursuit, to which the common and well-known principles of hydrostatics appeared to direct the way, though it had been of all others the most neglected: this was the obvious one that any body which is specifically, or bulk for bulk, lighter than common air, will rise and swim in it, and submit to the action of the wind; therefore, if any body could be found which was in any considerable degree lighter than air, by making it of a sufficient size, a person might attach himself to it, and float along with it. Another century, however, elapsed before this was accomplished.

Bishop Wilkins (who lived from 1614 to 1672) was an early disciple of this art. In his *Discovery of a New World, or That the World may be a Moon*, one of his propositions is, "That 'tis possible for some of our posterity to find out a conveyance to this other world," which can not seem more incredible to us than did the invention of ships:

> "So bold was he, who in a ship so frail
> First ventured on the treacherous waves to sail."

The bishop agrees with Kepler, that whenever the art is invented by which a man may be conveyed some twenty miles high, or thereabouts, it is not altogether improbable that some other art may enable him to fly to the moon; and supposing that he could fly as fast and as long as the swiftest bird, were he to keep on in a straight line, and fly 1000 miles a day, he would not arrive at the moon under 180 days, or half a year. As for the means of flying, Wilkins points to angels pictured with wings, which Mercury and Dædalus are feigned to have had; that if there be, as Marco Polo says, a roc in Madagascar "which can scoop up a horse and his rider, or an elephant," then a man may ride up to the moon, as Ganymede does upon an eagle. Or the bishop affirms it possible to make a Flying Chariot large enough to carry up several men, with their food and luggage, on the same principle by which Archytas made his wooden dove, and Regiomon-tanus his wooden eagle. The bishop also devotes a chapter of his *Mathematical Magic* to solving the difficulties that seem to oppose the possibility of a Flying Chariot, and concludes with suggesting the wings of the bat as preferable to those of a bird, gravely adding that the bat's wings are most easily imitated, and perhaps Nature intended by them to direct us in such experiments. Wilkins was also an early advocate of the "pleasant uses" of *heated air;* in his *Mathematical Magic* he minutely describes the moving of sails in a chimney, as in the smoke-jack; and he adds that Ctesibius "made by this kind of motion his representations of living creatures, whether birds or beasts."

Leaving these phantasies, we reach some practical illustrations of the art. In 1709, Gusman, a Portuguese friar, constructed a machine in the form of a bird, with tubes and bellows to supply the wings with air; he was rewarded with a liberal pension, but his machine failed. Gusman, however, was not discouraged, for in 1736 he constructed a wicker basket 7 feet in diameter, and covered with paper, which rose to the height of 200 feet in the air, the success of which experiment procured for the inventor the reputation of being a sorcerer.

M. Laurent's Bird Machine.

As air was considered the lightest of all things, there appeared little reason to believe that the discovery of flying would be made, when in 1755 Joseph Gallien, of Avignon, in a treatise, recommended the employment of a bag of cloth or leather, filled with air lighter than that

E

of the atmosphere. Eleven years later, in 1766, this desideratum was supplied by Mr. Cavendish announcing to the world that the gas now known as hydrogen, but at that time called inflammable air, was at least seven times lighter than common air. This important discovery led Dr. Black to suggest in his lectures that if a bladder sufficiently light and thin were filled with this air, it would form a mass much lighter than the same bulk of atmospheric air, and that it would float in the latter. Dr. Black, however, did not pursue the subject farther; and it rested for nearly twenty years, until Cavallo, reflecting on Dr. Black's remarks, in 1782 made several experiments to elevate a bag filled with hydrogen gas: he tried the largest and thinnest bladders, but they were found somewhat too heavy for the purpose. He also tried bags of the finest China paper, of such a size that, had it been possible to fill them with the gas, their ascent would have been certain; but the experiments failed, for, though common air would not pass through this paper, hydrogen gas passed through it like water through a sieve. In short, Cavallo was completely successful only in filling soap-bubbles with the gas, which was easily done by pressing small quantities of hydrogen out of a bladder, while a small pipe was immersed in a solution of soap and water; these bubbles rapidly ascended in the ambient air, and they may be considered as the first inflammable air-balloons that were ever exhibited. Cavallo read to the Royal Society the paper in which he gave an account of his experiments, on the 20th of June, 1782.

Here it should be observed that, although the art of flying had been diligently studied, or at least discussed, for centuries, the exceedingly simple contrivance we shall presently describe had not been tried, or even mentioned, by any of the projectors, some of whom were men of ingenuity. Nothing can set in a stronger light the antipathy of the earlier moderns to experimental research. And it is no small honor to the Montgolfiers, that the hint given by Lana, together with the every-day experiment of soap-bubbles, and the like, should have remained without results to their time.

"We consider" (says an able writer) "him the inventor of the balloon who raised a mass of solid substance to some considerable height in the atmosphere. But if we were to take the license which is so frequent, of disputing the right of an inventor on account of some experiments containing a principle common with his own, we might say that this machine has been invented from time immemorial in the ascent of soap-bubbles; or we might cite Candido Buono, who made one scale of a balance ascend by rarefying with a red-hot iron the air beneath it." We have seen how Cavendish discovered the gas seven-fold lighter than air; how Black took up its application, but then halted; and how Cavallo followed, but could not succeed in raising, by means of hydrogen, any thing heavier than a soap-bubble. We shall next show that, natural as it might appear to use hydrogen for the purpose, the experiment succeeded only with a very different agent. From this point practical aerostation commences.

In the last-mentioned year, 1782, but unknown to the English philosophers, two brothers, Stephen and Joseph de Montgolfier, paper manufacturers at Annonay, about thirty-six miles from Lyons, formed a scheme which led in a short time to the practice of aerostation on a large scale. They had both studied natural philosophy and chemistry, and their business gave them facilities for procuring large masses of light envelopes, so that we owe the invention of balloons to one of two accidents, either to that of philosophers being paper-makers, or to that of paper-makers being philosophers. Stephen Montgolfier is said to have derived the first idea from the accidental circumstance of the paper cover of a conical sugar-loaf which he had flung into the fire becoming inflated with smoke, and remaining suspended in the chimney. Struck with the notion of confining something lighter than air in a recipient as the means of making the latter ascend, the Montgolfiers tried this method at about the same period as M. Cavallo, by confining hydrogen in paper. They succeeded to some extent; but the gas so soon escaped through the paper that they abandoned the idea of any thing like perpetual elevation by means of it. They next thought that, as it was supposed the elevation of the clouds was caused by the presence of electric matter, and as it seemed to them from experiments that electrified bodies were diminished in weight, it might be possible to raise a surface of great extent, in proportion to its specific gravity, by means of electricity. After trying various methods, they applied fire underneath a balloon, *not to rarefy the inclosed air*, but " as well to increase the layer of electric fluid upon the vapor in the vessel as to divide the vapor into smaller molecules, and dilate the gas in which they are suspended." Thus they thought they were imitating a cloud by electrifying the gases and vapors contained in the atmosphere.

The first experiment was made at Avignon by Stephen Montgolfier. He prepared a bag of silk in the shape of a parallelopipedon; its capacity was about forty cubic feet, and he applied to its aperture burning paper, and inflated the bag with a kind of cloud, when the bag ascended rapidly to the ceiling of the room. This was referred to the electric theory, as above; but in the report made to the Academy of Sciences (December, 1783) by the commission appointed to investigate Montgolfier's invention, the inventors are spoken of as simply rarefying the air contained in the balloon, when they had probably arrived at the correct view of the subject. Their first public experiment was made at Annonay, June 5, 1783. At the appointed time, nothing was seen in the public *place* of the town but immense folds of paper, 100 feet in circumference, and fixed to a nearly spherical wooden frame, the whole weighing about 500 lbs., and containing 22,000 cubic feet, French measure. It was suspended, in a flaccid state, on a pole thirty-five feet high : straw and chopped hay were burnt underneath the opening at the bottom, the heated air from which entered ; the mass gradually assumed the form of a large globe, and ascended with such velocity, that in less than ten minutes it reached the elevation of 6000 feet. A breeze carried it in a horizontal direction to the distance of 7668 feet, when it fell gently on the ground. Machines on this principle were called *Montgolfiers*, to distinguish them from the hydrogen balloons which were made immediately afterward.

The news of this phenomenon flew to Paris, where it immediately produced an excitement almost unheard of before. That hydrogen could not have been used was

The first Montgolfier.

·evident from the description given, namely, that it was half as heavy as air. On August 23 the experiment was resumed at Versailles with hydrogen inclosed in lutestring which had been dipped in a solution of India-rubber. The gas was obtained in the usual manner by the action of diluted sulphuric acid on iron filings; but the machine was not filled until August 26, when it was allowed to rise 100 feet, to which height it was confined by ropes. Next day the balloon was conveyed to the Champ de Mars, where it was set free in the presence of an enormous crowd. It fell five leagues from Paris, after being about a quarter of an hour in the air.

Meanwhile Joseph Montgolfier arrived in Paris, where he exhibited one of his balloons on the 12th of September; and on the 19th, in front of the Palace of Versailles. The interest attached to the mere ascent of the *balloon alone* here ceases. Various repetitions of the experiment were made at Paris previously to the time when men trusted themselves to this conveyance. The first aerial voyagers were a sheep, a cock, and a duck, who were sent up on September 19, and came down safe. Human life was not, however, trusted to a balloon till the experiment of *holding the machine with ropes* had been made. In this manner M. Pilatre de Rozier ascended 100 feet on the 5th of October, and 324 feet on the 19th, in a spheroidal balloon seventy-five feet high.

De Rozier's Balloon.

The first persons who offered to *leave the earth entire-ly* were the Marquis d'Arlandes and M. de Rozier, in a *Montgolfier*, from the Château de la Muette, near Passy, November 21, 1783. Their balloon, magnificently dec-orated, was terminated below by a circular gallery for the aeronauts; inside a grate was suspended within their reach, so that they could, during the voyage, feed the fire in it with straw, a supply of which they took with them. The sky was loaded with heavy clouds, driven about by irregular winds. After a first trial which had nearly proved fatal to the aeronauts, the balloon was again filled, and a provision of straw taken up to supply the fire. The machine first mounted with a steady and majestic pace to more than 3000 feet, and traversed the whole extent of Paris, intercepting the body of the sun, and giving to the gazers on the towers of Nôtre Dame, for a few seconds, the spectacle of a total eclipse. When the balloon had reached so high that the objects on earth were not distinguishable, the Marquis d'Arlandes was anxious to descend; but his companion still kept feeding the fire. At last, on hearing some cracks from the top of the balloon, and observing holes burning in the sides, the marquis became alarmed, and applying wet sponges to stop the progress of the burning, he compelled M. de Rozier to desist. As they now descended too fast, M. d'Arlandes threw fresh straw on the fire, in order to gain such elevation as would enable them to clear the lofty buildings; and after a journey of twenty or twenty-five minutes, they safely alighted beyond the Boulevards, having described a track of six miles, and the balloon being quite empty and flattened.

The next voyage—the first made in a hydrogen balloon—was that of MM. Charles and Robert at sunset, on December 1, 1783, from the Tuileries. After coming down, M. Charles reascended alone, and was soon nearly two miles high: he saw the sun rise again, and he says, "I was the only illuminated object, all the rest of nature being plunged in shadow." A small balloon, launched by Montgolfier just before the ascent, was found to have run a totally opposite course, which first gave rise to the suspicion of different directions in the currents of air at different heights.

The third voyage—from Lyons, January 19, 1784—was made in the largest *Montgolfier* yet constructed (102 feet diameter, 126 feet high), by seven persons, among whom were J. Montgolfier and M. de Rozier. A rent in the balloon caused it to descend with great velocity, but no one was hurt.

The mania for aerial voyaging soon passed from France to Italy; and the Chevalier Paul Andreani constructed a spherical balloon about sixty-eight feet diameter, made of linen, lined with fine paper, and provided with a *brazière*, or fireplace. It was inflated in fifteen minutes; and on February 25, 1784, the chevalier and two assistants ascended from Monsucco, near Milan, and after remaining up twenty minutes, returned there in safety.

Three days previous to the above, February 22, a small balloon, launched by itself from Sandwich, crossed the

Channel, and was found nine miles from Lisle, having traveled above thirty miles an hour.

On April 25, 1784, MM. de Morveau and Bertrand ascended in a balloon 1300 feet at Dijon, and experimented with effect with oars.

On May 20, 1784, M. Montgolfier, two other gentlemen, and four ladies, ascended, the balloon being confined by ropes.

In September, 1784, the Duke of Orleans (Louis Philippe), accompanied by Messrs. Robert, ascended in a balloon furnished with oars and rudder; to this a small balloon was attached, for the purpose of being inflated with bellows, and thus supplying the means of descent without waste of the hydrogen gas. At 1400 feet high they encountered a thunder-storm and whirlwind; by throwing out ballast, they rose to 6000 feet, when the heat of the sun caused so great an expansion of the gas that a rupture in the balloon was feared. The duke pierced the silk with his sword in several places, and thus let out the gas; and having narrowly escaped falling into a lake, the aeronauts descended unhurt, after an excursion of five hours.

Hitherto the several balloon experiments had been confined to the Continent, but now reached our country.

The first balloon experiment in England was made by Count Zambeccari. On November 25, 1783, a balloon of oiled silk, richly gilt, and filled with hydrogen gas, ascended from the Artillery Ground, Moorfields; it was found forty-eight miles from London, near Petworth. At the end of the same year, Mr. Sadler sent up a hydrogen balloon from Oxford. About the same time (December 25th), Mr. Boulton, the partner of James Watt, constructed a balloon, to which a match and serpent were attached, that the gas might explode in the air. The object was to ascertain whether the reverberation of thunder is caused by echo or by successive explosions; but the point remained unsettled, owing to the shouting of the people.

Count Zambeccari may be considered as among the most unfortunate of the early voyagers. In an ascent from Ancona, he was driven for many hours over the Adriatic Sea, until picked up by a bark; and in another journey, from near Bologna, in his descent, the car, by the upsetting of a lamp and spirit of wine, took fire, and burnt the clothes of the count and his companion; the balloon fell into the Adriatic, twenty-five miles distant from the Italian coast. The half-burnt car sank, but Zambeccari held fast by the ropes of the balloon, though immersed in the water to his neck. By means of a bit of glass he detached a rope from the bag, and with it fastened his body to the machine. In this situation he floated on the water for some hours, the balloon being still inflated. At length, in the evening, the count was taken up by some fishermen, who, in attempting to seize the balloon, cut the ropes, when it rose and took its course toward the Turkish coast.

In Scotland, the earliest attempts at aerostation emanated from a chemist at Edinburgh, Mr. Scott, who, on March 12, 1784, let off from Heriot's Gardens a balloon, which was taken up twenty miles from Edinburgh. About the same time, various balloons were let off from other places in Scotland; one launched from the Observatory of Aberdeen went thirty-eight miles in half an hour.

The first Ascension on Horseback.

E 2

By a singular coincidence, on the above day, Philip Astley, the celebrated horse-rider, and founder of the Amphitheatre, launched an aerostatic globe (or balloon) in St. George's Fields; it was afterward found at Faversham, forty-seven miles distant.

In the same year Mr. J. Tytler, another chemist at Edinburgh, had constructed a balloon on the Montgolfier principle, and exhibited it as "the Edinburgh Fire-balloon." It was 40 feet high, and 30 feet in diameter. On August 27 he ascended with this balloon 350 feet; but, as no furnace was taken up with it to maintain the supply of heated air, Tytler soon descended, with the triple fame of being the first native of *Great Britain who achieved an aerial ascent;* of having accomplished the first aerial voyage in these realms; and, with one exception, the only person who ascended in Great Britain by the agency of atmospheric air rarefied by artificial heat. Nevertheless, the merit of the first aerial voyage in Great Britain was long ascribed to Lunardi, whereas he did not ascend till September 15th following. From an admission ticket in the British Museum, Tytler's balloon appears to have been constructed by M. W. Brodie; and the engraving represents it in the shape of a cask, hooped, of varnished linen in eight pieces; the car, provided with a pair of wings or sails, being suspended by eight large cords.

The exception above referred to was the ascent of Mr. Sneath, in a balloon of his own construction, from Bleak Hill, near Mansfield, on the night of May 24, 1837: after being in the air two hours, the balloon descended; but Mr. Sneath, fearing it might be destroyed if he quitted it, remained there till aid arrived in the morning.

On June 4, 1784, Madame Thible, the first female aeronaut, and possibly the only woman who has ascended in a fire-balloon, did so in a *Montgolfier*, from Lyons, in company with M. Fleuraud, in the presence of the court, and of Gustavus, King of Sweden, then traveling as Count Haga. Madame Thible's intrepidity was soon paralleled; for in the following year, June 29, 1785, the first English female aeronaut, Mrs. Sage, ascended in a balloon.

The first aerial voyage in England was made by Vincentio Lunardi, accompanied by a cat, a dog, and a pigeon:

he started from the Artillery Ground, and landed at Standon, near Ware.

On January 7, 1785, Messrs. Blanchard and Jefferies crossed the English Channel in an inflammable air-balloon constructed by the former gentleman. They rose from near Shakspeare's Cliff, at Dover; but the weight being too great for the power of the balloon, they rapidly descended. They threw out ballast from time to time, but without success; next they threw out a parcel of books, anchors, and cords, but ineffectually; and as the balloon approached the sea, the aeronauts threw away their clothes, and, fastening themselves to slings, prepared to cut away the boat as a last resource. Calais was now distinctly seen at a distance of about four miles in the direction of the wind; and the balloon rising quickly, they at length descended in safety in the forest of Guisnes.

The longest and most interesting voyage which was performed about this time was that of Messrs. Roberts and Hullin, from Paris, in a balloon filled with heated air. Its diameter was 27¾ feet, and its length 46¾ feet, and it was made to float with the longest part parallel to the horizon, with a boat nearly 17 feet long attached to a net that went over it as far as the middle. To the boat were annexed wings, or oars, in the form of an umbrella. At 12 o'clock they ascended, and descended at 40 minutes past 6, near Arras, in Artois. By working the oars they accelerated their course; but the current of air was uniform from the height of 600 to 4200 feet. In their voyage of 150 miles they heard two claps of thunder; the thermometer fell from 77° to 59°, and condensed the air in the balloon so as to make it descend very low. From experiments, they concluded that they were able, by the use of the two oars, to deviate from the direction of the wind about 22°.

On June 15, 1785, M. de Rozier and M. Romain ascended from Boulogne in a *Montgolfier*, with the intention of crossing the Channel. This machine was a sort of double balloon, one inflated with hydrogen gas; below it was suspended a fire-balloon, and between them were sails. In a short time the upper balloon was seen to be rapidly expanding, while the aeronauts tried to facilitate the escape of the gas. Soon afterward the whole apparatus appeared to be on fire, and the remains of the machine descended from the height of three quarters of a mile with the mangled bodies of the voyagers. In July following Major Money ascended in a balloon of his own construction from Norwich, which burst; he was precipitated into the German Ocean, where he remained five hours, clinging to the wreck of the balloon, by the aid of which he kept himself floating, till he was picked up by the *Argus* sloop-of-war off the coast of Yarmouth.

Testu-Brissy's Balloon.

The ascent of M. Testu from Paris, in June, 1786, lasted twelve hours. His balloon was furnished with wings and other steering apparatus; and when he had ascended 3000 feet, the distention of the balloon led him to descend in a corn-field in the plain of Montmorenci. His balloon was seized by the villagers, when he cut the cords and reascended, and was driven about through the night by a terrific thunder-storm, but descended at sunrise uninjured, seventy miles from Paris.

About this time attempts were made to render aeros-

tation useful in military operations.   A captive balloon
was held attached to a cord of sufficient length, so that
a person could ascend to a corresponding height, and obtain a bird's-eye view of the enemy's movements.   The
most successful result was obtained early in the French
Revolutionary war, when a balloon, prepared by the

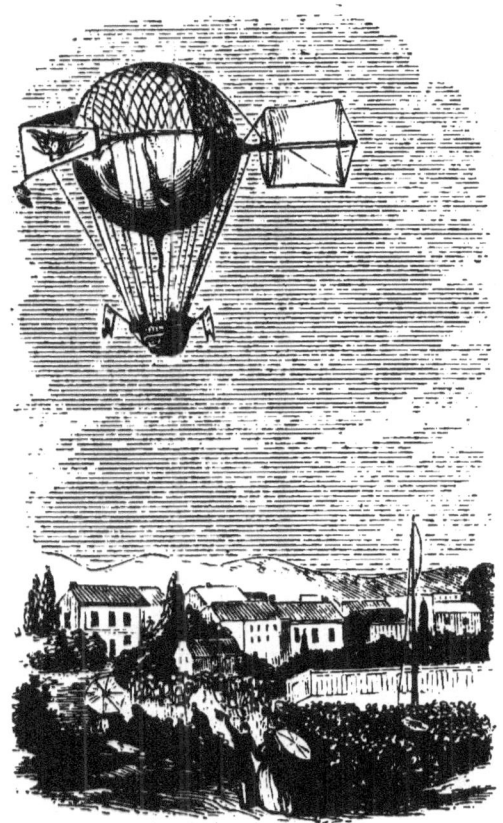

The French Academy's Balloon.

Aerostatic Institute in the Polytechnic School, and intrusted to the command of experienced officers, was distributed to each of the Republican armies.   The deci-
sive victory which General Jourdan gained in June, 1794,
over the Austrians at Fleurus, has been ascribed princi-

pally to the accurate information of the enemy's movements, before and during the battle, communicated by telegraphic signals from a balloon sent up to.a moderate height in the air. The aeronauts, headed by Guyton de Morveau, mounted twice in the course of the day, and continued about four hours each time hovering in the rear of the army, at an altitude of 1300 feet. In the second ascent, the enterprise being discovered by the enemy, a battery was opened against them; but they soon gained an elevation above the reach of the cannon.

In 1802 Garnerin visited England, and ascended in a balloon from Ranelagh Gardens, Chelsea, with a naval officer, when they reached Colchester in less than an hour. In July and September, Garnerin repeated his ascent; and in the latter month descended in a parachute in safety from a height at which he could scarcely be distinguished.

In 1807 Garnerin made a night ascent, and, rising with unusual rapidity, attained a great elevation. By some neglect, the apparatus for discharging the gas became unmanageable; the aeronaut was obliged to make an incision in the balloon, which then descended so rapidly that he cast out his ballast. The balloon, in this way, alternately rose and sank for eight hours; and the aeronaut was driven by a thunder-storm against the mountains, and landed at Mount Tonnere, 300 miles distant from the place of his ascent.

Among the most perilous ascents on record are those of Mr. Sadler from Bristol in 1810, and Dublin in 1812. In the latter voyage he was wafted across the Irish Channel, when, on his approach to the Welsh coast, the balloon descended nearly to the surface of the sea. By this time the sun was set, and the shades of evening began to close in. He threw out nearly all his ballast, and suddenly sprang upward to a great height; and by so doing, brought his horizon to *dip* below the sun, producing the whole phenomena of a western sunrise. Subsequently, descending in Wales, he of course witnessed a second sunset on the same evening (Sir John Herschel's *Outlines of Astronomy*). Mr. Sadler was long a famous aeronaut, and he was one of the earliest manufacturers of soda-water. His two sons, John and Windham, were also aeronauts: the latter was killed in 1824 by falling from a balloon.

Before the introduction of gas-lighting, the mode of inflating a balloon with hydrogen gas was by a slow chemical process from oil of vitriol and water, and sheet zinc, zinc filings, or iron filings; when, the water being decomposed, the vitriol causes the zinc or iron to attract the oxygen, and form with it an oxyd, while the hydrogen, the other component of water, is liberated. The hydrogen was made in casks, whence it was conveyed by hose into the balloon. Coal-gas was first substituted for hydrogen in 1821 by Charles Green, who, on the coronation-day of George IV., ascended from St. James's Park. The success of this experiment vastly increased the facilities and diminished the expenses of balloon ascents. This was Green's first aerial voyage; he subsequently made upward of 500. In 1836 a vast balloon was constructed in Vauxhall Gardens, at the cost of 2000 guineas. In this balloon Green ascended November 7, in the above year, with Mr. Monck Mason and Mr. Holland, and, crossing the British Channel, descended in eighteen hours at Weilburg in Nassau. In the same balloon, Green, September 10, 1838, with Mr. Rush, reached the greatest altitude ever attained—27,146 feet, or 5 miles 746 feet.

In 1838 a return was made to the heated-air system: there was constructed by subscription of a party of amateur aeronauts an egg-shaped Montgolfier balloon, the height of the York Column, and half the circumference of the dome of St. Paul's Cathedral. The furnace was dropped into the centre of the car, and the chimney was placed in the lower aperture of the balloon: the heat could be raised to 200° Fahr. in three minutes, and the bag filled with 170,000 cubic feet in eight minutes. On May 24, the balloon having been inflated upon a platform in the Surrey Zoological Gardens, an attempt to ascend failed from the furnace being too small, when the disappointed spectators tore the machine into pieces.

It has been at various times attempted to turn the balloon to scientific account, of which efforts the following are instances:

De Luc, the celebrated Genevese philosopher, made a scientific voyage in a balloon, taking up with him a barometer, which fell at the greatest altitude to 12 inches. Supposing the barometer to have stood at that time at 30 inches, it follows from this that he must have left below him in quantity exactly three fifths of the entire atmosphere, since 12 inches would be only two fifths of the complete column sustained in the barometric tube. His elevation at this moment was estimated to have been 20,000 feet; but it is certain that he had not attained a point amounting to more than a small fraction of the entire altitude of the atmosphere.

In 1804, MM. Gay-Lussac and Biot ascended at Paris to a height of 13,000 feet, provided with apparatus. The same year M. Gay-Lussac ascended alone 23,000 feet.* In the latter voyage he confirmed two important points: 1. That the magnetic force experiences no sensible variation, either in its inclination or its intensity, from the surface of the earth to the greatest height to which it is possible to ascend.   2. That in this interval the constitution of the atmosphere is entirely the same.   M. Gay-Lussac observed that the heat decreased nearly in arithmetical progression in proportion as he rose in the atmosphere, and that each degree of the depression of his centigrade thermometer corresponded to an elevation of about 85 toises 5 feet.

In 1806, Carlo Brioschi (died 1833), astronomer royal at Naples, ascended with Andreani, the first Italian aeronaut. Trying to rise higher than M. Gay-Lussac had done, they got into an atmosphere so rarefied as to burst the balloon. Its remnants checked the velocity of their descent, and this, with their falling on an open space, saved their lives; but Brioschi contracted a complaint which brought him to his grave.

On June 17, 1823, Captain Beaufoy, the able meteorological observer, ascended with Mr. Graham in his balloon, which, at the height of 6605 feet, became enveloped in clouds, above which was a vast expanse of frozen snow, with enormous mountain-like masses, burnished at every summit by the rays of the sun, which shone most brilliantly from a deep blue sky. The aeronauts rose 11,711 feet, at which height they heard the report of a gun, and could distinguish the metropolis. At the highest elevation, 19,000 feet, clouds were still visible, and the atmosphere was filled with fine crystals of snow : these aeronauts found no difficulty in breathing.

Four ascents were made in 1852 with Mr. Green, in his Nassau Balloon, by a Committee of the British Association, which reported to the Royal Society the meteorological observations obtained : the air collected in the ascents scarcely differed from that at the earth at the same time.

The aerial phenomena witnessed by Mr. E. Vivian, M.A., in a balloon ascent from the metropolis, were, the altitude of the horizon, which remained practically on a level with the eye at an elevation of two miles, causing the surface of the earth to appear concave instead of convex, and to recede during the rapid ascent, while the horizon and the balloon seemed to be stationary : the definite outlines and pure coloring of objects directly beneath, although reduced to micro-

---

* A like elevation has since been attained by MM. Barral and Bixio.—*Dr. Lardner*, 1855.

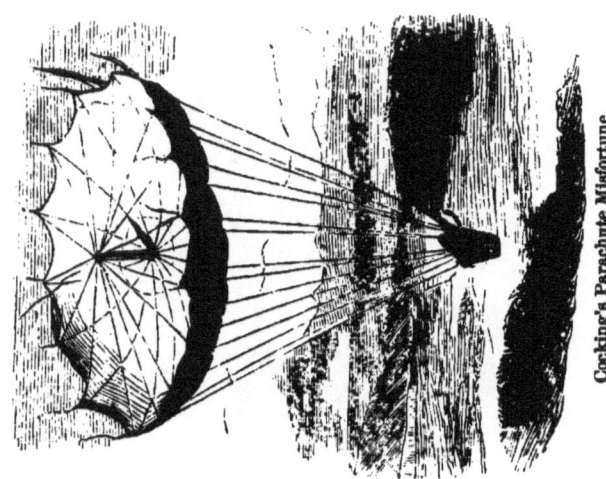

Cocking's Parachute Misfortune.

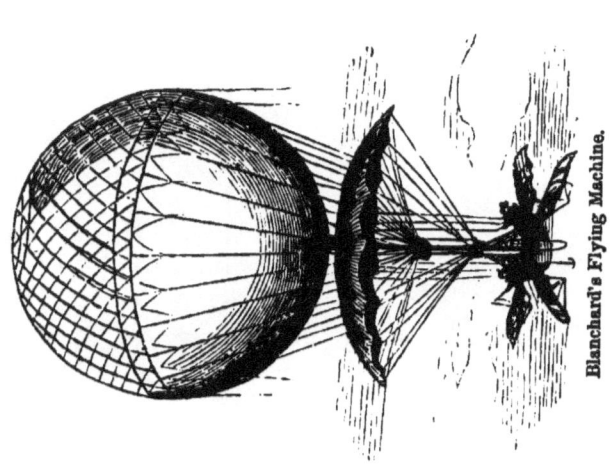

Blanchard's Flying Machine.

scopic proportions: the rich combination of rays bursting through clouds, and having the sun's disk for their focus, contrasted with shadows upon the earth which radiate from a vanishing point on the horizon, the narrow shadows of clouds and eminences, such as Harrow and Richmond, being projected several miles, as seen in the lunar mountains: the magnificent Alpine scenery of the upper surfaces of cloud, still illumined at high altitudes by the cold silvery ray, contrasted with the rich hues of clouds at lower levels, and the darkness of the earth after sunset.

In acoustics, several interesting phenomena were noticed. The sound of London rolled westward as far as its smoke, but was lost above the clouds, where the most intense silence prevailed, as also near the surface of the earth, showing that sound ascends.

It is now time to mention *Parachutes*, expedients by which an aeronaut is enabled to lower himself from a balloon to the earth. The Parachute resembles a vast umbrella, to the handle of which is attached a basket to support the aeronaut. When it is first detached from the balloon, it is shut up; but, as it descends, the air causes the folds to expand.

The idea of using a parachute to break the fall is not new. Two centuries ago two umbrellas were seen used as parachutes in Siam. In France, in 1783, M. le Normand used an umbrella as a parachute in jumping from a house to the ground.

Blanchard, in his first ascent from Paris in a hydrogen balloon, March 2, 1784, added wings and a rudder, but found them useless; and he first carried a *parachute*, or open umbrella, attached above the car, to break the fall, in case it separated from the balloon.

In October, 1797, M. Garnerin ascended from Paris, and when at the height of 2000 feet, disengaged from his balloon a parachute, in which he descended: at first the motion was slow and steady, then oscillatory, but he reached the earth in safety, as related at page 110.

A most disastrous descent was made July 24, 1837, by Mr. Cocking, in a parachute constructed by him, and attached to Mr. Green's Nassau balloon. The parachute resembled an *inverted* open umbrella; and when Cocking cut the connecting rope from the balloon, the parachute collapsed, he descended to the earth with great velocity, and was taken up dead, at Lee, near Blackheath, six miles from the scene of his ascent. The result had been nearly equally fatal to the persons in the car of Green's balloon, which shot up so rapidly that the gas was forced out, and for nearly five minutes they suffered great pain. Most luckily, they had provided a large silken bag full of atmospheric air, and furnished with two metal tubes; these they applied to their mouths, and were thus enabled to breathe: without such a precaution, suffocation would have been inevitable.

In September, 1838, Mr. Hampton ascended with a parachute, attached to a coal-gas balloon, from Cheltenham, to the height of 9000 feet; he then cut the connecting cord, when the balloon rose some hundred feet, and burst; Mr. Hampton safely descending in the parachute within thirteen minutes, the collapsed balloon having reached the earth before him.

We conclude with notices of a few of the more ingenious varieties of contrivances which have been made for navigating the air in the present century.

In 1843 Mr. Monck Mason proposed to propel balloons by the Archimedean Screw, so successfully applied to move vessels through water. He accordingly constructed a large egg-shaped balloon, placed upon a wooden frame in the form of a canoe, to the centre of which he suspended an oblong car. At the head of this car he placed, at the end of an iron axle, a portion of an Archimedean Screw; and at the stern of the car was a large oval-shaped rudder, to guide the balloon on either side, horizontally. In a model, Mr. Mason set the screw in motion by clock-work, which propelled the balloon round the room; but he left it to others to devise machinery for practical working.

A still bolder draft upon credulity was presented in 1843, with all the appliances which the graphic art could lend to design. This was the "Aerial Transit Machine," patented by a Mr. Henson, and to consist of a car attached to a huge rectangular wing-like frame, covered with oiled silk, or canvas; the machine to be propelled by a steam-engine in the car, working two vertical fan-wheels with oblique vanes; while a frame, like the tail of a bird, was to act as a rudder, and make the apparatus ascend or descend. But, as Mr. Henson had not provided for the buoyancy of all this machinery, the "Aerial Transit Machine" never rose but in the region of the brain of the speculative inventor!

PETIN'S PROJECTED GRAND FLYING MACHINE.

In 1844 there was constructed in Paris, by M. Marey
Monge, an immense balloon of sheets of copper the 200th
part of an inch thick, in extent about 1500 yards; the
sheets were soldered by bands, like the ribs of a melon;
the machine weighed 800 pounds, and was to be filled
with hydrogen gas. M. Monge submitted his project to
the French Academy. By substituting copper for silk,
he maintained that the aeronaut might remain in the air
for any length of time, and thus be enabled to study the
atmospheric currents; and by connecting the balloon by
a metal wire with the earth, Monge expected to conduct
the electric matter from the clouds, and thus prevent the
formation of hail, which is so destructive to agriculture.
However, the project entirely failed.

In 1850, M. Julien, a watchmaker of Paris, constructed a model balloon, in the form of a fish, which floated
against the wind by clock-work moving a pair of wings.
The model was of goldbeater's skin, filled with gas, and
was four yards long. Twenty years previously, Mr. Egg,
the celebrated gunmaker, constructed in a building erected for the purpose, at Knightsbridge Grove, a huge fish-
shaped balloon of goldbeater's skin, which could not be
navigated in the air, although the experiments with the
model were successful.

On May 24, 1850, Mr. H. Bell ascended from Kenning-
ton in an "Aerial Machine," shaped like an elongated
egg, which he propelled with a single screw, and steered
by an apparatus for nearly thirty miles, and descended
safely at High Laver, Essex. This is one of the few successes of steering.

In 1850 M. Petin designed at Paris "a System' of
Aerial Navigation," consisting of a vast framework 162
yards long, holding four balloons, each 90 feet diameter,
and four parachutes, at two levels, the whole worked by
two horizontal helices and wheel-work; the platform for
several persons.*

Bishop Wilkins has his followers in our time. In 1827
Colonel Viney patented a *Char Volant*, to be drawn by

* In 1859 there was constructed at New York an "Aerial Ship,"
the height of St. Paul's Cathedral, London, and provided with a life-
boat attached to the car; the aeronaut, T. S. C. Lowe, projecting to
cross the Atlantic Ocean in this vast machine.

Kites, occasionally tandem fashion; and in the Great Exhibition of 1851 there was shown a *Kite Carriage*, which many years previously had been experimented with on Durdham Downs, near Bristol. It was impelled by the air acting upon large kites, at the rate of twenty to twenty-five miles an hour. In 1857 Lord Carlingford patented an *Aerial Chariot*, of very light wood, with three wheels, two net-work wings, and a tail, the latter worked by "an aerial screw," turned by a winch acting on three multiplying wheels: possibly, adds the inventor, two eagles may be trained to draw the vehicle, like the Chariot of Jupiter!

The problem to be solved in aerial navigation is to move through the air in any desired direction. Until this be accomplished, the Balloon will remain a toy to amuse a crowd, and not productive of any gain to science. The accounts of the ascents made during the last thirty-nine years would fill a large volume; and the details of the catastrophes include the deaths of several of the aeronauts.

Besnier's Flying Machine.

Roger Bacon. From a scarce print by Ægidius Sadeler.

THE TRUE HISTORY OF FRIAR BACON.

FEW of the early workers in science have been so strangely misrepresented as Roger Bacon, the philosopher of the thirteenth century, but, until lately, more popularly known as the "Friar Bacon" of the story-books, and the legend of the Brazen Head, which he is said to have made to speak. Yet he was the author of upward of eighty scientific and philosophical treatises, and the reputed inventor of gunpowder and of spectacles. Tradition framed his character on the vulgar notions entertained in his day of the results of experimental science: he was regarded as a learned monk, searching for the philosophers' stone in his laboratory, aided only by infernal spirits; whereas he was the sagacious advocate of reform in education, reading, and reasoning; and, what was equally rare, the real inquirer into the phenomena of nature. Bacon died at Oxford in the year 1292, where existed, until nearly our own time, a traditionary memorial of "the Wonderful Doctor," as he

F

was styled by some of his contemporaries. On Grand-pont, or the Old Folly Bridge, at the southern entrance into Oxford, stood a building, called "Friar Bacon's Folly," from a belief that the philosopher was accustom-ed to ascend this building in the night and "study the stars." It was entirely demolished in 1778; and the bridge, of which Wood says "no record can resolve its precise beginning," was taken down in 1825, and rebuilt in modern style; but you have only to look across Christ Church Meadow to the pinnacled tower of Merton Col-lege to be reminded that this was the earliest home of science of a decidedly English school, and that for two centuries there was no other foundation, either in Oxford or Paris, which could at all come near it in the cultiva-tion of the sciences.*

Roger Bacon belonged to this distinguished founda-tion, although there is a doubt whether he was not of Brazen-nose College. He was born near Ilchester, in Somersetshire, in 1214, the year before the signing of Magna Charta. He was educated at Oxford, next stud-ied at Paris, and returned to Oxford with a doctor's de-gree, which was confirmed by the latter University. He next took the vows of a Franciscan in a convent possessed by that order at Oxford, from which time, 1240, he close-ly studied languages and experimental philosophy. His brethren soon grew envious of his success; the lectures

* An eloquent writer in the *Quarterly Review* thus draws the con-trast presented by a rapid transition from London to Oxford, such as may now be accomplished by railway almost within an hour: "From noise, glare, and brilliancy the traveler comes upon a very different scene—a mass of towers, pinnacles, and spires, rising in the bosom of a valley from groves which hide all buildings but such as are conse-crated to some wise and holy purpose. The same river which in the metropolis is covered with a forest of masts and ships, here gliding quietly through meadows, with scarcely a sail upon it; dark and an-cient edifices clustered together in forms full of richness and beauty, yet solid, as if to last for ever—such as become institutions raised, not for the vanity of the builder, but for the benefit of coming ages; streets, almost avenues, of edifices, which elsewhere would pass for palaces, but all of them dedicated to God; thoughtfulness, repose, and gravity in the countenance, and even dress, of their inhabitants; and, to mark the stir and the business of life, instead of the roar of car-riages, the sound of hourly bells calling men together to prayer." The one is a city in which wealth is created for man; and the other is one in which it has been lavished, and is still expended, for God.

which he gave in the University were prohibited, as well as the transmission of any of his writings beyond the walls of his convent. The charge made against him was that of practicing magic, which was then frequently brought against those who studied the sciences, and particularly chemistry. Yet, in his tract *De Nullitate Magiæ*, Bacon declares that experimental science enables us to investigate the practices of magic, not with the intent of confirming them, but that they may be avoided by the philosopher.

Meanwhile due allowance must be made for the times in which Bacon lived. Even his astrology and alchymy —those two great blots upon his character, as they are usually called—are, when considered by the side of a later age, irrational only because unproved, and neither impossible nor unworthy of the investigation of a philosopher in the absence of preceding experiments. According to Dr. Hutton's laborious inquiries, Bacon expended in twenty years' researches, some £2000, a very large sum for the time, supplied by some of the heads of the universities.

That Bacon was by far the truest philosopher of the Middle Ages is now generally admitted. He was fully acquainted with the works of Euclid, and he displayed great knowledge in the mixed mathematics. He is said also to have understood Greek.

To Pope Clement IV. we owe the production of Bacon's great work, the *Opus Majus*. Clement had previously been legate in England, where he had heard of Bacon's discoveries, and earnestly desired to see his writings; but the prohibition of the Franciscans prevented his wish being complied with. After his election as head of the Church, Bacon, conceiving that there would be no danger or impropriety in disobeying his immediate superiors at the command of the Pope, wrote to Clement, stating that he was now ready to send him whatever he wished for. The answer was a repetition of the former request, and Bacon accordingly drew up the *Opus Majus*, of which, it may be presumed, he had the materials ready. The book was accordingly sent, but was hardly received by Clement before he was seized with his last illness.

Bacon enjoyed freedom from open persecution until the year 1278, when, in his sixty-fourth year, he was summoned before a council of Franciscans at Paris, who condemned his writings, and committed him to close confinement; the particular ground of accusation being the charge of innovation, according to some, but, according to others, his writings upon astrology. During ten years Bacon tried to procure his enlargement, but without success: at last, however, he was set at liberty, through the intercession of some powerful nobles with the Pope, but who they were is not mentioned. Some say that Bacon died in prison; but the best authorities state that he returned to Oxford, where he wrote a compendium of theology, and died in 1292. He was buried in the church of the Franciscans at Oxford. The manuscripts which he had left behind him were immediately put under lock and key by the magic-fearing survivors of his order, until the documents are said to have been eaten by insects.

Mr. Hallam considers it hard to determine whether or not Roger Bacon is entitled to the honors of a discoverer in science. The two great points by which he is known are his reputed acquaintance with Gunpowder and the Telescope. In his *Opus Majus*, some detonating mixture, of which saltpetre is an ingredient, is spoken of as commonly known; and in his *De Secretis Operibus*, he expressly mentions sulphur, charcoal, and saltpetre as ingredients. But, independently of the claims of the Chinese and Indians, Marcus Græcus, who is mentioned by an Arabic physician of the ninth century, gives the recipe for gunpowder. The discovery has sometimes been given to Schwartz, the German monk; and the date 1320 annexed to it, which is much posterior to that claimed for Bacon.

Bacon's discovery of Optic Lenses has been established beyond a doubt; and he *conceived* the Telescope, though there is no proof that he carried his conception into practice, or *invented* the instrument. He truly describes a telescope; but if he had constructed one, he would have found that there are impediments to the indefinite increase of the magnifying power, and, still more, that a boy does not appear a giant, although he attributes these properties to the telescope. At the same time, Bacon asserted that a small army could be made to appear very large; and that the sun and moon could be made to descend, to all appearance, down below, and stand over the head of the enemy—ideas which, in after times, produced either the telescope or some modification of it, consisting in the magnifying of images produced by reflection, and that before the date either of Jansen or Galileo.

Whether the invention of Spectacles is due to Bacon, or whether they had been introduced just before he wrote, is doubtful. In his *Opus Majus* he writes: "This instrument, a plano-convex glass, or large segment of a sphere, is useful to old men, and to those who have weak eyes, for they may see the smallest letters sufficiently magnified;" whence we may conclude that the particular way of assisting decayed sight was known to him. The invention is commonly attributed to Alexander de Spina, a monk of Pisa, who died in 1313. Friar Jordan de Rivalto tells his audience, in a sermon published in 1305, that "it is not twenty years since the art of making spectacles was found out," thus placing the invention in 1285, seven years before Bacon's death. Among other inventions attributed to him is that of the introduction of the Arabic numerals into England; but this has been completely disproved.

"The mind of Roger Bacon," says Hallam, "was strangely compounded of almost prophetic gleams of the future course of science, and the best principles of the inductive philosophy, with a more than sacred credulity in the superstitions of his own time. Some have deemed him overrated by the nationality of the English. But, if we may have sometimes given him credit for the discoveries to which he had only borne testimony, there can be no doubt of the originality of his genius." He bears a singular resemblance to Lord Bacon, not only in the character of his philosophy, but in several coincidences of expression; and the latter has even been charged with having borrowed much from Roger Bacon, without having acknowledged his obligations.

There is little reason to suppose that Roger Bacon's writings were read much out of his own University. But

to those who will study them, there is, even at this day,
a combination of simplicity of style and independence of
thought altogether unusual in his time.   His *Opus Majus*
contains books on the necessity of advancing knowledge;
on the use of philosophy in theology; on the utility of
grammar and mathematics: in the latter of which he
runs through the various sciences of astronomy, chronol-
ogy, geography, and music.   The work also includes a
treatise on optics and experimental philosophy, besides
discussions upon the connection and causes of phenomena
—all treated in a manner greatly in advance of the learning of the thirteenth century—the dark age in which the
wisdom of Roger Bacon was as a light hidden beneath
a bushel measure.

" Bacon's Folly," Grandpont, Oxford.

THE DISCOVERIES OF LEONARDO DA VINCI.

VINCI has been well characterized as "one of the most accomplished men of an accomplished age, and for the extent of his knowledge in the arts and sciences yet unrivaled." Although he devoted himself enthusiastically to painting, he appears to have found time also to study sculpture, architecture, engineering, and mechanics generally; botany, anatomy, mathematics, and astronomy; and he was not only a student of these branches of knowledge, but a master.

"None of the writings of Leonardo," says Hallam, "were published till more than a century after his death; and, indeed, the most remarkable of them are still in manuscript. As Leonardo was born in 1452, we may presume his mind to have been in full expansion before 1500. His *Treatise on Painting* is known as a very early disquisition on the rules of art. But his greatest literary distinction is derived from those short fragments of his unpublished writings that appeared not many years since, and which, according at least to our estimate of the age in which he lived, are more like revelations of physical truth vouchsafed to a single mind than the superstructure of its reasoning upon any established basis. The discoveries which made Galileo, and Kepler, and Mæstlin, and Maurolycus, and Castelli, and other names illustrious; the system of Copernicus; the very theories of recent geologists, are anticipated by Da Vinci within the compass of a few pages; not, perhaps, in the most precise language, or on the most conclusive reasoning, but so as to strike us with something like the awe of preternatural knowledge. In an age of so much dogmatism, he first laid down the grand principle of Bacon, that experiment and observation must be the guides to just theory in the investigation of nature. If any other doubt could be harbored, not as to the right of Leonardo da

Vinci to stand as the first name of the fifteenth century, which is beyond all doubt, but as to his originality in so many discoveries, which probably no one man, especially in such circumstances, has ever made, it must be an hypothesis, not very untenable, that some parts of physical science had already attained a height which mere books do not record. The extraordinary works of ecclesiastical architecture in the Middle Ages, especially in the fifteenth century, lend some countenance to this opinion. Leonardo himself speaks of the Earth's Annual Motion, in a treatise that appears to have been written about 1510, as the opinion of many philosophers in his age."

Mr. Hallam adds, in a note, "The manuscripts of Leonardo da Vinci, now at Paris, are the justification of what has been said in the text." Our historian then quotes from a short account of the MSS. by Venturi, published at Paris in 1797, a few extracts, whence we select the following:

In Mechanics, Vinci was acquainted with, among other things, 1. The theory of applied forces obliquely to the power of the lever. 2. The respective resistance of beams. 3. The laws of friction, afterward given by Amontons. 4. The influence of the centre of gravity upon bodies at rest and in motion. 5. In optics, he described the Camera Obscura before Porta; he also taught aerial perspective, the nature of colored shadows, the movements of the iris, the effects of the duration of visible impressions, and many other phenomena of the eye which are not to be found in Vitellio. Lastly, Vinci stated all that Castelli, in an age after him, produced upon the motion of water, and thus gained the reputation of having been the first who applied the new doctrine of motion to hydraulics, on which subject he was long considered as the earliest writer of the experimental school.

Leonardo must therefore be placed at the head of the writers on the physico-mathematical sciences, and of the true method of study by the moderns. The first extract Venturi gives is entitled "On the descent of heavy bodies, combined with the rotation of the earth." He here assumes the latter, and conceives that a body falling to the earth from the top of a tower would have a compound motion in consequence of the terrestrial rotation.

Venturi thinks that the writings of Nicolas de Cusa had set men speculating concerning this before the time of Copernicus.

Vinci had very extraordinary lights as to mechanical motions. He says plainly that the time of descent on inclined planes of equal height is as their length; that a body descends along the arc of a circle sooner than down the chord; and that a body descending on an inclined plane will reascend with the same velocity as if it had fallen down the height. He frequently repeats that every body weighs in the direction of its movement, and weighs the more in the ratio of its velocity; by weight evidently meaning what we call force. He applies this to the centrifugal force of bodies in rotation: "Pendant tout ce temps elle pèse sur la direction de sa mouvement." Mr. Hallam then quotes another passage, and adds, that if it be not as luminously expressed as we should find it in the best modern books, it seems to contain the philosophical theory of motion as unequivocally as any of them.

Leonardo had a better notion of Geology than most of his contemporaries, and saw that the sea had covered the mountains which contained shells. He seems also to have had an idea of the elevation of the Continents, though he gives an unintelligible reason for it.

He explained the obscure light of the unilluminated part of the Moon by the reflection of the Earth, as Mæstlin did long after him.

Vinci understood Fortification well, and wrote upon it. "Since, in our time," he says, "artillery has four times the power it used to have, it is necessary that the fortification of towns should be strengthened in the same proportion." He was employed on several great works of engineering. So wonderful was the variety of power in this miracle of nature.[*]

* Hallam's *Introduction to the Literature of Europe*, fifth edition, vol. i., p. 222-225. The MSS., after Venturi had inspected them, were returned to Milan, where they are still preserved. It is said that Napoleon I. carried these and Petrarch's *Virgil* to his hotel himself, not allowing any one to touch them, exclaiming with delight, "Questi sono miei" ("These are mine"). When they were in the hands of Count Galeazzo Areonauti, James I. of England is said to have offered him 3000 Spanish doubloons for them (nearly £10,000); but this

His acquirements are told in his own words, in a letter to Ludovico il Moro, Duke of Milan, when he offered him his services: "Most illustrious Signor,—Having seen and sufficiently considered the specimens of authors who repute themselves inventors and makers of Instruments of War, and found them nothing out of the common way, I am willing, without derogating from the merit of another, to explain to your excellency the secrets which I possess; and I hope at fit opportunities to be able to give proof of my efficiency in all the following matters, which I will now only briefly mention:

"1. I have means of making bridges extremely light and portable, both for the pursuit of, or the retreat from, an enemy; and others that shall be very strong and fire-proof, and easy to fix or take up again. And I have means to burn and destroy those of the enemy.

2. In case of a siege, I can remove the water from the ditches; make scaling-ladders and all other necessary instruments for such an expedition.

3. If, through the height of the fortifications, or the strength of the position of any place, it can not be effectually bombarded, I have a means of destroying any such fortress, provided it be not built upon stone.

4. I can also make bombs most convenient and portable, which shall cause a great confusion and loss to the enemy.

5. I can arrive at any (place?) by means of excavations and crooked and narrow ways without any noise, even where it is required to pass under ditches or a river.

6. I can also construct covered wagons, which shall be proof against any force; and, entering into the midst of the enemy, will break any number of men, and make way for the infantry to follow without any hurt or impediment.

7. I can also, if necessary, make bombs, mortars, or field-pieces of beautiful and useful shapes quite out of the common method.

8. If bombs can not be brought to bear, I can make crossbows, balistæ, and other most efficient instruments; indeed, I can construct fit machines of offense for any emergency whatever.

9. For naval operations, I can also construct many instruments both of offense and defense; I can make vessels that shall be bomb-proof.

10. In times of peace, I think I can, as well as any other, make designs of buildings for public or for private purposes; I can also convey water from one place to another.

I will also undertake any work in Sculpture—in marble, in bronze, or in terra cotta; likewise in Painting, I can do what can be done as well as any man, be he who he may.

---

patriotic nobleman refused the money, and presented them to the Ambrosian Library.

I can execute the bronze horses to be erected to the memory and glory of your illustrious father, and the renowned house of Sforza.

And if some of the above things should appear to any one impracticable and impossible, I am prepared to make experiments in your park, or any other place in which it may please your excellency, to whom I most humbly recommend myself, etc."

There is no date to this letter; it was probably written about 1483, or perhaps earlier. The duke took Leonardo into his service. Why he chose to leave Florence is not known; he had made several propositions for the improvement of the city and the state, which were not listened to: one of his projects was to convert the River Arno, from Florence to Pisa, into a canal.

To the above may be added the evidence, discovered in 1841 among Da Vinci's manuscripts, of his knowledge of *steam power applied to warfare*, accompanied by pen and ink sketches of the apparatus of a "steam gun," which he designates the Architonnere, a machine of fine copper, which throws balls with a loud report and great force. One third of the instrument contains a charcoal fire, to heat the water, which being done, a screw at the top of the vessel must be made quite tight. All the water will then escape below into the heated portion of the instrument, and be immediately converted into a vapor so abundant and powerful that the machine will carry a ball a talent in weight. This invention Leonardo attributes to Archimedes.

THE STORY OF PARACELSUS.

This audacious Swiss charlatan and daring innovator was born at the close of the fifteenth, and died in the middle of the sixteenth century. His family was noble, though poor, and he was early initiated into the secrets of Astrology and Alchymy by his father, a physician, and by the Abbé Tritheim. He passed his youth in visiting mines, curing diseases, foretelling the future, and seeking the Philosophers' Stone. During a journey in Poland he was made prisoner by the Tartars, from whom he is said to have learned some arcana of Alchymy. He then went to Egypt, where he was initiated into farther mysteries. Thus equipped, he wandered through Europe, figuring among the doctors, astrologers, and quacks; picking up stray secrets from old women, gipsies, magicians, and headsmen. A peripatetic philosopher, not a book-worm, he read but little: he was never regularly educated, and had a horror of languages, insomuch that at one time he did not open a book for ten years together. But he talked and listened to all classes, and amassed a strange medley of knowledge, which he poured forth in his lectures with amazing facility. Alchymy at this period was fast falling into discredit, when Paracelsus undertook to revive and rehabilitate the study: his enthusiasm, his eloquence, and his audacity produced an impression, created a public for him, and therefore ruined him through his vanity.

In 1526 he returned to Switzerland. A lucky and striking cure fixed on him public attention, and led to his being appointed Professor of Physic and Surgery at Basle. He set himself in opposition against all traditions, declaring himself the rival of all doctors, past and present. His audience had no means of criticism. They took him at his own valuation; and the delighted students so thoroughly entered into his polemic against the schools, that they burned the writings of Hippocrates,

Galen, Avicenna, and Averroes, in the very court of the University. He lectured to students in his and their native language, instead of in the barbarous style and Latinity then universal. Some lucky cures confirmed his reputation; his failures, as usual in such cases, were passed over. Princes consulted and enriched him; professors corresponded with him. But Paracelsus reigned only a short time. Success ruined him : hitherto he had lived temperately, but now he took to drinking and debauchery. Success had raised him enemies, who drove him at last from his professorship; and he once more resumed the profession of a wandering empiric. In a few years he died, at the conclusion of a debauch, struck by apoplexy, in his forty-eighth year.

As a medical reformer, Paracelsus propounded a physiology which was novel, and in those days striking. The leading idea was an application of Astrology to Physiology. In the stars he placed the organ of the vital force. The Sun acts upon the heart and abdomen, the Moon upon the brain, Jupiter on the head and the liver, Saturn on the spleen, Mercury on the lungs, Venus on the loins, etc. Man, being a compound of body and spirit, can only act upon his spiritual part by means beyond the ordinary terrestrial phenomena. Dreams will reveal medicines; but the culmination of the medical art is in Magic; by it not only can life be restored, but health prolonged indefinitely; yet this boasted possessor of the Philosophers' Stone and the Elixir of Life died in poverty, at an early age.

Nevertheless, Paracelsus had genius enough to make posterity forget his errors and absurdities, as a glance at his discoveries will show. To him we owe the idea of employing poisons as medicines ; for he knew that, physiologically, there was a profound difference between a large dose and a moderate dose of the same substance. He also made known to Europe various preparations of antimony, mercury, iron, etc. He employed preparations of lead for diseases of the skin, and first used copper, arsenic, and sulphuric acid as medicaments. He knew that when oil of vitriol acts upon a metal there is an *air* disengaged, which air is a constituent of water. He knew, moreover, that air is indispensable to the respiration of animals and the combination of bodies; that is to say, he was on the threshold of the modern doctrine of combustion. Farther, he knew that digestion was a dissolution of the aliments, that putrefaction was only transformation, and that all which lives dies only to resuscitate under another form. He maintained that the virus of small-pox is a ferment, and that the fever which accompanies eruptions is a sort of boiling which separates the impure from the pure elements of the blood. By a bold generalization, he placed man at the head of the animal series, asserting that his organization was closely allied to that of animals, a position on which rests the whole science of Comparative Anatomy.*

"The vaunts of Paracelsus of the power of his chemical remedies and elixirs, and his open condemnation of the ancient pharmacy, backed as they were by many surprising cures, convinced all rational physicians that chemistry could furnish many excellent remedies, unknown till that time; and a number of valuable experiments began to be made by physicians and chemists desirous of discovering and describing new chemical remedies. The chemical and metallurgic arts, exercised by persons empirically acquainted with their secrets, began to be seriously studied, with a view to the acquisition of rational and useful knowledge."†

The original discoveries of Paracelsus, Brande considers to have been few and unimportant: his great merit lies in the boldness and audacity which he displayed in introducing chemical preparations into the *Materia Medica*, and in subduing the prejudices of the Galenical physicians against the productions of the laboratory. But, though we can fix upon no particular discovery on which to found his merits as a chemist, and though his writings are deficient in the acumen and knowledge displayed by several of his contemporaries and immediate successors, it is undeniable that he gave a most important turn to pharmaceutical chemistry; and calomel, with a variety of mercurial and antimonial preparations, as likewise opium, thenceforth came into general use.

* Condensed (with interpolations) from a paper on *Etudes Biographiques*, par P. A. Cap, in the *Saturday Review*, No. 100.
† Sir John Herschel's *Disc. Nat. Phil.*, p. 112.

Paracelsus performed most of his cures by mercury and opium, the use of which latter drug he had learned from Turkey. The physicians of his time were afraid of opium, as being "cold in the fourth degree." Tartar was likewise a great favorite of Paracelsus, who imposed on it that name "because it contains the water, the salt, the oil, and the acid, which burn the patient as hell does;" in short, a kind of counterbalance to his opium.

Mr. Hallam, in taking leave of the absurd and mendacious paradoxes of Paracelsus, sagely observes: "Literature is a garden of weeds as well as flowers, and Paracelsus forms a link in the history of opinion which should not be overlooked."

If he found hundreds of admirers during his life, he obtained thousands after his death. A sect of Paracelsists sprang up in France and Germany to perpetuate the extravagant doctrines of their founder upon all the sciences, and alchymy in particular.

John Napier, of Merchiston.   From a rare print by Delaram.

NAPIER'S SECRET INVENTIONS.

FEW of the results of speculative science have been so
soundly appreciated as the invention of Logarithms, by
John Napier, early in the seventeenth century.   His ingenious and contriving mind did not, however, rest satisfied with these pursuits; for a paper with his signature,
which is preserved in the Library at Lambeth Palace,
asserts him to be the author of certain "Secret Inventions, profitable and necessary in these days for the defense of this island, and withstanding of strangers, enemies to God's truth and religion."   Of these, the first is
stated to be "a Burning Mirror for burning ships by the
sun's beams," of which Napier professes himself able to
give to the world the "invention, proof, and perfect demonstration, geometrical and algebraical, with an evident
demonstration of their error who affirm this to be made
a parabolic section."   The second is a Mirror for producing the same effect by the beams of a material fire.
The third is a piece of Artillery, contrived so as to send

forth its shot, not in a single straight line, but in all directions, in such a manner as to destroy every thing in its neighborhood. Of this the writer asserts he can give "the invention and visible demonstration." The fourth and last of these formidable machines is described to be "a round Chariot in Metal," constructed so as both to secure the complete safety of those within it, and, moving about in all directions, to break the enemy's array "by continual charges of shot of the arquebuse through small holes." "These inventions," the paper concludes, "besides devices of sailing under water, and divers other devices and stratagems for harassing the enemies, by the grace of God and work of expert craftsmen, I hope to perform. John Napier of Merchiston, anno dom. 1596, June 2."

From this date it appears that Napier's head had been occupied with the contrivances here spoken of long before he made himself known through those scientific labors by which he is now chiefly remembered. Some of his announcements are so marvelous as to lead us to suppose that he intended in this paper rather to state what he conceived to be possible than what he had himself actually performed. Yet several of his expressions will not bear this interpretation, and others confirm what he asserts as to his having really constructed some of the machines he speaks of. Thus Sir Thomas Urquhart, in a strange work, *The Jewel*, first published in 1652, evidently alludes to the third invention as "an almost incomprehensible device;" adding, "it is this: he had the skill (as is commonly reported) to frame an engine (for invention not much unlike that of Archytas's dove) which, by virtue of some secret springs, inward resorts, with other implements and materials fit for the purpose, inclosed within the bowels thereof, had the power to clear a field of four miles in circumference of all the living creatures exceeding a foot of height that should be found thereon, how near soever they might be to one another; by which means he made it appear that he was able, with the help of this machine alone, to kill 30,000 Turks without the hazard of one Christian. Of this, it is said, upon a wager, he gave proof upon a large plain in Scotland, to the destruction of a great many heads of

cattle and flocks of sheep, whereof some were distant
from others half a mile on all sides, and some a whole
mile." Little faith is attached to this statement, that
Napier actually put the power of his machine to the
proof; but, taken in conjunction with Napier's own account, it seems to prove that he had imagined some such
contrivance, and even that his having done so was matter of general notoriety in his own day, and some time
after. It should be added, that although Sir Thomas
Urquhart was born in 1613, some years before Napier's
death, *The Jewel* was not published until 1652, some years
after the reputed inventor's decease. Urquhart informs
us that Napier, when requested on his death-bed to reveal the secret of this engine for destroying cattle, sheep,
and Turks, refused to do so, on the score of there being
too many instruments of mischief in the world already
for it to be the business of any good man to add to their
number.*

An able writer in the *Philosophical Magazine*, vol.
xviii., has collected several notices of achievements similar to those which the Scotch mathematician is asserted
to have performed. In regard to the mirror for setting
objects on fire at a great distance by the reflected rays
of the sun, he adduces the well-known story of the destruction of the fleet of Marcellus, at Syracuse, by the
burning-glasses of Archimedes: and the other (not so
often noticed), which the historian Zonaras records, of
Proclus having consumed by a similar apparatus the ships
of the Scythian leader Vitalian, when he besieged Constantinople in the beginning of the sixth century. Ma-
laba, another old chronicler, however, says that Proclus
operated on this occasion, not by burning-glasses, but by
burning sulphur showered upon the ships from machines.
The possibility of the mirror-burning feat was long dis-
believed; but Buffon, in 1747, by means of 400 plane
mirrors, actually melted lead and tin at a distance of fifty
yards, and set fire to wood at a still greater, and this in
March and April. With summer heat it was calculated
that the same effects might have been produced at 400

* There is a common report among the people at Gartness that
this machine is buried in the ground near the site of the old castle
said to have been occupied by Napier.

yards' distance, or more than ten times that to which, in all probability, Archimedes had to send his reflected rays. "It may be concluded, therefore, that there is nothing absolutely incredible in the account Napier gives of his first invention."*

Napier's second announcement is, however, more startling: he professes to have fired gunpowder by a single mirror; but the only record of the kind we possess is of gunpowder being lighted by heat from charcoal collected by one concave, and reflected from another. Napier's fourth invention, the chariot, bears some resemblance to one of the famous Marquis of Worcester's contrivances. Sailing under water, the object of Napier's last invention, was performed in his own day by the Dutch chemist Debrell, who is reported to have constructed a vessel for King James I., which he rowed under the water of the Thames. It carried twelve rowers, besides several passengers; the air breathed by whom, it is said, was made again respirable by means of a certain liquor, the composition of which Boyle asserts that he learned from the only person to whom it had been divulged by Debrell.

Another scheme of the inventor of Logarithms is the manuring of land with salt, as inferred from the following notice in Birrell's Diary, Oct. 23, 1598: "Ane proclamation of the Laird of Merkistoun, that he tuik upon hand to make the land muir profitable nor it wes before, be the sawing of salt upon it." The patent, or gift of office, as it is called, for this discovery, was granted upon condition that the patentee should publish his method in print, which he did, under the title of *The new Order of Gooding and Manuring all sorts of Field-land with common Salt.* This tract is now probably lost; but the above facts establish Napier's claim to an agricultural improvement which has been revived in our day, and considered of great value, while it proves that Napier directed his speculations occasionally to the improvement of the arts of common life, as well as to that of the abstract sciences.

Reverting to the Logarithms, we may observe that among the persons who had the merit of first apprecia-

* *Pursuit of Knowledge, etc.,* vol. ii., p. 61.

ting the value of Napier's invention was the learned Henry Briggs, reader of the Astronomy Lectures in Gresham College, who was "so surprised with admiration of them (the Logarithms) that he could have no quietness in himself until he had seen the noble person, the Lord Marchiston, whose only invention they were. When they met, almost one quarter of an hour was spent in each beholding the other, almost with admiration, before one word was spoke. At last Mr. Briggs began: 'My lord, I have undertaken this long journey purposely to see your person, and to know by what engine of wit or ingenuity you came first to think of this most excellent help into astronomy, viz., the Logarithms; but, my lord, being by you found out, I wonder nobody else found it out before, when now known it is so easy.'"

Before his invention of Logarithms, Napier devised a method of performing multiplication and division by means of small rods, having the digits inscribed upon them according to such an arrangement, that when placed alongside of each other in the manner directed—in order, for instance, to multiply any two lines of figures—the several lines of the product presented themselves, and had only to be transcribed and added up to give the proper result. These rods, or bones, are thus alluded to by Butler in his *Hudibras*, where he recounts the "ruminaging of Sidrophel:"

       "A moon-dial, with *Napier's bones.*"

A set of the bones used by Napier is preserved in his family. Sir Walter Scott, in his *Fortunes of Nigel*, makes Davie Ramsay swear by "the bones of the immortal Napier," the novelist having an indistinct remembrance of what these "bones" were.

LORD BACON'S "NEW PHILOSOPHY."

THE claim of this wonderful man to rank as a discoverer in science will scarcely be allowed by those who question the title of his predecessor, and, in some respects, prototype, Roger Bacon, to that distinguished honor. Nevertheless, Francis Bacon, Lord Verulam, "by his hours of leisure, by time hardly missed from the laborious study and practice of the law, and from the assiduities of a courtier's life," became the father of modern science, and will be justly looked upon in all future ages as the great reformer of philosophy. His own actual contributions to the stock of physical truths were small; and his observations and experiments in physical science, viewed beside the results obtained by his immediate successors, do not appear to great advantage; nor can we compare them at all with the brilliant discoveries of his contemporary, Galileo. It is only when viewed in reference to the *general* state of knowledge in his own times that Bacon's recorded experiments and observations can be fairly estimated. To glance at these characteristics of his philosophic mind, and at the effect of his labors, rather than detail the labors themselves, is all that can here be attempted.

Francis Bacon was born in York House, on the south side of the Strand, in 1561. His health was very delicate; and to this circumstance may be partly attributed that gravity of carriage, and that love of sedentary pursuits, which distinguished him from other boys. We are told that while still a mere child he stole away from his playfellows to a vault in St. James's Fields for the purpose of investigating the cause of a singular echo which he had observed there. It is certain that at only twelve years of age he busied himself with very ingenious speculations on the art of legerdemain; a subject which, as Professor Dugald Stewart has most justly observed, merits much more attention from philosophers than it has ever received.

In his thirteenth year Bacon was sent to Trinity College, Cambridge, where he studied with diligence and success. Dr. Rawley, his chaplain and biographer, relates that "while he was commorant at the University, about sixteen years of age (as his lordship hath been pleased to impart unto myself), he first fell into the dislike of the philosophy of Aristotle—not for the worthlessness of the author, to whom he would ever ascribe all high attributes, but for the untruthfulness of the way —being a philosophy (as his lordship used to say) only strong for disputations and contentions, but barren of the production of works for the life of man; in which mind he continued to his dying day." Thus early Bacon is said to have planned that great intellectual revolution with which his name is inseparably connected.

In his great work on the *Instauration of the Sciences*, he first made a survey of knowledge as it then existed. In its second part, the *Novum Organum*, in the first book, the main object of science is pointed out, its true end being "to enrich human life with new discoveries and wealth." In the second book, Bacon explains the mode of studying nature which he proposed for the advancement of science. The last division includes the use of instruments in aiding the senses, in subjecting objects to alteration for the purpose of observing them better, and in the production of that alliance of knowledge and power which has in our day crowded every part of civilized life with the most useful inventions. The great merit of Bacon undoubtedly consists in the systematic method which he laid down for prosecuting philosophical investigation; and at the present day, those especially who busy themselves with physical pursuits would often do well to recur to the severe and rigorous principles of the *Organum*. Experience and observation are the guides through the Baconian philosophy, by which its author so largely contributed to the existing knowledge in matters of fact. Of his far-seeing anticipation we quote an instance. Bacon, after remarking that every change and every motion requires time, has the following very curious anticipation of facts which appeared then doubtful, but which subsequent discovery has ascertained :

The consideration of these things produced in me a doubt altogether astonishing, namely, whether the face of the serene and starry heavens be seen at the instant it really exists, or not till some time later; and whether there be not with respect to the heavenly bodies a true time and an apparent time, no less than a true place and an apparent place, as astronomers say, on account of parallax. For it seems incredible that the species or rays of the celestial bodies can pass through the immense interval between them and us in an instant, or that they do not even require some considerable portion of time.

"The measurement of the velocity of light," Professor Playfair subjoins, "and the wonderful consequences arising from it, are the best commentaries on this passage, and the highest eulogy on its author."

It must not be forgotten how much is due for the foundation of the Royal Society to Lord Bacon, who died only thirty-six years before its incorporation. In his *Novum Organum*, rejecting syllogism as a mere instrument of disputation, and putting no trust in the hypothetical system of ancient philosophy, he recommends the more slow but satisfactory method of induction, which subjects natural objects to the test of observation and experience, and subdues nature by experiment and inquiry; and "it will be seen how rigidly the early Fellows of the Royal Society followed Bacon's advice." It is, however, in his *New Atlantis* that we have the plan of such an institution more distinctly set forth; and Sprat considered that there should have been no other preface to his account of the Royal Society than some of Bacon's writings.

After the glory of Bacon had set forever, and his name had become tarnished with infamy, he was stripped of his offices, banished from the court, heavily fined, and imprisoned; but then, discharged and his sentence commuted, his ruined fortunes were never repaired; and the record of his frauds, deceits, impostures, bribes, corruptions, and other malpractices, is one of the blackest pages in history. He passed the remainder of his days in the society of the few friends whom adversity had left him. Scientific pursuits were his consolation, and at last caused his death. The father of experimental philosophy was the martyr of an experiment. It had occurred to him that snow might be used with advantage for the purpose of preventing animal substances from putrefying. On a very cold day, early in the spring of the year 1626, he alighted from his coach near Highgate in order to try the experiment. He went into a cottage, bought a fowl, and with his own hands stuffed it with snow. While thus engaged he felt a sudden chill, and was soon so much indisposed that it was impossible for him to return to Gray's Inn. The Earl of Arundel, with whom he was well acquainted, had a house at Highgate. To that house Bacon was carried. The earl was absent; but the servants who were in charge of the place showed great respect and attention to the illustrious guest. Here, after an illness of about a week, he expired, early on the morning of Easter Day, 1626. His mind appears to have retained its strength and liveliness to the end. He did not forget the fowl which had caused his death. In the last letter that he ever wrote, with fingers which, as he said, could not steadily hold a pen, he did not omit to mention that the experiment of the snow had succeeded "excellently well." In this letter Bacon calls himself the "martyr of science," and compares himself to Pliny the elder, whose death was caused by his over-zealous observation of Vesuvius.

In his will, Lord Bacon "expressed, with singular brevity, energy, dignity, and pathos, a mournful consciousness that his actions had not been such as to entitle him to the esteem of those under whose observation his life had been passed, and, at the same time, a proud confidence that his writings had secured for him a high and permanent place among the benefactors of mankind. So at least we understand those striking words which have been often quoted, but which we must quote once more: 'For my name and memory, I leave it to men's charitable speeches, and to foreign nations and to the next age.'

"His confidence was just. From the day of his death his fame has been constantly and steadily progressing; and we have no doubt that his name will be named with reverence to the latest ages, and to the remotest ends of the civilized world."

The great practical value of the benefits which have resulted from the Baconian philosophy has been thus eloquently illustrated by Lord Macaulay:

Ask a follower of Bacon what the New Philosophy, as it was called in the reign of Charles II., has effected for mankind, and his answer is ready: "It has lengthened life; it has mitigated pain; it has extinguished diseases; it has increased the fertility of the soil; it has given new securities to the mariner; it has furnished new arms to the warrior; it has spanned great rivers and estuaries with bridges of form unknown to our fathers; it has guided the thunderbolt innocuously from heaven to earth; it has lighted up the night with the splendor of the day; it has extended the range of human vision; it has multiplied the power of human muscles; it has accelerated motion; it has annihilated distance; it has facilitated intercourse, correspondence, all friendly offices, all dispatch of business; it has enabled man to descend to the depths of the sea, to soar into the air, to penetrate securely into the noxious recesses of the earth, to traverse the land in cars which whirl along without horses, and the ocean in ships which run ten knots an hour against the wind. These are but a part of its fruits, and of its first-fruits; for it is a philosophy which never rests, which has never attained, which is never perfect. Its law is progress. A point which yesterday was invisible is its goal to-day, and will be its starting-post to-morrow."

The same brilliant writer denominates the two leading principles of the Baconian philosophy to be *utility* and *progress*, of which there can not be more direct evidence than in the fact that the writings of Lord Bacon have been more extensively read in England during the last forty years than in the two hundred years which preceded.

G

INVENTIONS OF PRINCE RUPERT.

THIS ill-starred soldier of fortune, born in 1619, and nephew of King Charles I., was a man of distinguished talent and bravery, but lacked "the better part of valor" —discretion. His checkered fortunes are prominent in the records of the Civil Wars; and we have here to glance at his later life, when the impetuosity of the soldier had subsided into the calmness of the philosopher; and it is to the prince's peculiar readiness for the change of employment and pursuits that we trace the peaceable close of his busy life.

After his reconciliation with Charles II., Rupert took up his residence with the Elector of Mentz; and here, says Mr. Eliot Warburton, in the first leisure of his manhood, his mind reverted with a sense of luxury to the philosophical pursuits in which even his youth had taken pleasure. He now found new sources of unexhausted interest in the forge, the laboratory, and the painter's studio.

It was during this lull in the stormy life of Rupert that he discovered or improved upon his art of Mezzotinto. So long ago as 1637, when immured in the castle of Lintz, he had exercised his active genius in some etchings that still remain, and bear that date. He was long said to have discovered the art of Engraving in Mezzotinto, stated to have been suggested to him by observing a soldier one morning rubbing off from the barrel of his musket the rust which it had contracted by being exposed to the night-dew. The prince perceived, on examination, that the dew had left on the surface of the steel a collection of very minute holes, so as to form the resemblance of a dark engraving, parts of which had been here and there already rubbed away by the soldier. He immediately conceived the idea that it would be practicable to find a way of covering a plate of copper in the same manner with little holes, which, being inked, and laid upon paper, would undoubtedly produce a black impression; while by scraping away in different degrees such parts of the surface as might be required, the paper would be left white where there were no holes. Pursuing this thought, he at last, after a variety of experiments, invented a kind of steel roller, covered with points, or salient teeth, which, being pressed against the copper plate, indented it in the manner he wished; and then the roughness thus occasioned had only to be scraped down, where necessary, in order to produce any gradation of shade that might be desired. This anecdote obtained currency from its being related by Lord Orford, in his famous work upon the Arts, as well as from the avidity with which origins of the arts are commonly set down as the results of accident.

The discovery of Mezzotinto has likewise been claimed for Sir Christopher Wren; but his communication to the Royal Society upon the subject is of date four years subsequent to the date of the earliest of the mezzotinto plates engraved by Prince Rupert.

The real inventor of this art was Louis von Siegen, a lieutenant colonel in the service of the Landgrave of Hesse Cassel, from whom Prince Rupert learned the secret when in Holland, and brought it with him to England, when he came over a second time in the suite of Charles II. Some curious and very rare prints, purchased on the Continent, and now deposited in the British Museum, place the claims of Von Siegen beyond doubt. In this collection is a portrait of the Princess Amelia Elizabeth of Hesse, dated 1643, which is *fifteen years anterior* to the earliest of Prince Rupert's dates: there is another portrait of the same date; and another by Von Siegen bears the most conclusive evidence of its having been produced in the very infancy of the art; besides which is the fact that Von Siegen frequently attached the words "*primus inventor*" to his plates. There are also works by Fürstenburg, dated 1656.

Prince Rupert's plates, however, evince a more matured knowledge of the power of Mezzotinto than those of its inventor, Von Siegen; and Rupert by himself, or with the assistance of Wallerant Vaillaint, an artist whom he retained in his suite, is thought to have improved the

mechanical mode of laying the mezzotinto ground; but
this observation does not apply to the principle of the art.

The prince was, in the fullest sense, a working invent-
or: he labored heartily at his own forge, and applied
himself to the practical as well as the theoretical details
of science. The writer of his funeral ode describes him
as forging "the thunderbolts of war his hands so well
could throw." The *Transactions of the Royal Society*
record his mode of fabricating a gunpowder of ten times
the ordinary strength at that time used; likewise a mode
of blowing up rocks in mines, or under water; "an instrument to cast platforms into perspective;" an hydraulic éngine; a mode of making hail-shot; and an improvement in the naval quadrant. Among his mechanical labors are also to be reckoned his improvement in the locks
of fire-arms, and his guns for discharging several bullets
very rapidly. Among his chemical discoveries was the
composition now called Prince's Metal, of which candle-
sticks and small kitchen pestles and mortars are made:
this is an alloy of copper and zinc, which contains more
copper than brass does, and is prepared by adding this
metal to the alloy. To the list of the prince's inventions
must be added a mode of rendering black-lead fusible,
and rechanging it into its original state. To him also
has been attributed the toy that bears his name as "Rupert's Drop," that curious bubble of glass which has long
amused children and puzzled philosophers.

The prince also discovered a method of boring guns,
which was afterward carried into executien in Romney
Marsh by a speculator; but some secret contrivance of
annealing the metal was not understood except by Rupert, and the matter died with him. His mode of tem-
pering the Kirby fish-hooks was among his lesser discoveries.

Nor must Rupert's pursuits in Glass-making be forgotten. The prince had at Chelsea an experimental
glass-house, which adjoined Chelsea College; for we find
by the Council Minutes of the Royal Society that the
college and lands "might have been well disposed of
(before 1682) but for the annoyance of Prince Rupert's
glass-house, which adjoined it." Sir Jonas Moore wrote
to the prince, at the request of the council, urging him to

PRINCE RUPERT'S LABORATORY IN WINDSOR CASTLE: VISIT OF CHARLES II.

"consider the Society, on account of the mischief that his glass-house was doing to the college" (Weld's *History of the Royal Society*, vol. i., p. 279).

After the Restoration Rupert was received with honor by the king; and Mr. Warburton tells us that the prince "established a seclusion for himself in the high tower in Windsor Castle,* which he soon furnished after his own peculiar taste. In one set of apartments forges, laboratory instruments, retorts, and crucibles, with all sorts of metals, fluids, and crude ores, lay strewed around in the luxurious confusion of a bachelor's domain; in other rooms, armor and arms of all sorts, from that which had blunted the Damascus blade of the Holy War to those which had lately clashed at Marston Moor and Naseby. In another was a library stored with strange books, a list of which may still be seen in the *Harleian Miscellany*."

* Rupert's residence in Windsor Castle, of which Charles II. appointed him governor, was in the keep or round tower.

"PRINCE RUPERT'S DROPS."

THESE philosophical toys, to which we have just alluded, take their English name from having been first made known in England by Prince Rupert, and not from his having invented them, as commonly supposed.

Their origin has been much disputed. Beckmann considers it more than probable that these Drops, and the singular property which they possess, have been known from time immemorial. All glass, when suddenly cooled, becomes brittle, and breaks on the least scratch. On this account, as far back as the history of the art can be traced, a cooling furnace was always constructed close to the fusing furnace. A drop of fused glass falling into water might easily have given rise to the invention of these Drops; at any rate, this might have been the case in rubbing off what is called the navel—that piece of glass which remains adhering to the pipe when any article has been blown, and which the workman must rub off.

It is, however, certain that these Drops were not known to experimental philosophers before the middle of the seventeenth century. Monconys, who traveled in the year 1656, was present when some experiments were made at Paris before a learned society, which assembled at the house of Mommor, the well-known patron of Gassendi; and in the same year he saw similar experiments made by several scientific persons in London. Beckmann then shows it to be probable that these Drops were sent to Paris from Stockholm by Chanut, who was then French embassador at the Swedish court. About fifteen years before, that is, in 1641, the first glass-houses were established in Sweden, and in all probability by Germans; and it is possible that when the blowing of glass was first seen, glass drops may have excited attention, which they had not met with in Germany, where glass-houses had been long established. It can nevertheless be proved that the Drops were known to the

German glass-blowers much earlier. In 1695, Schulen-burg, of the Cathedral school of Bremen, published a German dissertation on glass drops and their properties, in which he says that he was informed by glass-blowers worthy of credit that these Drops had been made more than seventy years before at the Mecklenburg glass-houses—that is to say, about the year 1625.

Professor Reyher, of Kiel, states that Henry Sievers, teacher of mathematics at Hamburg, had assured him that such glass drops were given to his father by a glass-maker so early as the year 1637; and that his father had exhibited them in a company of friends, who were much astonished at their effects. Reyher adds that he himself had seen at Leyden, in 1656, the first of these glass drops which had been made at Amsterdam, where he afterward purchased some of the same kind; but in 1666 he procured for a trifling sum a great many of them from the glass-houses in the neighborhood of Kiel. - It is worthy of remark, that Huet, who paid considerable attention to the history of inventions, says that the first glass drops, which he had seen also in the society held at the house of Mommor, were brought to France from Germany. According, however, to Anthony le Grand (*Historia Naturalis*, 1680), they came from Prussia. The French call these "glass tears" *larmes Bataviques*, from the statement of the first being made in Holland.; but we incline to think that as the Drops were the result of a common operation in glass-houses, their property may have been as commonly known among glass-makers, but not so early observed by philosophers.

The drops were first brought to England in 1660, and in the proceedings of the Royal Society occurs this entry:

Aug. 14. Sir Robert Moray brought in glass drops, an account of which was ordered to be registered.

Accordingly, the first volume of the register book con-tains a very long account of them and their manufacture. They were so well known when *Hudibras* was written as to be used by Butler in popular illustration. In part ii., canto 2, we have,

"Honor is like that glassy bubble
That finds philosophers such trouble;

Whose least part crack'd, the whole does fly,
And wits are crack'd to find out why."

The Drop appears to have been first brought from the Continent by Prince Rupert, and hence associated with his name.   M. Rohalt, in his *Physics*, calls the Drop a kind of miracle in nature, and says:

" Ed. Clarke lately discovered and brought it hither from Holland, and which has traveled through all the universities in Europe, where it has raised the curiosity and confounded the reason of the greatest part of the philosophers."

He accounts for it as follows: "The Drop, when taken hot from the fire, is suddenly immersed in some appropriate liquor (cold water, he thinks, will break it*), by which means the pores on the outside are closed, and the substance of the glass condensed; while the inside not cooling so fast, the pores are left wider and wider from the surface to the middle, so that the air, being let in, and finding no passage, bursts it to pieces.   To prove the truth of this explication, he observes, that if you break off the very point of it the drop will not burst, because that part being very slender it was cooled all at once, the pores were equally closed, and there is no passage for the air into the wider pores below.   If you heat the drop again in the fire, and let it cool gradually, the outer pores will be opened, and made as large as the inner; and then, in whatever part you break it, there will be no bursting.   He gave three of the drops to three several jewelers, to be drilled or filed; but when they had worked them a little way—that is, beyond the pores which were closed—they all burst to powder."

The Drop is thus described in the *Philosophical Transactions*, vol. xlvi., p. 175:

" The bubble is in form somewhat pear-shaped, or like a leech; it is formed by dropping highly-refined green glass, when melted, into cold water.   Its end is so hard that it can scarcely be broken on an anvil; but if the smallest particle of its taper end is broken off, the whole flies at once into atoms and disappears.   The theory of this phenomenon is, that its particles, when in fusion, are in a state of repulsion; but on being dropped into

* Here he is mistaken.

the water its superficies is annealed, and the particles return into the power of each other's attraction, the inner
particles, still in a state of repulsion, being confined within their outward covering."

Though simple in structure, these Drops are difficult
to make. They may be purchased of philosophical instrument makers, and at a few toy-shops; but we remember Rupert's Drops, or "hand-crackers," as they
were called, common at fairs, as well as "candle-bombs"
(a little water in glass hermetically sealed), which are
mentioned by Hook in his *Micrographia*, 1665, but were
known in Germany before that date.

SIR SAMUEL MORLAND AND HIS INVENTIONS.

Among the records of the ingenious men of the seventeenth century, the life of Sir Samuel Morland is entitled to special regard, for the glimpses which his mechanical inventions afford us of the science of the period, as well as for the circumstances of his checkered career.

Samuel Morland was born in Berkshire about the year 1625. He received his education at Winchester School and Cambridge; he remained at the University ten years, but never took a degree. Soon after leaving college he was sent on the famous embassy to the Queen of Sweden, in company with Whitelocke, who, in his journal, calls him "a very civil man, and an excellent scholar." On his return Morland became assistant to Thurloe, the secretary of Oliver Cromwell; and he is said to have been privy to Sir Richard Willis's plot against King Charles, which he overheard while feigning sleep in Thurloe's chambers in Lincoln's Inn, and which he divulged to the king, who made him a knight, and soon afterward a baronet. Morland had already shown his genius for mechanical science; and on the Restoration, Charles made him *Master of Mechanics* to his majesty. In 1677 he took a house at Vauxhall, where he formed a large collection of mechanical contrivances. Morland subsequently removed to a house near the Thames, at Hammersmith, where he died in 1695, having spent his last three years very wretchedly. Poverty and loss of sight compelled him to rely, almost solely, upon the charity of Archbishop Tenison.

John Evelyn, in his *Diary*, describes, 25th of October, 1695, his visit with the archbishop to Morland, "who was entirely blind—a very mortifying sight." Evelyn says: "He showed us his invention of writing, which was very ingenious; also his wooden calendar, which instructed him all by feeling; and other pretty and useful inventions of mills, pumps, etc.; and the pump he had erected, that serves water to his garden, and to passengers, with an inscription, and brings from a filthy part of the Thames near it a most perfect and pure water."

The inscription to which Evelyn refers was a stone tablet fixed in the wall, and is still preserved. The following is a copy of it: " Sir Samuel Morland's Well, the use of which he freely gives to all persons; hoping that none who shall come after him will adventure to incur God's displeasure by denying a cup of cold water (provided at another's cost, and not their own) to either neighbor, stranger, passenger, or poor thirsty beggar. July 8, 1695."

Morland, shortly before his death, as a penance for his past life, buried in the ground, six feet deep, £200 worth of music-books, being, as he said, love-songs and vanity.

From some correspondence between Morland and Dr. John Pell, preserved in the British Museum, it appears that Sir Samuel, as early as 1666, had intended to publish a work on the quadrature of the curvilinear spaces, and had actually printed two sheets of the work, when, by the advice of Dr. Pell, he laid it aside. About this time also Morland invented his Arithmetical Machine, which he describes in a small work. Its operations are conducted by means of dial-plates and small indices, movable with a steel pin. By these means the four fundamental rules of arithmetic are very readily worked, and, to use the author's own words, " without charging the memory, disturbing the mind, or exposing the operations to any uncertainty." His " Perpetual Almanac" is given at the end, which was often printed separately.

We are indebted to Morland for the Speaking Trumpet in its present form. The ancient contrivances of this kind resembled hearing rather than speaking trumpets. Some have considered the great horn, described in an old manuscript in the Vatican Library as having been used by Alexander the Great to assemble his army, to be the oldest speaking trumpet on record; but the description does not expressly state that Alexander *spoke* through the horn.

Sir Samuel's claim to the credit of the invention is warmly contested by Athanasius Kircher. Morland, in

1671, describes his invention in a pamphlet of eight pages.   He first made a trumpet of glass in 1670, by which he was heard speaking at a very considerable distance, when it considerably multiplied the voice.   The next he made was of brass, about 4½ feet long, 12 inches in diameter at the large end, and only 2 inches at the small end; to which was affixed a mouth-piece, "made somewhat after the manner of bellows," to move with the mouth, and thereby prevent the escape of the breath.   This was tried in St. James's Park, and rendered the voice audible at a distance of near half a mile.   The third instrument was of copper, recurved in the form of a common trumpet; its total length was 16 feet 8 inches, the large end 19 inches, and the small end 2 inches in diameter: with this the voice was heard about a mile and a half.   Morland made other trumpets: with one of the largest, tried at Deal Castle, the voice was conducted a distance of between two and three miles over the sea.   He very excusably exaggerated the "manifold uses" of his instrument, and even said that it might be improved so as to carry the voice for the distance of ten miles!   Kircher asserted that he had published the description of a speaking trumpet several years before Morland's pamphlet appeared; but his invention more resembles a hearing trumpet, and he does not appear to have tried a proper speaking trumpet till about 1673.   There is one of Sir Samuel's original trumpets preserved in Trinity College Library, Cambridge, about six feet long, in bad condition. In an advertisement of 1671, it is stated that Morland's "tubes" were sold by Moses Pitt, a bookseller in St. Paul's Church-yard, at the price of £2 5s.   The invention excited much general interest at the time, so Butler makes Hudibras say,

> "I heard a formidable voice,
> Loud as the Stentophonic noise."

Morland was long claimed to be the inventor of the fire-engine; but, as early as 1590, Cyprian Lucar described a rude fire-engine, precisely like a huge squirt.   Evelyn also mentions a fire-engine, invented by Greatorix in 1656, ten years before he saw the "quench-fires," as Morland's engines were called.*

* Morland's invention reminds us that in the vestry of the church

The principal objects of Sir Samuel's study were water-engines, pumps, etc., which he carried to high perfection: his pumps brought water from Blackmore Park, near Winkfield, to the top of Windsor Castle. There is in the Harleian Collection of MSS., in the British Museum, a short tract on the steam-engine, in which "the Principles of the New Force of Fire," invented by Morland in 1682, are thus explained:

"Water being converted into vapor by the force of fire, these vapors shortly require a larger space (about 2000 times) than the water before occupied, and, rather than be constantly confined, would split a cannon; but, being duly regulated, according to the rules of status, and by science reduced to measure, weight, and balance, then they bear their load peacefully (like good horses), and thus become of great use to mankind, particularly for raising water, according to the following table, which shows the number of pounds that may be raised 1800 times per hour to a height of six inches by cylinders half filled with water, as well as the different diameters and depths of the said cylinders."

Then follows his table of the effects of different-sized cylinders. This indicates a perfect knowledge of the subject; and, to Morland's great credit also, let it not be forgotten that he has correctly stated the increase of volume that water occupies in a state of vapor, which must have been the result of experiment: his researches, however, seem to have had little influence on the progress of the practical application of steam.

In one of his letters to Archbishop Tenison, dated 28th of July, 1688, and preserved in Lambeth Palace, it appears that Morland then had an intention of publishing the first six books of Euclid, for the use of public schools.

Morland is said to have written a Treatise on the Barometer: he is also said to have invented the capstan to heave up anchors; but he must have been rather the improver than the inventor of that machine.

Morland's house at Vauxhall was built upon the site

of St. Dionis Backchurch, Fenchurch Street, are preserved four large syringes, at one time the only engines used in London for the extinction of fires: they are about 2 feet 2 inches long, and were attached by straps to the bodies of the firemen.

of Vauxhall Gardens, which appear to have benefited
from his inventive genius. Aubrey tells us that Sir Samuel " built a fine room at Vaux-hall, anno 1667, the inside
all of Looking-glass, and Fountains very pleasant to be-
hold, which is much visited by Strangers; it stands in
the middle of the Garden." In 1675 he obtained a lease
of Vauxhall House; and about the year 1794 there was
removed from the premises a lead pump inscribed S. M.,
1694. The room mentioned by Aubrey is believed to
have stood where the orchestra was afterward built; and
it was probably erected by Morland for the entertain-
ment of Charles II., when he visited this place with his
ladies. The gardens were planted with trees, and laid
out in walks for Sir Samuel, as we see them in a plan of
1681. Their embellishments have, from the earliest date
to our time, consisted of whimsical proofs of skill in mechanics, such as Morland indulged in. The *rococo* orchestra, which was only removed on the clearance of the
Gardens in the autumn of 1859, had plastic ornaments of
a composition resembling plaster of Paris, but known
only to the architect who designed it. The model pictures in the Gardens, too, had their mechanism, as artificial cascades, a water-mill, and a bridge with a mail-
coach and a Greenwich long-stage passing over; an animated cottage scene, with figures drinking and smoking
by machinery, were in existence in 1820; and bushes and
subterranean musical sounds were among the attractions
—all partaking of Morland's taste, which in the present
day is termed *polytechnic ;* so that the King's *Master of
Mechanics* may have originally set the fashion of the curiosities of Vauxhall Gardens, which existed a century
and a half after Morland's death.

Edward, Marquis of Worcester. From the family picture by Vandyck.

THE MARQUIS OF WORCESTER'S "CENTU-
RY OF INVENTIONS."

As the tourist passes by the right of the Abergavenny
or great road from Monmouthshire into Wales, he will
scarcely fail to notice the picturesque remains of Raglan
Castle, "the most perfect decorated strong-hold of which
this country can boast—a romance in stone and lime."
Its historic interest can be traced through five centuries;
but its culminating point was during the time of Henry,
fifth Earl and first Marquis of Worcester, who, in his
eighty-sixth year, made here a desperate struggle in
favor of King Charles I., Raglan being the last castle
throughout this broad realm which defied the power of
Cromwell. In 1642 the marquis raised and supported
an army of 1500 foot and near 500 horse soldiers, which
he placed under the command of his son, Lord Herbert,
who, succeeding his father, became better known as the
Marquis of Worcester, who left in manuscript the *Cen-*

tury of Inventions. During the civil commotions Charles made several visits to Raglan, and on these occasions particularly distinguished the young Lord Herbert, whom his majesty subsequently invested with the command of a large body of troops. His bravery and devotedness to the royal cause led to his being commissioned by the king in Ireland, failing in which the marquis embarked for France. Meanwhile Raglan was surrendered to the Parliamentary forces: we do not hear of the young marquis until 1654, when we find him attached to the suite of Charles II., who then resided at the court of France; and in the following year he was dispatched by the exiled monarch to London for the purpose of procuring private intelligence and supplies of money, of which the king was in the greatest need. Worcester was, however, speedily discovered, and committed a close prisoner in the Tower, where he remained in captivity for several years: he was set free at the Restoration. Of his lordship's private life we find few records. He probably found leisure for the scientific pursuits to which he was much attached during his sojourn in France, where he wrote the first manuscript of his *Century of Inventions,* the notes of which he appears to have lost; but he rewrote them, it is said after his committal to the Tower. This we infer from the manuscript now in the possession of the Beaufort family, which opens thus:

"A
CENTURY
OF THE
NAMES AND SCANTLINGS
OF SUCH
INVENTIONS
As at present I can call to mind to have tried and perfected; which (my former notes being lost) I have, at the instance of a powerful friend, endeavored now, in the year 1655, to set these down in such a way as may sufficiently instruct me to put any of them in practice.

Artis et Naturæ proles."

At the period of the usurpation, Worcester House, in the Strand,* the London residence of the marquis, was sold by Parliament; but at the Restoration it reverted to his lordship, who leased the house to the great Lord

* Afterward called Beaufort House, upon the site of the present Beaufort Buildings.

Clarendon, who resided here until the erection of his new house at the top of St. James's Street.

In 1663 appeared the first edition of the marquis's *Century of Inventions;* and on April 3, in the same year, a bill was brought into Parliament for granting to Worcester and his successors the whole of the profits that might arise from the use of an engine described in the last article in the *Century.* Lord Orford describes this bill to have passed on the simple affirmation of the discovery that he (the marquis) had made; but the journals of the Lords and Commons for 1663–4 show there were no less than seven meetings of committees on the subject, composed of some of the most learned men in the House, who, after considerable amendments, finally passed the bill on the 12th of May.

There is anecdotic evidence of at least the latter portion of the *Century* being written by the author while confined in the Tower. It is said that he was preparing some food in his apartment, when the cover of the vessel, having been closely fitted, was, by the expansion of the steam, suddenly forced off, and driven up the chimney. This circumstance, attracting the marquis's attention, led him to a train of thought which terminated in the completion of the above invention, which he denominated a " Water-commanding Engine."

Lord Worcester's engine was shown in operation; and when Cosmo de' Medici, Grand-Duke of Tuscany, visited England in 1656 (at which time the marquis was a close prisoner in the Tower), his invention was exhibited at Lambeth, as thus recorded in the Grand-Duke's Diary:

His Highness went "beyond the Palace of the Archbishop of Canterbury to see an hydraulic machine, invented by my Lord Somerset, Marquis of Worcester. It raises water more than forty geometrical feet by the power of one man only, and in a very short space of time will draw up four vessels of water through a tube or channel not more than a span in width."

Precisely four years after the bill was brought into Parliament for securing the above invention, viz., upon April 3, 1667, the marquis died in retirement near London, and his remains were conveyed with funeral solemnity to the vault of the Beaufort family in Raglan Church.

Worcester has been illiberally described as a "fantastic projector," and his *Century** as "an amazing piece of folly." But Mr. Partington, in his edition of the work published in 1825, has, throughout an able series of notes, fully demonstrated not only the practicability of applying the major part of the hundred inventions there described, but the absolute application of many of them, though under other names, to some of the most useful purposes of life. It is surely injustice and ingratitude to apply the name of a "fantastic projector" to the man who first discovered a mode of applying steam as a mechanical agent—an invention alone sufficient to immortalize the age in which he lived.

Many of Worcester's contrivances have since been brought into general use: among them may especially be mentioned stenography, telegraphs, floating baths, speaking statues, carriages from which horses can be disengaged if unruly, combination locks, secret escutcheons for locks, candle-moulds, etc.

We have not space to do more than quote the table of the Inventions, which will convey some idea of their great variety:

No.
1. Seals abundantly significant.
2. Private and particular to each owner.
3. A one-line cipher.
4. Reduced to a point.
5. Varied significantly to all the twenty-four letters.
6. A mute and perfect discourse by colors.
7. To hold the same by night.
8. To level cannons by night.
9. A ship-destroying engine.

No.
10. How to be fastened from aloof and under water.
11. How to prevent both.
12. An unsinkable ship.
13. False-destroying decks.
14. Multiplied strength in little room.
15. A boat driving against wind and tide.
16. A sea-sailing fort.
17. A pleasant floating garden.
18. An hour-glass fountain.
19. A coach-saving engine.

* The second edition of the *Century* was published in 1746; the third in 1767; the fourth, which may be considered as the best edition, is a reprint from the first, and is furnished with an Appendix, "containing an Historical Account of the Fire-engine for raising Water." It is dated Kyo, near Lancaster, June 18, 1778. The fifth is a reprint from the Glasgow copy, "by W. Bailey, Proprietor of the Speaking Figure, now showing, by permission of the Right Hon. the Lord Mayor, at No. 41, within Bishopsgate," 1786. The sixth edition was confined to 100 copies, and dated London, 1813.

No.
20. A balance water-work.
21. A bucket-fountain.
22. An ebbing and flowing river.
23. An ebbing and flowing castle clock.
24. A strength - increasing spring.
25. A double-drawing engine for weights.
26. A to-and-fro lever.
27. A most easy level draught.
28. A portable bridge.
29. A movable fortification.
30. A rising bulwark.
31. An approaching blind.
32. A universal character.
33. A needle alphabet.
34. A knotted - string alphabet.
35. A fringe alphabet.
36. A bracelet alphabet.
37. A pinked-glove alphabet.
38. A sieve alphabet.
39. A lantern alphabet.
40. An alphabet by the smell.
41. " " taste.
42. " " touch.
43. A variation of all and each of these.
44. A key-pistol.
45. A most conceited tinder-box.
46. An artificial bird.
47. An hour water-ball.
48. A screwed ascent of stairs.*
49. A tobacco-tongs engine.
50. A pocket-ladder.
51. A rule of gradation.
52. A mystical jangling of bells.
53. A hollowing of a water-screw.
54. A transparent water-screw.
55. A double water-screw.

No.
56. An advantageous change of centres.
57. A constant water flowing and ebbing motion.
58. An often-discharging pistol.
59. An especial way for carabines.
60. A flask charger.
61. A way for muskets.
62. A way for a harquebus, a crock, or ship-musket.
63. For sakers and minyons.
64. For the biggest cannon.
65. For a whole side of ship-muskets.
66. For guarding several avenues to a town.
67. For musketoons on horse-back.
68. A fire water-work.
69. A triangle key.
70. A rose key.
71. A square key with a turning screw.
72. An escutcheon for all locks.
73. A transmittible gallery.
74. A conceited door.
75. A discourse woven on tape or ribbon.
76. To write in the dark.
77. A flying man.
78. A continually-going watch.
79. A total locking of cabinet boxes.
80. Light pistol-barrels.
81. A comb conveyance for letters.
82. A knife, spoon, or fork conveyance.
83. A rasping mill.
84. An arithmetical instrument.
85. An untoothsome pear.
86. An imprisoning chair.

* Most probably the geometrical staircase now in general use, with the addition of a small flight of stairs in the centre, in lieu of the common hand-rail, which, being surrounded by a partition of boards, would serve as a private communication to the upper stories.—*Partington.*

No.
87. A candle-mould.
88. A coining engine.
88. A brazen head.
89. Primero gloves.
90. A dicing box.
91. An artificial ring-horse.
92. A gravel engine.
93. A ship-raising engine.
94. A pocket engine to open any door.
95. A double cross-bow.
96. A way for sea-banks.
97. A perspective instrument.
98. An engine so contrived that working the *primum mobile* forward or backward, upward or downward, circularly or corner-wise, to and fro, straight, upright or downright, yet the pretend-ed operation continueth and ad-vanceth; none of the motions above-mentioned hindering, much less stopping, the other; but unan-imously, and with harmony agree-ing, they all augment and con-tribute strength unto the intended work and operation; and therefore I call this a *semi-omnipotent engine*, and do intend that a model there-of be buried with me.
99. How to make one pound weight to raise one hundred as high as one pound falleth, and yet the hundred pounds descending doth what nothing less than one hundred pounds can effect.
100. Upon so potent a help as these two last-mentioned inven-tions, a water-work is, by many years' experience and labor, so advantageously by me contrived, that a child's force bringeth up an hundred feet high an incredible quantity of water, even two feet diameter. And I may boldly call it *the most stupendous work in the whole world*, not only with little charge to drain all sorts of mines, and furnish cities with water, though never so high seated, as well as to keep them sweet, run-ning through several streets, and so performing the work of scav-engers, as well as furnishing the inhabitants with sufficient water for their private occasions; but likewise supplying the rivers with sufficient to maintain and make navigable from town to town, and for the bettering of lands all the way it runs; with many more ad-vantageous, and yet greater effects of profit, admiration, and conse-quence; so that deservedly I deem this invention to crown my labors, to reward my expenses, and make my thoughts acquiesce in way of farther inventions. This making up the whole Century, and pre-venting any farther trouble to the reader for the present, meaning to leave to posterity a book, wherein, under each of these heads, the means to put in execution and visible trial all and every of these inventions, with the shape and form of all things belonging to them, shall be printed by brass plates. Besides many omitted, and some of three sorts willingly not set down, as not fit to be di-vulged, lest ill use may be made thereof, but to show that such things are also within my knowl-edge, I will here in myne owne cypher sett downe one of each, not to be concealed when duty and affection obligeth me.

The last three inventions, says Mr. Partington, may justly be considered as the most important of the whole *Century;* and when united with the 68th article, they appear to suggest nearly all the data essential for the construction of a modern steam-engine. The 68th article is as follows:

An admirable and most forcible way to drive up water by fire, not by drawing or sucking it upward, for that must be, as the philosopher calleth it, *infra sphæram activitatis*, which is but at such a distance. But this way hath no bounder, if the vessels be strong enough ; for I have taken a piece of a whole cannon, whereof the end was burnt, and filled it three quarters full, stopping and screwing up the broken end, as also the touch-hole ; and making a constant fire under it, within twenty-four hours it burst, and made a great crack ; so that having found a way to make my vessels so that they are strengthened by the force within them, and the one to fill after the other, have seen the water run like a constant fountain stream forty feet high ; one vessel of water, rarefied by fire, driveth up forty of cold water ; and a man that tends the work is but to turn two cocks, that one vessel of water being consumed, another begins to force and refill with cold water, and so successively, etc.

The marquis has also furnished us with a "Definition" of the above engine, which is exceedingly rare, as the only copy known to be extant is preserved in the British Museum. It is printed on a single sheet, without date, and appears to have been written for the purpose of procuring subscriptions for a Water Company then about to be established. The invention is described as

"A stupendous, or a water-commanding engine, boundless for height or quantity, requiring no external, nor even additional help or force to be set or continued in motion but what intrinsically is afforded from its own operation, nor yet the twentieth part thereof. And the engine consisteth of the following particulars :

"A perfect counterpoise for what quantity soever of water.

"A perfect countervail for what height soever it is to be brought unto.

"A *primum mobile*, commanding both height and quantity, regulator-wise.

"A vicegerent, or countervail, supplying the place and performing the full force of man, wind, beast, or mill.

"A helm, or stern, with bit and reins, wherewith any child may guide, order, and control the whole operation.

"A particular magazine for water, according to the intended quantity or height of water.

"An aqueduct, capable of any intended quantity or height of water.

"A place for the original fountain or river to run into, and naturally of its own accord, incorporate itself with the rising water, and at the very bottom of the aqueduct, though never so big or high.

"By divine Providence and heavenly inspiration, this is my stupendous water-commanding engine, boundless for height and quantity.

"Whosoever is master of weight is master of force ; whosoever is master of water is master of both, and, consequently, to him all forcible actions and achievements are easie."

Among the documents in the possession of the Duke of Beaufort is the following impressive memorial of the success of the engine, and the pious gratitude of the inventor:

The Lord Marquesse of Worcester's ejaculatory and extemporary thanksgiving Prayer, when first with his corporal eyes he did see finished a perfect trial of his Water-commanding Engine, delightful and useful to whomsoever hath in recommendation either knowledge, profit, or pleasure.

Oh infinitely omnipotent God! whose mercies are fathomlesse, and whose knowledge is immence and inexhaustible, next to my creation and redemption I render thee most humble thanks from the very bottom of my heart and bowels for thy vouchsafing me (the meanest in understanding) an insight in soe great a secret of nature, beneficent to all mankind, as this my water-commanding engine. Suffer me not to be puffed upp, O Lord, by the knowing of it, and many more rare and unheard off, yea, unparalleled inventions, tryals, and experiments; but humble my haughty heart by the true knowledge of myne own ignorant, weake, and unworthy nature: proane to all evill, O most mercifull Father my creator, most compassionating Sonne my redeemer, and Holyest of Spiritts, the sanctifier, three diuine persons and one God, grant me a further concurring grace with fortitude to take hould of thy goodnesse, to the end that whatever I doe, unanimously and courageously to serve my kind and country, to disabuse, rectifie, and convert my undeserved, yet wilfully incredulous enemyes, to reimburse thankfully my creditors, to reimmunerate my benefactors, to reinhearten my distressed family, and with complaisance to gratifie my suffering and confiding friends, may, voyde of vanity or selfe ends, be only directed to thy honor and glory everlastingly. Amen.

As the pensive tourist strays amid the desolate courts and roofless halls of Raglan, or views from its battlements the golden glories of sunset, he may reflect upon the vicissitudes of the noble owners of this "famous castle fine;" and should the visitor extend his walk to the burial-place of the Beauforts in Raglan Church, he will there see the arched stone vault which enshrines the remains of Edward, Marquis of Worcester.

Of his greatest invention no record has been preserved beyond the articles to which reference has been made in the present *précis* of his labors; but in our day Professor Millington has designed an engine on similar principles, and which, with a few alterations, might be made available for the purposes recommended by our author.

In the *Transactions of the Society of Arts*, vol. iii.,

p. 6, is recommended to the attention of every mechanic the *Century,* " which, on account of the seeming improbability of discovering many things mentioned therein, has been too much neglected; but when it is considered that some of the contrivances, apparently not the least abstruse, have by close application been found to answer all that the marquis says of them, and that the first hint of that most powerful machine, the Steam-engine, is given in that work, it is unnecessary to enlarge on the utility of it."

H

GEORGE GRAHAM AND HIS IMPROVE-
MENT OF THE WATCH.

THE improvement of Clocks by the application of the pendulum was not more essential than the improvement in Watches by the application of the balance-spring. The honor of this invention is claimed by three very eminent men—Huyghens, a Dutchman; the Abbé Hautefeuille, a Frenchman; and our own countryman, Dr. Hooke. The balance-spring was soon universally applied, and even watches on the old construction were altered to receive it.

It was now found that the old vertical escapement (still used in common watches) did not produce sufficient accuracy. Hooke, Huyghens, Hautefeuille, and Tompion, therefore, introduced new principles; but as neither succeeded, probably from imperfect execution, the old crown-wheel was again adopted.

The talent and perseverance of these great men were not, however, lost, as each of their principles has since been successfully applied. Huyghens's escapement is used in producing the motion of the well-known bottle-jack; Hautefeuille's escapement appeared about sixty years ago, under the name of the patent (rack) lever; and Hooke's idea has since been fully developed in the duplex escapement.

The first real improvement in escapements was made by Graham. This is called the horizontal or cylinder escapement; it was introduced in the beginning of the last century, and has been successfully applied up to the present time.

George Graham was born at Horsgill, in Kirklington, Cumberland, in 1675, of parents belonging to the Society of Friends. At the age of thirteen he was apprenticed to Mr. Tompion, the celebrated watchmaker, who kept shop at the corner of Water Lane, Fleet Street. Graham soon evinced inventive fitness for the art he had chosen, conjoined with straightforward character and high principle—qualities which endeared him to his master, who treated him like his own offspring. By his inventive skill and careful work Graham became an excellent watchmaker and mechanician; and, by obtaining a sound knowledge of practical astronomy, he perfected several means for the nice measurement of time, and invented astronomical instruments of first-rate precision and accuracy. This was an era in the history of clockwork. The expansion and contraction of metal had been known above fifty years; and although the use of the clock for astronomical purposes demanded some compensation for the lengthening and shortening of the pendulum by heat and cold, art had not supplied this desideratum, until, in the year 1715, Graham, by substituting a jar of mercury for the pendulum ball, succeeded in retaining the point of suspension and the centre of oscillation at the same distance from each other. To guard against breakage of this pendulum, Graham provided the opposite expansions of different metals as a compensation by the dead-beat escapement, with which, and a gridiron, or mercurial pendulum, having a heavy ball moving in a very small arc of vibration, time-keepers are made whose average variation is less than a quarter of a second daily.

These inventions still continue to be employed, in all their early simplicity, in the construction of the best astronomical clocks of the present day. Graham's horizontal escapement is still extensively used in the Swiss and Geneva watches; but in the better sort of those of English manufacture it has been superseded by the duplex, and recently by the lever, which is nothing more than the application of Graham's dead-beat escapement to the watch, though patents have been taken out by various persons who have claimed the invention.*

The excellence of Graham's work is attested by the south mural quadrant, which was made under his inspection, and divided by his own hand, for Dr. Halley, at the Royal Observatory, Greenwich. He also invented a sector, with which Dr. Bradley discovered two new motions in the fixed stars; and Graham supplied with instruments the French academicians in their voyage to the North Pole, to ascertain the figure of the earth. Gra-

* Additions to Beckmann's *Hist. Inventions*, etc., vol. i., 4th edit.

ham's watches were highly prized. It is related that
when the French mathematician, Maupertuis, was made
prisoner at the battle of Molwitz, and taken to Vienna,
the Grand-Duke of Tuscany, afterward emperor, treated
him with much kindness, and asked him whether he re-
gretted the loss of any particular portion of his property
which the hussars had taken from him. Being much
pressed, the philosopher acknowledged that he wished to
save a watch by Graham, of which he had made use in
his astronomical observations. · The duke also had one
by the same maker, but enriched with diamonds: "See,"
said the duke, taking the watch from his pocket, "it was
but a joke; they have brought it to me, and I now return it."

Julien le Roy, the celebrated French horologist, also
bore testimony to the perfection of Graham's watches.
In 1728 he procured one, said to be the first horizontal
watch seen in Paris: it was presented to Maupertuis after having been fully proved by Le Roy. Graham distinguished himself as a Fellow of the Royal Society: he
was also one of the discoverers of the very remarkable
fact of the contemporaneous occurrence of magnetic dis-
turbances over large portions of the earth's surface. This
discovery was made on the 5th of April, 1741, by the pre-
concerted observations of Celsius at Upsala and Graham
in London.* The investigation of this phenomenon has
since been pursued with great success, especially by the
establishment of magnetic observatories, first proposed
by the illustrious Humboldt.

Desaguliers believes Graham, about the year 1700, to
have first invented a movement for exhibiting the motion of the earth about the sun at the same time that the
moon revolved round the earth. This machine being in
the hands of the instrument-maker, to be sent with some
other instruments to Prince Eugene, he copied it, and
made the first for the Earl of Orrery, and then several
others, with additions of his own. Sir Richard Steele,
who knew nothing of Graham's machine, in one of his

---

* Its rediscovery in the present century is due to a series of corre-
sponding observations undertaken by Arago in Paris and Kupffer in
Kasan, in the years 1825 and 1826.—WELD's *History of the Royal
Society,* vol. ii., p. 438.

lucubrations, thinking to do justice to the first encourager as well as to the first inventor of such a curious instrument, called it, after the earl, an *Orrery*, and gave Mr. J. Rowley the praise due to Mr. Graham.*

We find, however, earlier mention of an Orrery than the above, in the Journal of Dr. Rowland Davies, Dean of Ross (printed for the Camden Society, in 1857). The entry, under 1689, is as follows:

December 14*th*. In the evening Mr. Milbourn came and sat with me, and showed me an account of an automaton projected and made by Mr. Watson, of Coventry, whereby all the stars' motions and planets were exactly represented in clockwork, and all the problems and observations in astronomy therein fully answered.

Graham continued his useful labors for the benefit of science till his death at his house in Fleet Street in 1751. He was buried in the nave of Westminster Abbey, in the same grave with his friend and master Tompion; and over their remains was placed a slab, with the following inscription:

"Here lies ye body of Thomas Tompion, who died November 20th, 1713, aged 75. Also Geo. Graham, watchmaker, and F.R.S., whose curious inventions do honor to ye British genius, whose accurate performances are ye standard of mechanic skill. He died ye 16th of November, 1751, in ye 78th of his age."

---

* The machine has since retained the name, and its invention has often been attributed to Lord Orrery, from its being named after his lordship. Orreries have been constructed by several ingenious persons. There died lately in Scotland Mr. John Fulton, a native of Fenwick. He was a self-taught artist, and constructed a beautiful Orrery, which was greatly admired in the principal towns of England and Scotland, where it was exhibited. Hence the maker was called "Fulton of the Orrery." He was a working shoemaker in his native village, of scanty means and education, yet by dint of application during his leisure hours he executed the above instrument with the greatest accuracy and finish. He afterward removed to London, and was employed in the establishment of Mr. Bate, the well-known mathematical instrument-maker in the Poultry, where his ingenuity and skill were fully demonstrated in making theodolites for the Pacha of Egypt, and balances for the Royal Mint. Fulton also applied himself, almost unaided, to the study of languages: he became a good French scholar, a proficient in the German language, a student of Greek, with a considerable knowledge of Italian. His modesty, his unassuming manners, his generosity, his patience, his perseverance, and his piety, obtained for him a high place in the estimation of his friends. His health failed him through excessive application, and a lingering illness brought him to a comparatively early death.

But this memorial no longer exists, it having been taken up, in 1839, by order of the dean.   Mr. Adam Thomson, in his interesting volume on *Time and Time-keepers*, 1842, says : " Watchmakers, the writer among the number, until prevented by recent restrictions, were in the habit of making frequent pilgrimages to the sacred spot : from the inscription and the place they felt proud of their occupation, and many a secret wish to excel has arisen while silently contemplating the resting-place of the two men whose memory they so much revered. Their memory may last, but the slab is gone.   Who would suppose that on a small lozenge-shaped bit of marble was all that was left to indicate where lie the bodies of 'the Father of Clockmaking,' Thomas Tompion, and 'Honest George Graham;' greater benefactors to mankind than thousands whose sculptured urns impudently emblazon merits that never existed?"   Graham was a man of strict integrity, and of kind and generous nature.   Many pleasant anecdotes are related of his aids to science in communicating to others in the same path the results of his own experiments.   In money-matters he was liberal and open-handed ; and rather than invest his savings, he kept them in the house, ever ready to relieve the necessities of deserving applicants.   These are traits of loving-kindness which require no monumental marble to perpetuate their memory.

There is not, probably, any example of human skill which demands higher qualifications than a perfect watch. And Berthoud does not exaggerate when he tells us that " to become a good watchmaker it is necessary to be an arithmetician, in order to find the revolutions of each wheel ; a geometrician, to determine the curve of the teeth ; a mechanician, to find the forces that must be applied ; and an artist, to be able to put into execution the principles and rules which these sciences prescribe : he must know how fluids resist bodies in motion, the effects of heat and cold on different metals, and, in addition to these acquirements, he must be endowed by nature with a happy genius."

JOHN HARRISON AND THE LONGITUDE WATCH.

The method of ascertaining Longitude by means of the Watch is briefly as follows. If a navigator has a chronometer showing him the exact time at Greenwich, the instant that the sun comes to his meridian it is twelve o'clock, and the difference between this time and the hour marked by the chronometer gives him his Longitude; or, when the time is known at which any particular star passes the meridian at Greenwich, if the navigator marks the instant at which the star comes to his meridian, the difference between this time and the time it would appear at Greenwich is the difference in Longitude.

This problem had, however, been but inaccurately solved for want of good watches. Huyghens is supposed to have been the first who thought of constructing timekeepers for this purpose; but at that period, 1664, sufficient attention had not been paid to the effects produced on metals by the variations of temperature in different climates, and he unfortunately failed in his experiments. Maritime nations had already promised rewards to any one who should make the discovery. In 1598, Philip III. of Spain offered a prize of 1000 crowns; the Dutch followed this example; the Duke of Orleans, Regent of France, offered in the name of the king 100,000 livres; and the French Academy awarded annually a prize to those who made the most useful discoveries connected with the subject. The English, being the greatest navigators, were most interested; and in 1714, Parliament appointed a committee to consider the question, foremost of whom was Sir Isaac Newton, who at once suggested the discovery of the Longitude by the dial of an accurate time-keeper; and, upon their recommendation, the Legislature of Queen Anne, in 1714, passed an act granting £10,000 if the method found discovered the longitude to a degree, or 60 geographical miles, £15,000 if to 40 miles, and £20,000 if to 30 miles, to be determined by a voyage from a port in Great Britain to any port in America. At length, after the golden promises of sovereigns, and the researches of the greatest philosophers of the age had for nearly a century and a half failed in the great discovery, it was made by a self-taught genius, who was bred a village carpenter, and never acquired any acquaintance with literature.

John Harrison was born at Faulby, near Pontefract, in Yorkshire, in 1693; he was the son of a carpenter, which occupation he followed for several years: yet he very early manifested a taste for mathematical science, said to have been first awakened by a copy of some lectures of Saunderson the blind mathematician, that accidentally fell into his hands. He was also fond of mechanical pursuits; and before he was twenty-one he had made two wooden clocks by himself, and without having received any instruction in the art. His residence in view of the sea is said to have led him to devote himself to the construction of marine time-pieces, and in 1728 he first came up to London to prosecute this object; in 1736 he completed the first chronometer used at sea: it neither varied from change of temperature nor the motion of the vessel. Having obtained certificates of its excellence from Halley, Graham, and others, this time-keeper was placed on board a ship of war going to Lisbon, the captain of which attested that Harrison had corrected an error of about a degree and a half upon their return to the English Channel. The Parliamentary Commissioners now presented Harrison with £500, to enable him to proceed with his experiments. In 1739 he produced a smaller chronometer, which promised to give the longitude with even greater accuracy. In 1741 he finished another smaller than either, which the Fellows of the Royal Society considered to be more simple, and less likely to be deranged; and in 1749 Harrison received the Society's gold medal.

Having much improved and corrected this third chronometer, Harrison claimed a trial of it; and the commissioners accordingly, in 1761, sent out his son William in a king's ship to Jamaica. After eighteen days' navigation, the vessel was supposed to be 13° 50′ west of Portsmouth, while the watch, marking 15° 19' was condemned as useless. Harrison, however, maintained that, if Portland Island were correctly marked on the chart, it would be seen on the following day; in this he persisted so strongly that the captain was induced to continue in the same course, and accordingly the island was discovered the next day. This raised Harrison and his watch in the estimation of the crew, who otherwise would not have been able to procure the necessary stores during the remainder of the voyage. In like manner, Harrison was enabled by his watch to announce all the islands in the order in which they would fall in with them. On his arrival at Port Royal, after a voyage of eighty-one days, the chronometer was found to be about five seconds slow; and on his return to Portsmouth, after a voyage of five months, it had kept time within about one minute five seconds, which gives an error of about eighteen miles. This was much within the limits of thirty miles prescribed by the act of 1714, and Harrison claimed the reward; but several objections being taken to the proofs, William Harrison made a second voyage, which left no farther doubt of Harrison's claim, his chronometer having determined the position of Barbadoes within the limits prescribed by the act. The sum of £20,000 was then awarded to him—£10,000 immediately on his explaining the principle of construction, the other half on its being ascertained that the chronometer could be made by others. Liberal as this reward appears, it must be remembered that Harrison devoted upward of forty years before his inventions were perfected, or their general merit fully established. The most important of these improvements are the gridiron pendulum and the expansion balance-wheel; the one serving to equalize the movements of a clock, and the other those of a watch, under all changes of temperature, and both depending upon the unequal stretching, under change of temperature, of two different metals, which are so employed to form the rod of the pendulum and the circumference of the wheel, that the contraction of the one exactly counterbalances the expansion of the other.* Another of Harrison's impor-

* An interesting account of the trial of Harrison's Watch at the

tant inventions is the *going fusee*, by which a watch can be wound up without interrupting its movement. This curious machine, as well as the other time-keepers of Harrison, is still preserved at the Royal Observatory, Greenwich. Being discovered there in a very dilapidated state several years ago, it was put in repair at the expense of Messrs. Arnold and Dent. Excepting the escapement-wheel, all the wheels were of wood—merely flat disks with wooden teeth. The pinions also were of wood. Mr. Dent states that the arrangements for obviating friction were so admirable, that on the removal of part of the escapement the train of wheels ran down with great velocity, although they had not revolved for more than a century before.

Harrison died at his house in Red Lion Square in 1776, in his eighty-third year. On mechanics, and subjects connected with that science, he could converse clearly; but he found great difficulty in expressing his sentiments in writing, as is evident from a work which he left on the construction of time-pieces. Still, his labors present a remarkable instance of what natural genius can accomplish in one particular line without cultivation.

It should, however, be added, that the complexity of Harrison's time-keeper, and its high price, £400, left to be invented, for practical purposes, an instrument of greater simplicity, in the time-keeper of John Arnold, for which he and his son received the government reward of £3000.* In this machine each part performs unchecked the office assigned to it; and its extreme vari-

Royal Observatory, Greenwich, and the general method of rating Marine Chronometers, will be found in the *Curiosities of Science*, by the author of the present work, p. 229-232.

* Arnold is celebrated for the manufacture of the smallest repeating watch ever known: it was made for George III., to whom it was presented on his birthday, June 4, 1764. Although less than six tenths of an inch in diameter, it repeated the hours, quarters, and half quarters, and contained the first ruby cylinder ever made. It is the size of a silver twopenny piece, and its weight that of a sixpence. So novel was its construction, that Arnold not only designed and executed the work himself, but had to manufacture the greater part of the tools employed in its construction. The king presented 500 guineas to Mr. Arnold for this curious watch: and the Emperor of Russia afterward offered the maker 1000 guineas for a duplicate of it, which he declined.

ation in twelve months has been 57 hundredths only. It is therefore highly honorable to the English artists, that by their ingenuity and skill they have accomplished the great object which had occupied the attention of the learned of Europe for nearly 300 years, namely, the means of discovering the Longitude at Sea.

In 1793 a committee of the House of Commons gave to Thomas Mudge, a London watchmaker (or to his son for him), in opposition to the opinion of the Board of Longitude, a reward of £3000 for inventing a remontoire escapement for chronometers, "not worth a farthing," says Mr. E. B. Denison, "and, as indeed it turned out, worth a great deal less to his son, who proceeded to make the chronometer." Mr. Denison maintains that Thomas Earnshaw brought the chronometer to the state in which it has remained for the last eighty years, with scarcely any alteration: the chronometers of inferior artists were always beaten by his whenever they came into competition, and these artists afterward copied Earnshaw's inventions, and did their best to prevent his being rewarded for them.

The English chronometers, on the whole, enjoy a reputation superior to those of any other nation; nevertheless, the latter have attained high excellence. One of the New York chronometers supplied to the Grinnell Arctic Expedition was subjected to all sorts of exposure to which such instruments are liable in a Polar winter, but was so exquisitely provided with adjustments and compensations for the very great extremes of temperature to which it had been subjected, that it was returned with *a change in its daily rate, during a year and a half, of only the eighteen thousandth part of one second of time.* It should be borne in mind that the temperature registered during the winter in Wellington Straits was actually 46° below zero.*

* Alderman Carter, elected Lord-mayor of London in 1859, is a very successful chronometer-maker, having received several government rewards.

DR. WILLIAM HARVEY AND THE CIRCULATION OF THE BLOOD.

THE discovery which has given an imperishable glory to the name of Harvey places him in the highest rank of natural philosophers. "The same services which Newton afterward rendered to optics and astronomy by his theories of light and gravitation, Harvey conferred upon anatomy and medicine by his true doctrine of the Circulation of the Blood."[*]

The early life of Harvey, and the opportunities of his education, led him step by step in the brilliant career of his investigation, till it was finally crowned with success. He was descended from a respectable family in the county of Kent, and was born at Folkestone on the 1st of April, 1578, in a house of fair stone, which Harvey left by will, together with some land adjoining, to Caius College, Cambridge. At ten years of age he was sent to the Grammar School in Canterbury, and having there laid a proper foundation of classical learning, was removed to Gonville and Caius College, Cambridge, and admitted as a pensioner in May, 1593. After spending five years at the University, he went abroad for the acquisition of medical knowledge; and, traveling through France and Germany, fixed himself, in his twenty-third year, at Padua University. Here he attended with the utmost diligence the lectures of Fabricius ab Aquapendente, the Professor of Anatomy. He taught the existence of valves in all the veins of the body; and from that moment Harvey endeavored to discover the use of these valves, his success in which inquiry was the foundation of his after fame. He took his doctor's degree at Padua in 1602, when he was only twenty-four years of age; in the same year he returned to England, again graduated at Cambridge, and settled in the practice of his profession in London. In 1604 he was admitted of

[*] Pettigrew's *Life of Harvey*.

the College of Physicians; and in 1615, when thirty-seven years old, he was appointed reader of the anatomical and surgical lectures at the College. He now seriously prosecuted his researches on the Circulation of the Blood, and it was in the course of these lectures that he first publicly announced his new doctrines; but many years of experimental verification elapsed before he ventured to commit these doctrines to the press. Nevertheless, there is historical evidence to prove that, although Harvey discovered the *fact* of the Circulation of the Blood, he did not discover the *course* nor the *causes* of the circulation. He knew that the blood was carried from the heart through the arteries to the tissues, and from the tissues, through the veins and lungs, back again to the place whence it started. But he knew not *how* the blood passed from arteries to veins; he knew not *why* the blood thus moved. In our day, science is in possession of the exact *course* of the circulation; but the exact *causes* are still under question. We know that the circulating system consists of heart, arteries, capillaries, veins, and lymphatics. Harvey knew not the capillaries and lymphatics, so that his knowledge of the course taken by the blood was necessarily incomplete. To form an estimate of what Harvey actually discovered, we will first take a rapid view of the Circulation.

"The heart, as the great centre, shall be our point of departure. It is composed of four cavities: two ante-chambers, or *auricles*, and two chambers, or *ventricles*. Into the right auricle the blood is poured by the veins; it passes thence into the right ventricle, and is driven therefrom by a strong contraction along the *pulmonary artery* into the lungs. Here it comes in contact with the oxygen of the atmosphere, and changes from venous into arterial blood. It now passes along the *pulmonary veins* into the left auricle of the heart, thence into the left ventricle, from which it is driven by a powerful contraction into the arteries. The pulsing torrent rushes through the arteries to the various tissues, where it passes into the net-work of capillary vessels. Having served the purposes of nutrition, the blood continues its course along these capillaries into the veins. Here the stream is joined by that of the lymphatics, which, like the roots of a plant in the earth, absorb lymph from the organs in which they arise. This confluence of streams hurries on till the blood is emptied into the right auricle, from which it originally started, and thus is the circle completed."*

The story of Harvey's discovery is one of the most interesting and instructive in the whole range of science. Its episodes extend over not less than seventeen centuries; and the two centuries that have elapsed since Harvey's discovery have not sufficed entirely to complete it. Three capital errors, for sixteen centuries, masked the fact of circulation. The first was, that the arteries did not contain blood. The second error was, that the two chambers of the heart communicated with each other by means of holes in the septum dividing them. The third error was, that the veins carried the blood to the various parts of the body. The first of these errors was in part set aside by Galen's proving that the arteries did carry blood; but the composition of the atmosphere being unknown in his days, it remained for modern science to prove that atmospheric air is not contained in the arteries, but only the oxygen thereof, with a slight amount of nitrogen and a certain amount of carbonic acid gas. The second assertion, of the holes in the septum, was disproved in 1543 by Vesalius, the father of modern anatomy. The third error, that of the veins carrying the blood *to* the tissues, was disproved by Michael Servetus showing that the two bloods, venous and arterial, pass one into the other in the lungs, or by the pulmonary circulation. This he showed in a work which was burned by the theologians; and Servetus himself was subsequently burned for speculations of another kind. Two copies of Servetus's work still exist: one, reddened and partly consumed by the flames, is in the Imperial Library of Paris. Nothing can be less equivocal than its description of the passage of the blood from the heart to the lungs, where it is agitated, prepared, changes its color, and is poured from the pulmonary artery into the pulmonary vein. Still, this was but a lucky guess, without influence, and soon forgotten. Six years afterward Realdo Colombo rediscovered the pul-

* *Blackwood's Edinburgh Magazine*, No. 514.

monary circulation; and then Cæsalpinus, the great botanist, unaware of what Colombo had written, announced
the same discovery, and was the first to pronounce the
phrase " Circulation of the Blood."

But nearly every thing remained for Harvey to discover. So far from any one having had a clear idea of
the true theory, no one had even accurately conceived
the true theory of *pulmonary* circulation; for, although
Servetus, Colombo, and Cæsalpinus knew that the blood
passed through the lungs, they fancied only so much
passed as was necessary for the reception of the " vital
spirits;" a quantity which their predecessors fancied
took its course through the perforated septum of the
heart. But they had no conception of the entire mass
of blood traversing the lungs.

The finding that the veins had valves, opening and
closing like doors, brought the discovery of the Circulation within compass. It was made in 1574 by Fabricius,
under whom Harvey studied at Padua. These valves,
preventing any flow from the heart, but admitting the
flow to the heart, ought to have suggested to their discoverer the true interpretation of their use; but five-
and-forty years elapsed before any one arose who had
the sagacity to perceive the real value of this anatomical
structure in respect to the blood-currents. Meanwhile,
although every thing that had been discovered or surmised respecting the Circulation was familiar to every
anatomist of the great Paduan school in which Harvey
studied, nevertheless, when he promulgated his theory,
it was vehemently opposed. No one except Harvey
had, for nearly half a century, seen the significance of
the fact; and he not only conceived a clear idea of the
process, but described it minutely and accurately. He
noticed the successive contractions of each auricle and
ventricle, which forced the blood into the ventricle when
the auricle contracted, and forced it from the ventricle
into the lungs when the ventricle contracted—a process
repeated on the left side with the aerated blood. And
at each passage of the blood from one cavity to another,
there were the valves, or " little doors," opening to let
the current pass, and closing to prevent its reflux. He
described the course of the blood along the arteries,

which he attributed to the pulsations of the heart; and in this, instead of Galen's "pulsific virtue," he recognized the cause of the blood's movement.

The overthrow of ancient authority was now completed. Men dared no longer swear by Galen; they swore by Harvey, who had discovered the greatest fact in the animal economy—a fact totally unknown and unsuspected by Galen, or any other ancient. The opposition to the new system was loud and vehement, but it has been greatly exaggerated by historians. It is true that the Faculty rejected the doctrine, but eminent men accepted it. If Guy Patin was caustic in opposition, Molière laughed at Guy Patin's prejudice, and Boileau ridiculed the Faculty. Some anatomists accepted the doctrine, and the great Descartes warmly espoused it. Swammerdam and Malpighi, two of the greatest names of the century, speak of Harvey with reverence; and soon no one spoke of him in any other tone. Among his admirers was the writer of certain verses, "To the Incomparable Dr. Harvey, on his Book of the Motion of the Heart and Blood," in which these lines occur:

> "There didst thou trace the blood, and first behold
> What dreams mistaken sages coined of old.
> For till thy Pegasus the fountain brake,
> The crimson blood was but a crimson lake,
> Which first from thee did tyde and motion gaine,
> And veins became its channel, not its chaine.
> With Drake and Ca'endish hence thy bays are curl'd,
> Famed circulator of the lesser world."

But the epithet *circulator*, in its Latin invidious signification (*quack*), was applied to Harvey by many in derision. To an intimate friend he complained that, after his book of the Circulation came out, he fell considerably in his practice, and it was believed by the vulgar that he was crack-brained. Nevertheless, about twenty-five years after the publication of his system, it was received in all the universities of the world; and Hobbes has observed that Harvey was the only man, perhaps, who ever saw his own doctrines established in his lifetime.

The *course* of the Circulation was not, however, known to Harvey, nor, with the means at his disposal, could he have traced it. The Microscope was needed; and the first to employ this instrument in such researches was Malpighi, who, four years after Harvey's death, in 1661, detected those capillaries which form the channel of communication between the arteries and veins. Nevertheless, in 1668, Leuwenhoeck describes them as if they had been previously quite unknown: this was in the tail of a tadpole. "A sight presented itself," says Leuwenhoeck, "more delightful than any that my eyes had ever beheld, for here I discovered more than fifty circulations of the blood in different places. I saw that not only the blood in many places was conveyed, through exceedingly minute vessels, from the middle of the tail toward the edges, but that each of these vessels had a curve or turning, and carried the blood back toward the middle of the tail, in order to be conveyed to the heart. Hereby it appeared plainly to me that the blood-vessels I now saw in this animal, and which bear the names of arteries and veins, *are, in fact, one and the same*—that is to say, that they are properly termed arteries so long as they convey the blood to the farthest extremities of its vessels, and veins when they bring it back toward the heart." Thus, then, was the demonstration of the course of the blood completed; and we must confess that it is with surprise we find all historians overlooking the great gap in the doctrine which had been left by Harvey—a gap only filled up by Malpighi and Leuwenhoeck in their discovery of those capillaries forming the true passage of arterial to venous blood.

Harvey was appointed physician to Charles I., and was in the habit of exhibiting to him and the most enlightened persons of his court the motion of the heart, and the other phenomena upon which his doctrines were founded.* During the Civil War he traveled with the king; and, while staying a short time in Oxford, was made by him Master of Merton College, and received the degree of Doctor of Medicine. He held the mastership only for a few months, when he was displaced by the Parliamentary party, his house was plundered, and several unpublished works, of which we have only notices in his other writings, were destroyed. He and his brother, who was

* Mr. Hannah has painted this scene with excellent effect: it has been engraved.

a Turkey merchant, drank coffee before coffee-houses came into fashion in London. His visits to his patients he made on horseback, with a footcloth, his man following on foot, in the same way in which the judges were then accustomed to ride to Westminster Hall. In 1654 he was elected President of the College of Physicians; but, from age and infirmities, he declined the office. He died June 3, 1657, in the eightieth year of his age, and was buried at Hampstead, in Essex, where he is lapped in lead, and on his breast in large letters was to be read, DR. WILLIAM HARVEY. There is a fine portrait of Harvey by Jansen in the library of the College of Physicians; here also are preserved some of the nerves and blood-vessels used by Harvey in his lectures on the Circulation, which he delivered in the house of the College, then in Amen Corner: he built also a Museum in the adjoining garden, upon the site of the present Stationers' Hall. The old College buildings were destroyed in the Great Fire. The Harveian Oration (in Latin) is delivered annually by a Fellow, usually on June 25.

The real *cause* of the Circulation, however, remains to be established. That the heart pumps blood incessantly into the arteries, and thus drives the stream onward with great force, there is no doubt. This, however, is not the sole agent. Professor Draper supplies the answer by an hypothesis grounded on a well-known physical law, namely, that if two fluids communicate in a capillary tube which have different degrees of affinity for the walls of that tube, the fluid having the highest affinity for the tube will drive the other fluid before it. The two fluids in the blood-vessels are arterial and venous, and the greater affinity of the arterial blood for the venous tissues causes it to drive the venous blood onward.

In conclusion, Professor Draper's hypothesis is briefly this: The arterial blood has an affinity for the tissues, which causes it to press forward in the capillaries; and no sooner is that affinity satisfied, than the blood becomes venous, and is pressed forward by the advancing column. In the lungs venous blood presses forward to satisfy its affinity for the oxygen which is in the air; having satisfied this, and become arterial, it is pressed on by the advancing column.

The reader who is desirous of pursuing this subject more in detail is referred to an able exposition in *Blackwood's Edinburgh Magazine*, No. 514, p. 148–164 — "Circulation of the Blood, its Course and History;" the writer availing himself of M. Flourens' *Histoire de la Découverte de la Circulation du Sang*, 1854, and completing his labors by a masterly reasoning upon this very curious and intricate subject. To this paper we are mainly indebted for the facts and new views in the present article.

DR. JENNER AND HIS DISCOVERY OF VACCINATION.

FEW of the many thousand ills which human flesh is heir to have spread such devastation among the family of man as Small-pox. Its universality has ranged from the untold tribes of savages to the silken baron of civilization, and its ravages on life and beauty have been shown in many a sad tale of domestic suffering. To stay the destroying hand of such a scourge, which by some has been identified with the Plague of Athens, was reserved for a genial spirit of our time—such a benefactor to his species was Edward Jenner, the discoverer of Vaccination.

The great fact can, however, be traced half a century before Jenner's time. In the Journal of John Byrom, F.R.S., under June 3, 1725, it is recorded that,

"At a meeting of the Royal Society, Sir Isaac Newton presiding, Dr. Jurin read a case of Small-pox, where a girl who had been inoculated and had been vaccinated was tried and had them not again; but another [a] boy, caught the small-pox from this girl, and had the confluent kind and died."

This case occurred at Hanover. The inoculation of the girl seems to have failed entirely; it was suspected that she had not taken the true small-pox: doubts, however, were removed, as a boy, who daily saw the girl, fell ill and died, "having had a very bad small-pox of the confluent sort." This is the first use of the word *Vaccination*, or, more familiarly, Cow-pox, which is an eruption arising from an insertion into the system of matter obtained from the eruption on the teats and udders of cows, and especially in Gloucestershire: it is also frequently denominated *vaccine matter ;* and the whole affair, inoculation and its consequences, is called Vaccination, from the Latin *vacca*, a cow.

It is admitted that Jenner's merit lay in the scientific application of his knowledge of the fact that the chapped hands of milkers of cows sometimes proved a preventive of small-pox, and from those of them whom he endeavored to inoculate resisting the infection. These results were probably known far beyond Jenner's range, and long before his time; for we have respectable testimony* of their having come within the observation of a Cheshire gentleman, who had been informed of them shortly after settling on his estate in Prestbury parish in or about 1740. This does not in the least detract from Jenner's merit, but shows that to his genius for observation, analogy, and experiment we are indebted for this application of a simple fact, only incidentally remarked by others, but by Jenner rendered the stepping-stone to his great discovery, or, in other words, extending its benefits from a single parish in Gloucestershire to the whole world.

We agree with a contemporary that, "among all the names which ought to be consecrated by the gratitude of mankind, that of Jenner stands pre-eminent. It would be difficult, we are inclined to say impossible, to select from the catalogue of benefactors to human nature an individual who has contributed so largely to the preservation of life and to the alleviation of suffering. Into whatever corner of the world the blessing of printed knowledge has penetrated, there also will the name of Jenner be familiar; but the fruits of his discovery have ripened in barbarous soils, where books have never been opened, and where the savage does not pause to inquire from what source he has derived relief. No improvement in the physical sciences can bear a parallel with that which ministers in every part of the globe to the prevention of deformity, and, in a great proportion, to the exemption from actual destruction."†

The ravages which the Small-pox formerly committed are scarcely conceived or recollected by the present generation. An instance of death occurring after vaccination is now eagerly seized and commented upon; yet seventy years have not elapsed since this disease might fairly be termed the scourge of mankind, and an enemy more extensive and more insidious than even the plague.

* *Notes and Queries*, No. 278.
† Mr. Pettigrew, in his *Lives of British Physicians and Surgeons*, 1830.

A family blighted in its fairest hopes through this terrible visitation was an every-day spectacle: the imperial house of Austria lost eleven of its offspring in fifty years.* This instance is mentioned because it is historical; but in the obscure and unrecorded scenes of life this pest was often a still more merciless intruder.†

Edward Jenner was the third son of the Vicar of Berkeley, in Gloucestershire, where he was born May 17th, 1749. Before he was nine years of age he showed a growing taste for natural history in forming a collection of the nests of the dormouse; and when at school at Cirencester he was fond of searching for fossils, which abound in that neighborhood. He was articled to a surgeon at Sudbury, near Bristol, and at the end of his apprenticeship came to London, and studied under John Hunter, with whom he resided as pupil for two years, and formed a lasting friendship with that great man. In 1773 he returned to his native village, and commenced practice as a surgeon and apothecary with great success. Nevertheless, he abstracted from the fatigues of country practice sufficient time to form a museum of specimens of comparative anatomy and natural history. He was much liked, was a man of lively and simple humor, and loved to tell his observation of Nature in homely verse; and in 1788 he communicated to the Royal Society his curious paper on the Cuckoo. At the same time, he carried to London a drawing of the casual disease, as seen on the hands of the milkers, and showed it to Sir Everard Home and to others. John Hunter had alluded frequently to the fact in his lectures; Dr. Adams had heard of the Cow-pox both from Hunter and Cline, and mentions it in his *Treatise on Poisons*, published in 1795, three years previous to Jenner's own publication. Still no one had the courage or penetration to prosecute the inquiry except Jenner.

Jenner now resolved to confine his practice to medicine, and obtained, in 1792, a degree of M.D. from the University of St. Andrew's.

---

* The grandfather of Maria Theresa died of it, wrapped, by order of the faculty, in twenty yards of scarlet broadcloth.

† In the Russian empire, small-pox is said to have swept away two millions in a single year.—*Woodville on Small-pox.*

We now arrive at the great event of Jenner's life. While pursuing his professional education in the house of his master at Sudbury, a young countrywoman applied for advice; and the subject of small-pox being casually mentioned, she remarked that she could not take the small-pox because she had had cow-pox; and he then learned that it was a popular notion in that district that milkers who had been infected with a peculiar eruption which sometimes occurred on the udder of the cow were completely secure against the small-pox. The medical gentlemen of the district told Jenner that the security which it gave was not perfect; and Sir George Baker, the physician, treated it as a popular error. But Jenner thought otherwise; and, although John Hunter and other eminent surgeons disregarded the subject, Jenner pursued it. He found at Berkeley that some persons, to whom it was impossible to give small-pox by inoculation, had had cow-pox, but that others who had had cow-pox yet received small-pox. This led to the doctor's discovery that the cow was subject to a certain eruption which had the power of guarding from small-pox; and, next, that it might be possible to propagate the cow-pox, and with it security from the small-pox, first from the cow to the human body, and thence from one person to another. Here, then, was an important discovery, that matter from the cow, intentionally inserted into the body, gave a slighter ailment than when received otherwise, and yet had the same effect of completely preventing small-pox. But of what advantage was it for mankind that the cows of Gloucestershire possessed a matter thus singularly powerful? How were persons living at a distance to derive benefit from this great discovery? Dr. Jenner, having inoculated several persons from a cow, took the matter from the human vesicles thus produced, and inoculated others, and others from them again, thus making it pass in succession through many individuals, and all with the same good effect in preventing small-pox.

An opportunity occurred of making a trial of the latter on May 14th, 1796 (a day still commemorated by the annual festival at Berlin), when a boy, aged eight years, was vaccinated with matter from the hands of a milkmaid; the experiment succeeded, and he was inoculated for small-pox on the 1st of July following without the least effect. Dr. Jenner then extended his experiments, and in 1798 published his first memoir on the subject. He had originally intended to communicate his results to the Royal Society, but was admonished not to do so, lest it should injure the character which he had previously acquired among scientific persons by his paper on the natural history of the Cuckoo. In the above work Dr. Jenner announces the security against small-pox afforded by the true cow-pox, and also traces the origin of that disease in the cow to a similar affection of the heel of the horse.

The method, however, met with much opposition, un-til, in the following year, thirty-three leading physicians and forty eminent surgeons of London signed an earnest expression of their confidence in the efficacy of the cow-pox. The royal family of England exerted themselves to encourage Jenner: the Duke of Clarence, the Duke of York, the king, the Prince of Wales, and the queen, bestowed great attention upon Jenner. The incalculable utility of cow-pox was at last evinced, and observation and experience furnished evidence enough to satisfy the Baillies and Heberdens, the Monros and Gregorys of Britain, as well as the physicians of Europe, India, and America. The new practice now began to supersede the old plan pursued by the Small-pox Hospital, which had been founded for inoculation. The two systems were each pursued until 1808, when the Hospital governors discontinued small-pox.

A Committee of Parliament was now appointed to consider the claims of Jenner upon the gratitude of his country. It was clearly proved that he had converted into scientific demonstration a tradition of the peasantry. Two parliamentary grants of £10,000 and £20,000 were voted to him. In 1808 the National Vaccine Establishment was formed by government, and placed under his direction. Honors were profusely showered upon him by various foreign princes, as well as by the principal learned bodies of Europe.

Dr. Jenner passed the remainder of his years principally at Berkeley and at Cheltenham, continuing to the last his inquiries on the great object of his life. He died at Berkeley in February, 1823, at the green old age of seventy-four: his remains lie in the chancel of the parish church of Berkeley. A marble statue by Sievier has been erected to his memory in the nave of Gloucester Cathedral; and another statue of him has been placed in a public building at Cheltenham. Five medals have been struck in honor of Jenner: three by the German nation, one by the Surgeons of the British Navy, and the fifth by the London Medical Society.

No monument of Jenner has been placed in Westminster Abbey, whose proudest inmates would be honored by such companionship. It was, however, at length determined to honor this good and great man by placing his statue in the metropolis. A subscription for this purpose was originated in England,* but nearly half the amount (£340) was collected by the Philadelphia Committee. The statue—which was inaugurated in Trafalgar Square, May 17, 1858, the hundred and ninth anniversary of Dr. Jenner's birth—was modeled by Mr. Calder Marshall, R.A., and is cast in bronze. As a composition it is successful, the sitting position and the reflective attitude being very characteristic of Jenner's placid and amiable nature. The doctor wears his university gown, and is seated in a classic chair, which is ornamented with the wand of Æsculapius. The pedestal is of gray granite, and is simply inscribed JENNER.

Dr. Jenner was endowed with a rare quality of mind, which it may be both interesting and beneficial to sketch. A singular originality of thought was his leading characteristic. He appeared to have naturally inherited what in others is the result of protracted study. He seemed to think from originality of perception alone, and not from induction. He arrived by a glance at inferences which would have occupied the laborious conclusions of most men. In human and animal pathology, in comparative anatomy, and in geology, he perceived facts and formed theories instantaneously, and with a spirit of inventive penetration which distanced the slower approaches of more learned men. But, if his powers of mind were

* The German nation had already struck three medals of Jenner; and in England, the prince consort, to his honor, subscribed liberally to the statue fund.

I

singularly great, the qualities which accompanied them were still more felicitous. He possessed the most singular amenity of disposition with the highest feeling, the rarest simplicity united to the highest genius. In the great distinction and the superior society to which his discovery introduced him, the native cast of his character was unchanged. Among the great monarchs of Europe, who, when in great Britain, solicited his acquaintance, he was the unaltered Dr. Jenner of his birthplace. In the other moral points of his character, affection, friendship, beneficence, and liberality were pre-eminent. In religion, his belief was equally remote from laxity and fanaticism; and he observed to an intimate friend not long before his death that he wondered not that the people were ungrateful to him for his discovery, but he was surprised that they were ungrateful to God for the benefits of which he was the humble means.

Statue of Dr. Jenner, in Trafalgar Square, London.

EULER'S POWERS OF CALCULATION.

LEONARD EULER, one of the most distinguished mathematicians of the eighteenth century, was born at Basle in 1707, and was educated in the University of that city. In 1730 he obtained the Professorship of Natural Philosophy in the Academy of St. Petersburg. In 1735, a very intricate problem in mathematics having been propounded by the Academy, he completed the solution of it in three days; but the exertion of his mind had been so violent that it threw him into a fever, which endangered his life, and deprived him of the use of one of his eyes. In 1741, by invitation of Frederick the Great, Euler went to Berlin, where the Princess of Anhalt, the king's niece, received from him instructions in the well-known facts in the physical sciences; and on his return to St. Petersburg in 1766, Euler published his celebrated work, *Letters to a German Princess,* in which he discusses with clearness the most important truths in mechanics, optics, sound, and physical astronomy. This work has been translated into most of the languages of Europe. Euler had previously published several isolated treatises and some hundred memoirs on mathematics. During his residence at Berlin the king often employed him in calculations relative to the Mint and other subjects of finance; in the conducting of the waters of San Souci, and in the inspection of canals and other public works. By invitation from the Empress Catharine, Euler returned to St. Petersburg to end his days. Shortly afterward he lost the sight of his other eye, having been for a considerable time obliged to perform his calculations with large characters traced with chalk upon a slate. His pupils and his children copied his calculations, and wrote all his memoirs from his dictation. To one of his servants, who was quite ignorant of mathematical knowledge, he dictated his *Elements of Algebra,* a work of great merit, and translated into English and many other languages.

Euler now acquired the rare faculty of carrying on in his mind the most complicated analytical and arithmetical calculations; and his powers of memory wonderfully increased even in his old age. M. d'Alembert, when he saw him at Berlin, was astonished at some examples of Euler's calculating powers which occurred in their conversation. To instruct his grandchildren in the extraction of roots, Euler formed a table of the first six powers of all numbers from 1 to 100, and he recollected them with the utmost accuracy. Two of his pupils having computed to the 17th term a complicated converging series, their results differed one unit in the 50th chapter; and an appeal being made to Euler, he went over the calculation in his mind, and his decision was found correct. His principal amusement, after he had lost his sight, was to make artificial loadstones, and to give lessons in mathematics to one of his grandchildren who evinced a taste for science.

In 1771 a dreadful fire broke out at St. Petersburg, and reached the house of Euler; when Peter Grimen, a native of Basle, having learned the danger in which his illustrious countryman was placed, rushed through the flames to Euler's apartment, and brought him away on his shoulders. His library and his furniture were consumed, but his manuscripts were saved by the exertions of Count Orloff.

Euler underwent the operation of couching, which happily restored his sight; but, either from the negligence of his surgeon, or from his being too eager to avail himself of his new organs, he again lost it, and suffered much severe pain from the relapse. His love of science, however, continued unabated. On September 7th, 1783, after having amused himself with calculating upon a slate the law of the ascensional motion of balloons, which at that time occupied the attention of philosophers, he dined with his relation, M'Lexell, and spoke of the planet Herschel (then recently discovered), and of the calculations by which its orbit was determined. A short time afterward, as he was playing with one of his grandchildren, his pipe fell from his hand; he was struck with apoplexy, and expired, in the seventy-ninth year of his age.

Euler's knowledge was not limited to mathematics and

the physical sciences. He had carefully studied anatomy, and botany, and he was deeply versed in ancient literature. He could repeat the Æneid of Virgil from the beginning to the end, and he could even tell the first and last lines in every page of the edition which he used. In one of his works there is a learned memoir on a question in mechanics, of which, as he himself informs us, a verse of the Æneid gave him the first idea. He amused himself with questions of pure curiosity, such as the knight's move in chess so as to cover all the squares. His various researches have gone far toward creating the geometry of situation, a subject still imperfectly known. The following is one of the questions which Euler has generalized: "At Königsburg, in Prussia, the river divides into two branches, with an island in the middle, connected by seven bridges with the adjoining shores: it was proposed to determine how a man should travel so as to pass over each bridge once, and once only."

MR. GEORGE BIDDER AND MENTAL CALCULATION.

THE boyhood of Mr. Bidder will be remembered among the few records we possess of the higher class of mental calculators. The youth has now matured as an eminent engineer; and in 1856 Mr. Bidder delivered to the Institution of Civil Engineers two addresses, conveying that process of reasoning, or action of the mind, by which, when a boy, he trained himself in Mental Arithmetic, and thus laid the basis of that professional skill which he has exercised so beneficially in his great engineering works.

Mr. Bidder is convinced that Mental Calculation can be taught to children, and be acquired with greater facility and less irksomeness than ordinary arithmetic. Still, the eminent mental calculators have been extremely few during the last two centuries, among whom Jedediah Buxton and Zerah Colborne were the most remarkable; but even their powers have not been usefully employed, in consequence of their not having subsequently had the opportunity of receiving a mathematical education. It has been commonly thought that Mental Calculation is an art naturally ingrafted upon peculiarly constituted minds; it has also been attributed to the possession of great powers of memory; and it has been generally imagined that Mr. Bidder himself has been indebted to unusual powers of memory and a naturally mathematical turn of mind for the celebrity he has acquired. Now Mr. Bidder emphatically declares this not to have been the case; he has sought every opportunity of comparing himself with boys and men who possess this faculty, and, except so far as being carefully trained and practiced in the cultivation and use of figures, he has not found that his memory was more than ordinarily retentive. In fact, while at school and at college, he had some difficulty in maintaining a decently respectable position in the mathematical class.

Mr. Bidder enunciates as a principle that there is not any royal or short road to Mental Calculation. All the rules which he employed were invented by him, and are only methods of so arranging calculation as to facilitate the power of registration; in fact, he thus arrived at a sort of natural algebra, using actual numbers in the place of symbols.

He believes that when he began to deal with numbers he had not learned to read, and certainly long after that time he was taught the symbolical numbers from the face of a watch. His earliest recollection is that of counting up to 10, then up to 100, and afterward to 1000; then, by intuitive process, he taught himself the method of abbreviating the labor of counting—arriving, in fact, at the natural multiplication of numbers into each other, attributing to each a separate and individual value.

In this manner the actual value of every number up to 1000 was impressed upon his memory, and he then proceeded onward, *seriatim*, up to a million. It was his practice to count numbers practically by peas, marbles, or shots; to compose rectangles of various values, and, by counting them, the multiplication table was ultimately the result of actual experience and test; and thus he attained an intimate acquaintance with numbers multiplied with each other by a tangible process, divested of that formidable character under which it was generally brought before the young student.

In this way he learned to multiply up to two places of figures before he knew the symbolical characters of the figures, or the meaning of the word "multiply;" as, instead of the term "multiplying 27 by 73," he only understood the expression "27 times 73."

All the varieties of numbers up to a million being represented by six different designations, or varieties of numbers, viz., units, tens, hundreds, thousands, tens of thousands, and hundreds of thousands, their permutations were only eighteen in number. A boy, therefore, who knows his multiplication table up to 10 times 10 registers 50 facts in his mind, and with the permutations above mentioned has only to store 68 facts. The ordinary multiplication table of 12 times 12 gives him 72 facts to store, or 4 additional facts. The machinery, therefore, necessary to enable him to multiply to 6 places of figures consists of 4 facts less than that required to enable him to carry the multiplication table in his mind.

The application of this, when fairly acquired, may be thus illustrated; for example, multiplying 173 by 397, the following process is performed mentally:

```
100×397=39,700
 70×300=21,000=60,700
 70× 90      ..=  6,300=67,000
 70×  7      ..      ..=   490=67,490
  3×300      ..      ..      ..=   900=68,390
  3× 90      ..      ..      ..      ..=   270=68,660
  3×  7      ..      ..      ..      ..      ..=    21=68,681
```

The last result in each operation being alone registered by the memory, all previous results being obliterated.

To show the aptitude of the mind by practice, he will know at a glance

That	..	..	$400 \times 173 = 69,200$
And then	..	..	$3 \times 173 = 519$

The difference being $68,681$, as above.

In Addition and Subtraction the same principle, as already explained for Multiplication, is adhered to, viz., that of commencing with the left-hand side, or the large numbers, and adding successively, keeping one result only in the mind.

Division is, as in ordinary arithmetic, much more difficult than Multiplication, as it must be a tentative process, and is only carried out by a series, more or less, of guesses; but no doubt, in this respect, the training arrived at by Mental Arithmetic gives the power of guessing to a greater extent than is usually attained, and affords a corresponding facility in the process. Supposing, for instance, it is necessary to divide 25,696 by 176, the following will be the process: 100 must be the first figure of the factor: 100 times 176 are known at once to be 17,600; subtracting that from 25,696, there remains 8096. It is perceived that 40 is the next number in the factor; 40 times $176 = 7040$: there then remains 1056; that, it is immediately perceived, gives a remaining factor of 6, making in all 146. Thus only one result is retained in the mind at a time; but, as contrasted with multiplication, it is necessary to keep registered in the mind two results which are always changing, viz., the remainder of the number to be divided, and the numbers of the factor as they are determined; but if it is known, as in the present instance, that 176 is the exact factor, without any remainder, having got the first factor, 100, which is perceived at a glance, it is known that there are only four numbers which, multiplied by 76, can produce a result terminating in 96, viz., 21, 46, 71, and 96, and therefore the immediate inference is that it must be 46, as 121 must be too little, and 171 must be too much, therefore 146 must be the factor. Thus, the only facility afforded by Mental Calculation is the greater power of guessing at every step toward the result.

Mr. Bidder recommends, as the true course in teaching arithmetic, that, before any knowledge of figures is symbolically acquired, the process of counting up to ten should be mastered, then up to 100, and subsequently to 1000; then the multiplication table, up to 10 times 10, should be taught practically, by the use of peas, marbles, or shots, or any bodies of uniform dimensions, by placing them in rectangles or squares.

Having thus induced the student to teach himself the multiplication table, nothing will be more easy than to teach him to multiply 10 by 17, which will be $10 \times 10 + 10 \times 7$; having accomplished this, the multiplication of 17×13 easily follows, being $10 \times 17 + 3 \times 10 + 3 \times 7$. This being executed, it only remains for him to practice multiplication up to two places of figures. Concurrently with this should be taught the permutations of 100, 1000, etc., into each other, and thus will be laid the basis of Mental Calculation for whatever extent, the individual student relying upon his own resources for framing his rules for any other branch of arithmetic. In order to do this, however, his mind must be stored with a certain number of facts, which must be completely at his command; and advantage should be taken of the mode of giving him an insight into natural algebra and geometry. With this view, the training should be extended; and there would be no difficulty in conveying to young minds the knowledge of certain leading facts connected with the sciences long before they were capable of comprehending the beautiful trains of reasoning by which their truths were established. There is no difficulty in impressing permanently an appreciation of the relative proportion of the diameter to the circumference of a circle; of the beautiful property of the square of the hypothenuse of a right-angled triangle being equal to the squares of the two sides containing the right angle; or of the equality of the areas of triangles on the same base, contained between the same parallel; and many others which must occur to all geometricians.

The same with respect to the properties of several series of numbers; for instance,

$$1+3+5, \text{ etc., or } 1+2+3, \text{ etc., or } (1)+(1\times6)+(1+3\times6)+(1+6\times6), \text{ etc.}$$

Mr. Bidder suggests that his mode of proceeding presents advantages of much greater importance than even the teaching of figures, namely, the cultivation of the reasoning powers in general. He would through this means introduce a boy to natural geometry and algebra. By placing shots or any small symmetrical objects on the circumference and diameter of a circle, he will be able, by actual observation, to satisfy himself of their relative proportions. He may simultaneously be taught the relation of the area of the circle to the area of the square. Advantage may also be taken of this mode to develop many other ideas connected with geometry; as, for instance, that all the angles subtended from the same chord in the circle are equal. This may be shown by having a small angle cut in pasteboard, and fitted to every possible position in which two lines can be drawn within the circle upon the same chord. He may also be taught that the rectangles of the portions of any two lines intersecting a circle are equal. If the learner once acquire a feeling for the beauty of the properties of figures—surmising that he has any natural taste for arithmetic—the discovery of these facts by his own efforts may incite him to farther investigations, and enable him to trace out his own path in the science.

"As nearly as I can recollect," says Mr. Bidder, "it was at about the age of six years that I was first introduced to the science of figures. My father was a working-man, and my elder brother pursued the same calling. My first and only instructor in figures was that elder brother; the instruction he gave me commenced by teaching me to count up to 10. Having accomplished this, he induced me to go on to 100, and there he stopped. Having acquired a certain knowledge of numbers by counting up to 100, I amused myself by repeating the process, and found that by stopping at 10, and repeating that every time, I counted up to 100 much quicker than by going straight through the series. I counted up to 10, then to 10 again=20, 3 times 10=30, 4 times 10= 40, and so on. This may appear to you a simple process, but I attach the utmost importance to it, because it made me perfectly familiar with numbers up to 100; they became, as it were, my friends, and I knew all their relations and acquaintances. You must bear in mind that at this time I did not know one written or printed figure from another, and my knowledge of language was so restricted that I did not know there was such a word as 'multiply;' but, having acquired the power of counting up to 100 by 10 and by 5, I set about, in my own way, to acquire the multiplication table. This I arrived at by getting peas or marbles, and at last I obtained a treasure in a small bag of shot. I used to arrange them into squares of 8 on each side, and then, on counting them throughout, I found that the whole number amounted to 64; and that fact, once established, has remained there undisturbed until this day, and I dare say it will remain so to the end of my days. It was in this way that I acquired the whole multiplication table up to 10 times 10, beyond which I never went; it was all I required.

"At the period referred to there resided in a house opposite to my father's an aged blacksmith, a kind old man, who, not having any children, had taken a nephew as his apprentice. With this old gentleman I struck up an early acquaintance, and was allowed the privilege of running about his workshop. As my strength increased, I was raised to the dignity of being permitted to blow the bellows for him; and on winter evenings I was allowed to perch myself on his forge-hearth, listening to his stories. On one of these occasions, somebody by chance mentioned a sum—whether it was 9 times 9, or what it was, I do not now recollect; but, whatever it was, I gave the answer correctly. This occasioned some little astonishment; they then asked me other questions, which I answered with equal facility. They then went on to ask me up to 2 places of figures: 13 times 17, for instance. That was rather beyond me at the time; but I had been accustomed to reason on figures, and I said, 13 times 17 means 10 times 10 plus 10 times 7, plus 10 times 3 and 3 times 7. I said 10 times 10 are 100, 10 times 7 are 70, 10 times 3 are 30, and 3 times 7 are 21; which, added together, give the result 221. Of course I did not do it then as rapidly as afterward; but I gave the answer correctly, as was verified by the old gentleman's nephew, who began chalking it up to see if I was right. As a natural consequence, this increased my fame still more, and, what was better, it eventually caused halfpence to flow into my pocket, which, I need not say, had the effect of attaching me still more to the science of arithmetic; and thus by degrees I got on, until the multiple arrived at thousands. Then, of course, my powers of numeration had to be increased, and it was explained to me that 10 hundreds meant 1000. Numeration beyond that point is very simple in its features: 1000 rapidly gets up to 10,000 and 20,000, as it is simply 10 or 20 repeated over again, with thousands at the end, instead of nothing. So, by degrees, I became familiar with the numeration table up to a million. From 2 places of figures I got to 3 places; then to 4 places of figures, which took me up, of course, to tens of millions; then I ventured to 5 and 6 places of figures, which I could eventually treat with great facility; and on one occasion I went through the task of multiplying 12 places of figures by 12 figures, but it was a great and distressing effort."*

* Mr. Bidder's Addresses, *in extenso*, have been edited and published by Mr. Charles Manby, F.R.S.

CALCULATING MACHINES.

The employment of shells and pebbles for performing separate arithmetical operations was common before computers by the pen had attained proficiency for that purpose. The Roman Abacus was the oldest instrument of this kind: it was employed in the south of Europe till the end of the fifteenth century, and in England to a later period. It consisted of counters, movable in parallel grooves, or on parallel wires in a frame, and having the different denominations, units, tens, hundreds, etc., according to the grooves in which they were placed. In China, where the whole system is decimal, this instrument, called Schwampan, is used with great rapidity. From the merchants of China, at the great fair of Novogorod, the Muscovites are thought to have first learned the utility of the Abacus, since it is, at the present day, the common mode of reckoning in the shops of Moscow, the Russian money being in decimals. This is the simplest form of Calculating Machine with which we are acquainted.

"Napier's bones," described at page 140, is another instrument for arithmetical calculations; and Saunderson, the blind mathematician, invented a machine by which he was enabled to make computations.

Blaise Pascal, when scarcely nineteen years of age, devised a machine for performing arithmetical operations; its construction, however, was a much more troublesome task than its contrivance, and Pascal not only injured his constitution, but wasted the most valuable portion of his life in his attempts to bring it to perfection. A clockmaker in Rouen, to whom he had described his earliest model, made one of his own accord, which, though utterly unfit for its purpose, was placed in the cabinet of curiosities at Rouen, and annoyed Pascal so much that he dismissed all the workmen in his service, under the apprehension that other imperfect models might be made of the new machine they were employed to construct. Some time afterward, the Chancellor Seguier, having seen Pascal's first model, encouraged him to proceed, and obtained for him, in May, 1649, the exclusive privilege of constructing it; and he then gave up all his time to the machine. The first model which he executed proved unsatisfactory both in its form and materials. After successive improvements, he made a second, and then a third, which went by springs, and was very simple in its construction. This machine Pascal actually used several times in the presence of many of his friends; but defects gradually presented themselves; and he executed more than fifty models—all of them different; some of wood, others of ivory and ebony, and others of copper—before he completed the machine.

From this remarkable invention Pascal doubtless expected much more reputation than posterity has awarded. This over-estimate of its merits, founded on the length of time and the mental energy which it had exhausted, is strongly exhibited in a letter which he wrote to Christina, Queen of Sweden, in 1650, accompanying one of the machines. The tone of this letter is frank and manly; " for, though only in his twenty-seventh year, Pascal had witnessed, and even experienced, the truth, that nations who vaunt most loudly their superiority in science and learning have ever been the most guilty in neglecting and even starving their cultivators. The French monarch had, indeed, given him the exclusive privilege of his invention — the right of expending his time, his money, and his health in perfecting a machine for the benefit of France and the world; but, like a British patent bearing the Great Seal of England, it was not worth the wax which the royal insignia so needlessly adorned."*

Pascal's machine was an assemblage of wheels and cylinders: on the convex surfaces of the latter were the numbers with which the operations were to be performed, and attached to the axles of the cylinders were teethed wheels, which were turned by pointers, the additions being performed by means of the numbers in the lower series of numbers on the cylinders, and the subtractions

* *North British Review*, No. 2.

by the upper series. This machine excited a considerable sensation throughout Europe, and many attempts were made to improve its construction and extend its power. De l'Epine, Boitissendeau, and Grillet in France, S. Morland* and Gersten in England, and Poleni in Italy, applied to this task all their mathematical and mechanical skill, but none of them seem to have devised or constructed a machine superior to that of Pascal. The celebrated Leibnitz, however, is believed to have made two models of a Calculating Machine which surpassed Pascal's both in ingenuity and power; but its complicated structure, and the great expense and labor which the actual execution of it required, discouraged its inventor, and Leibnitz could not be prevailed upon to publish any detailed account of its mechanism: perhaps all that is known of it is that by wheel-work the operations of multiplication and division could be performed without the successive additions or subtractions which would be required if Pascal's machine were used.

The obvious value of these machines is for the obtaining numerical tables with the positive certainty of their being wholly exempt from errors; and, without numerical tables, astronomers, navigators, engineers, actuaries, and, indeed, laborers in every department of science and the useful arts, could have made but little progress in their several vocations.

The construction of a Calculating Machine which truly deserves that name was reserved for our distinguished countryman, Mr. Babbage. While all previous contrivances performed only particular arithmetical operations under a sort of copartnery between the man and the machine, in which the latter played a very humble part, the extraordinary invention of Mr. Babbage actually substitutes mechanism in the place of man. A problem is given to the machine, and it solves it by computing a long series of numbers following some given law. In this manner it calculates astronomical, logarithmic, and navigation tables, as well as tables of the powers and products of numbers. It can integrate, too, innumerable equations of finite differences; and, in addition to these

* See the notice of Morland's Arithmetical Machine at pages 157–8 of the present work.

functions, it does its work cheaply and quickly; it *corrects whatever errors are accidentally committed, and it prints all its calculations !*

The earliest allusion to this grand invention of the age occurs in a letter from Mr. Babbage to Sir Humphrey Davy, dated July 3, 1822, in which he gives some account of a small model of his engine for calculating differences (hence Mr. Babbage prefers to call it a Difference Engine†), which "produced figures at the rate of forty-four a minute, and performed with rapidity and precision all those calculations for which it was designed;" and Sir H. Davy witnessed and expressed his admiration of the performances of this engine. In the following year, upon the recommendation of a committee of the Royal Society, Mr. Babbage, at the desire of the government, undertook to superintend the construction of such an engine. He gave his mental labor gratuitously: drawings of the most delicate nature were made, tools were formed expressly to meet mechanical difficulties, and workmen educated in the construction of the machine. Mr. Babbage bestowed his whole time upon the subject for many years; and about £17,000 had been expended, when a dispute arose with the manager of the mechanical department, who withdrew, taking with him all the valuable tools that had been used in the work (which he had a legal right to do), and the works were suspended. Mr. Babbage now devised, upon a principle of an entirely new kind, an Analytical Engine of far simpler construction, to execute with greater rapidity the calculations for which the Difference Engine was intended, and which should contain a hundred variables, or numbers susceptible of changing, and each consisting of twenty-five figures. The government, however, abandoned the completion of the work, and in 1843 the portion of the Engine, as it existed, was placed in the Museum of King's College. It was capable of calculating to five figures and two orders of differences; but it is now out of order, and no portion of the printing machinery exists.

Throughout the long series of years which Mr. Babbage devoted to this great work, he did not receive one shilling for his invention, his time, or his services, while

* *North British Review*, No. 2.   † See Frontispiece.

he declined offers of great emolument, the acceptance of which would have interfered with his labors upon the Difference Engine.   Yet, with unwearied zeal, he has since occupied every working and almost every waking hour in the contrivance and the construction of the Analytical Engine, carrying on the drawings and experiments for this new machine at his own expense : the mechanical notations for the purpose cover 400 or 500 large sheets of paper, the original sketches extend to five volumes, and there are upward of 100 large drawings.   The following is a summary of the powers of the engine :

It will perform the several operations of simple arithmetic on any numbers whatever.   It can combine the quantities algebraically or arithmetically in an unlimited variety of relations.   It can use algebraic signs according to their proper laws, and develop the consequences of those laws.   It can arbitrarily substitute any formula for any others, effacing the first from the columns on which it is represented, and making the second appear in its stead.   And, lastly, it can effect processes of differentiation and integration on functions in which the operations take place by successive steps.   It is farther stated that the engine is particularly fitted for the operations of the combinatory analysis for computing the numbers of Bernouilli, etc.

The Difference Engine was elaborately described in the *Edinburgh Review*, July, 1834 ; from reading which Mr. George Scheutz, at that time the editor of a technological journal in Stockholm, was so fascinated with the subject that he set about constructing a machine for the same purpose as that of Mr. Babbage, namely, that of calculating and simultaneously printing numerical tables; but, after satisfying himself of the practicability of the scheme by constructing models of wood, pasteboard, and wire, he relinquished the design.   Three years afterward, in 1837, his son, Mr. Edward Scheutz, then a student in the Royal Technological Institute at Stockholm, being provided with a work-room in his father's house, as well as a lathe and other necessary tools, constructed a working model in metal, and succeeded in demonstrating the application of the scheme to practical purposes.   The father now applied to the government for aid, but was refused.   The father and son then worked together; but the severe economy they had been compelled to use in the purchase of materials and tools, and probably the absence in Sweden of those precious but expensive machine-tools which constitute the power of modern workshops, rendered this new model unsatisfactory in its operations, although perfectly correct in principle. Exhausted by these sacrifices, yet convinced that with better workmanship a more perfect instrument was within their reach, Mr. Scheutz applied for assistance to the Diet of Sweden; but the conditions on which they reluctantly consented to advance 5000 rix-dollars (about £280) were so stringent that the Messrs. Scheutz were compelled to renounce the work, and the model remained shut up in its case during the ensuing seven years.

The inventors then renewed their application to the Diet, and, by the assistance of some members of the Swedish Academy, the state being secured from loss, a limited amount was raised, and the Messrs. Scheutz, after working night and day, completed the machine before the end of October, 1853. The Diet now granted a reward of 5000 rix-dollars to the inventors, thus raising their total grant to 10,000 rix-dollars (about £560). The new engine performed its work so perfectly as to require no alteration whatever.

The size of Messrs. Scheutz's machine, when placed on its proper stand and protected by its cover, is about that of a small square pianoforte. The calculating portion consists of a series of fifteen upright steel axes, passing down the middle of five horizontal rows of silver-coated numbering rings, fifteen in each row, each ring being supported by and turning concentrically on its own small brass shelf, having within it a hole rather less than the largest diameter of the ring. Round the cylindrical surface of each ring are engraved the ordinary numerals from 0 to 9, one of which, in each position of the ring, appears in front, so that the successive numbers shown in any horizontal row of rings may be read from left to right, as in ordinary writing. The machine not only calculates the series of numbers, but it impresses each result on a piece of lead, from which a *cliché* in typemetal is taken, thus producing a stereotype plate from which printed copies may be obtained free from any error of composing, etc. The mechanism is peculiarly simple. The machine *calculates* to sixteen figures, but *prints* to eight only. By taking out certain wheels and inserting others, the machine can be readily caused to produce its results in £ *s. d.*, degrees, minutes, and seconds, or any other series of subdivisions which may be thought desirable. The machine performs its operations, when once set to the law on which the required table depends, by simply turning a handle, without any farther attention, the power required for the purpose being extremely small, not more than a child of ten years old could supply. The calculations are made, and the results impressed on the lead, at the rate of about 250 figures every ten minutes, the machine being worked slowly.

In the winter of 1854 the inventors brought their machine to London, where Mr. William Gravatt, F. R. S., civil engineer, showed and explained the invention to the Royal Society. It was next placed in the Great Exhibition at Paris, where Mr. Gravatt again kindly worked and explained the machine to many scientific gentlemen; and a jury unanimously awarded to it a gold medal. "The Emperor Napoleon" (says Mr. Babbage), "true to the inspirations of his own genius and to the policy of his dynasty, caused the Swedish Engine to be deposited in the Imperial Observatory of Paris, and to be placed at the disposal of the Members of the Board of Longitude."

In 1856 Mr. E. Scheutz revisited London; and the machine was brought from France, and was set to work in an apartment of Mr. Gravatt's house. It was subsequently purchased for the Dudley Observatory at Albany, U. S., by Mr. Rathbone, a merchant of that city.

Mr. Babbage, in some observations which he gracefully addressed to the Royal Society in 1856, says:

Mr. Scheutz's engine consists of two parts, the Calculating and the Printing; the former being again divided into two, the Adding and the Carrying parts.

With respect to the Adding, its structure is entirely different from my own, nor does it even resemble any one of those in my drawings.

The very ingenious mechanism for carrying the tens is also quite different from my own.

The Printing part will, on inspection, be pronounced altogether unlike that represented in my drawings, which, it must also be remembered, were entirely unknown to Mr. Scheutz.

The contrivance by which the computed results are conveyed to the printing apparatus is the same in both our engines; and it is well known in the striking part of the common eight-day clock, which is called "the snail."

A small volume of *Specimens of Tables calculated, stereo-moulded, and printed by Machinery*, by Messrs. Scheutz, is dedicated to Mr. Babbage, in recognition of the generous assistance he has afforded to the ingenious laborers in a similar field to that in which he has so long toiled. The remarkable and unique feature of the book itself is, that the tables and calculations are all printed from stereotyped plates produced directly from the machine, and without the use of any movable type.

One of Messrs. Scheutz's Difference Engines, made by Messrs. Donkin for the English government, is now worked in the Registrar General's Office in Somerset House.

Several other varieties of Calculating Machines have been produced within the last twenty years, but neither of them can be said to equal in the circumstances of its production the interest attached to the above engines.

"THE STARRY GALILEO." INVENTION OF THE TELESCOPE.

THERE is no instrument or machine of human invention (says Sir David Brewster) so recondite in its theory and so startling in its results as the Telescope. All others embody ideas and principles with which we are familiar; and, however complex their construction, or vast their power, or valuable their products, they are all limited in their application to terrestrial and sublunary purposes. The mighty steam-engine has its germ in the simple boiler in which the peasant prepares his food. The huge ship is but the expansion of the floating leaf freighted with its cargo of atmospheric dust; and the flying balloon is but the infant's soap-bubble lightly laden and overgrown. But the Telescope, even in its most elementary form, embodies a novel and gigantic idea, without an analogue in nature, and without a prototype in experience. It enables us to see what would forever be invisible. It displays to us the being and nature of bodies which we can neither see, nor touch, nor taste, nor smell. It exhibits forms and combinations of matter whose final cause reason fails to discover, and whose very existence even the wildest imagination never ventured to conceive. Like all other instruments, it is applicable to terrestrial purposes; but, unlike them all, it has its noblest application in the grandest and the remotest works of creation. The Telescope was never invented.* It was a divine gift which God gave to man,

* Among the individual claims to the invention, none appears to have made so near an approach as our celebrated countryman, Roger Bacon. In the following passage, extracted from his *Opus Majus*, he describes the phenomena depending on the refraction of light by lenses with so much truth, that we should almost feel justified in ascribing to him some share in the invention both of the telescope and microscope:

"Greater things than these may be performed by refracted vision. For it is easy to understand by the canons above mentioned that the in the last era of his cycle, to place before him and beside him new worlds and systems of worlds—to foreshadow the future sovereignties of his vast empire, the bright abodes of disembodied spirits, and the final dwellings of saints that have suffered, and of sages that have been truly wise.   With such evidences of His power and such manifestations of His glory, can we disavow His embassador, disdain His message, or disobey His commands?*

It was in the month of April or May, 1609, that a rumor, creeping through Europe by the tardy messengers of former days, at length found its way to Venice, where Galileo was on a visit to a friend, that a Dutchman had presented to Prince Maurice of Nassau an optical instrument which possessed the singular property of causing distant objects to appear nearer to the observer.   This Dutchman was Hans, or John Lippershey, who, as has been clearly proved by the late Professor Moll, of Utrecht, was in possession of a telescope made by himself so early as October, 1608.   A few days afterward this report was confirmed in a letter from James Badorere, at Paris, to Galileo, who immediately applied himself to the consideration of the subject.   On the first night after his return to Padua, he found, in the doctrines of refraction, the principle which he sought.   Having

greatest things may appear exceeding small, and the contrary.   For we can give such figures to transparent bodies, and dispose them in such order with respect to the eye and the objects, that the rays will be refracted and bent toward any place we please, so that we shall see the object near at hand or at a distance, under any angle we please; and thus from an incredible distance we may read the smallest letters, and may number the smallest particles of dust and sand, by reason of the greatness of the angle by which we may see them; and, on the contrary, we may not be able to see the greatest bodies close to us, by reason of the smallness of the angle under which they appear; for distance does not affect this kind of vision except by accident, but the magnitude of the angle does so.   And thus a boy may appear to be a giant, and a man as big as a mountain, forasmuch as we may see the man under as great an angle as the mountain, and as near as we please; and thus a small army may appear a very great one, and though very far off, yet very near to us, and the contrary.   Thus, also, the sun, moon, and stars may be made to descend hither in appearance, and to be visible over the heads of our enemies, and many things of the like sort, which persons unacquainted with such things would refuse to believe."

* *North British Review*, No. 3.

procured two spectacle-glasses, both of which were plane on one side, while one of them had its other side convex, and the other its second side concave, he placed one at each end of a leaden tube a few inches long, and, having applied his eye to the concave glass, he saw objects pretty large and pretty near him.   This little instrument, which magnified only three times, and which he held between his fingers or laid in his hand, he carried to Venice, where it excited the most intense interest.   Crowds of the principal citizens flocked to his house to see the magical toy; and after nearly a month had been spent in gratifying this epidemical curiosity, Galileo was led to understand from Leonardo Deodati, the Doge of Venice, that the senate would be highly gratified by obtaining possession of so extraordinary an instrument.   Galileo instantly complied with the wishes of his patrons, who acknowledged the present by a mandate conferring upon him for life his professorship at Padua, and raising his salary from 520 to 1000 florins.

These details are related, upon the authority of Viviani's *Life of Galileo*, by Sir David Brewster, who unhesitatingly asserts that a method of magnifying distant objects was known to Baptista Porta and others; but it seems equally certain that an *instrument* for producing these effects was first constructed in Holland, and from that kingdom Galileo derived the knowledge of its existence.   In considering the contending claims, it has been generally overlooked that *a single convex lens*, whose focal length exceeds the distance at which we examine minute objects, performs the part of a telescope, when an eye, placed behind it, sees distinctly the inverted image which it forms.   A lens twenty feet in focal length will in this manner magnify thirty times; and it was by the same principle that Sir William Herschel discovered a new satellite of Saturn, by placing his eye behind the focus of the mirror of his forty-feet telescope. The instrument presented to Prince Maurice, and which the Marquis Spinola found in the shop of John Lippershey, the spectacle-maker of Middleburg, must have been an astronomical telescope, consisting of two convex lenses. Upon this supposition, it differed from that which Galileo constructed, and the Italian philosopher will be justly entitled to the credit of having invented that form of telescope which still bears his name, while we must accord to the Dutch optician the honor of having previously invented the astronomical telescope.

. The interest which the exhibition of the telescope excited at Venice did not soon subside; Sirturi describes it as amounting to phrensy. When he himself had succeeded in making one of these instruments, he ascended the tower of St. Mark, where he might use it without molestation. He was recognized, however, by a crowd in the street; and such was the eagerness of their curiosity, that they took possession of the wondrous tube, and detained the impatient philosopher for several hours, till they had successively witnessed its effects. Desirous of obtaining the same gratification for their friends, they endeavored to learn the name of the inn at which Sirturi lodged; but he, overhearing their inquiries, quitted Venice early next morning.

The opticians speedily availed themselves of this wonderful invention. Galileo's tube, or the double eye-glass, or the cylinder, or the trunk, as it was called—for Demisiano had not then given it the appellation of *Telescope*—was manufactured in great numbers, and in a very inferior manner. The instruments were purchased merely as philosophical toys, and were carried by travelers into every corner of Europe. The art of grinding and polishing lenses was at this time very imperfect. Galileo, and those whom he instructed, were alone capable of making tolerable instruments. In 1634 a good telescope could not be procured in Paris, Venice, or Amsterdam; and even in 1637 there was not one in Holland which could show Jupiter's disk well defined.

After Galileo had completed his first instrument, which magnified only *three* times, he executed a larger and better one, with a power of about *eight*. "At length," as he himself remarks, "sparing neither labor nor expense," he constructed a telescope so excellent that it bore a magnifying power of more than *thirty* times.*

Thus was Galileo equipped for a survey of the heavens. The first celestial object to which he directed his telescope was the Moon, which, to use his own words, appeared as

* *North British Review*, No. 3.

near as if it had been distant only two semidiameters of
the earth.   It displayed to him her mountain ranges and
her glens, her continents and her highlands, now lying
in darkness, now brilliant with sunshine, and undergoing
all those variations of light and shadow which the surface of our own globe presents to the Alpine traveler or
to the aeronaut.   The four satellites of Jupiter, illuminating their planet, and suffering eclipses in his shadow like
our own moon; the spots on the sun's disk, proving his
rotation round his axis in twenty-five days; the crescent
phases of Venus; and the triple form, or the imperfectly
developed ring of Saturn, were the other discoveries in
the solar system which rewarded the diligence of Galileo.
In the starry heavens, too, thousands of new worlds were
discovered by his telescope; and the Pleiades alone,
which to the unassisted eye exhibits only *seven* stars,
displayed to Galileo no fewer than *forty*.*

It was then that, to his unutterable astonishment, Galileo saw, as a celebrated French astronomer (M. Biot) has
expressed it, " what no mortal before that moment had
seen—the surface of the moon, like another earth, ridged
by high mountains and furrowed by deep valleys; Venus,
as well as it, presenting phases demonstrative of a spherical form; Jupiter surrounded by four satellites, which
accompanied him in his orbit; the milky way; the nebulæ; finally, the whole heavens sown over with an infinite multitude of stars too small to be discerned by the
naked eye."   Milton, who had seen Galileo, described,
nearly half a century after the invention, some of the
wonders thus laid open by the telescope:

> "The moon, whose orb
> Through optic glass the Tuscan artist views
> At evening from the top of Fesolé,
> Or in Valdarno, to descry new lands,
> Rivers, or mountains, in her spotty globe."

"There are" (says Everett, the American orator) " occasions in life in which a great mind lives years of rapt
enjoyment in a moment.   I can fancy the emotions of
Galileo when, first raising the newly-constructed telescope to the heavens, he saw fulfilled the grand prophecy
of Copernicus, and beheld the planet Venus crescent like

* *Martyrs of Science.* By Sir David Brewster, K. H. 4th edit., 1858.

the moon. It was such another moment as that when 'the immortal printers of Mentz and Strasburg received the first copy of the Bible into their hands, the work of their divine art; like that when Columbus, through the gray dawn of the 12th of October, 1492 (Copernicus, at the age of eighteen, was then a student at Cracow), beheld the shores of San Salvador; like that when the law of gravitation first revealed itself to the intellect of Newton; like that when Franklin saw by the stiffening fibres of the hempen cord of his kite that he held the lightning in his grasp; like that when Leverrier received back from Berlin the tidings that the predicted planet was found."

"The starry Galileo, with his woes," is enshrined among "the Martyrs of Science." His noblest discoveries were the derision of his contemporaries, and were even denounced as crimes which merited the vengeance of Heaven. He was the victim of cruel persecution, and spent some of his latest hours within the walls of a prison; and, though the Almighty granted him, as it were, a new sight, to discern unknown worlds in the obscurity of space, yet the eyes which were allowed to witness such wonders were themselves doomed to be closed in darkness. Sir David Brewster eloquently says:

"The discovery of the moon's Libration was the result of the last telescopic observations of Galileo. Although his right eye had for some years lost its power, yet his general vision was sufficiently perfect to enable him to carry on his usual researches. In 1636, however, this affection of his eye became more serious, and in 1637 his left eye was attacked with the same disease: the disease turned out to be in the cornea, and every attempt to restore its transparency was fruitless. In a few months the white cloud covered the whole aperture of the pupil, and Galileo became totally blind. This sudden and unexpected calamity had almost overwhelmed Galileo and his friends. In writing to a correspondent, he exclaims: 'Alas! your dear friend and servant has become totally and irreparably blind. Those heavens, this earth, this universe, which by wonderful observation I had enlarged a thousand times beyond the belief of past ages, are henceforth shrunk into the narrow space which I myself

K

occupy.   So it pleases God ; it shall, therefore, please me
also.'   Galileo's friend, Father Castelli, deplores the calamity in the same tone of pathetic sublimity: 'The
noblest eye,' says he, ' which nature ever made, is darkened; an eye so privileged, and gifted with such rare
powers, that it may truly be said to have seen more than
the eyes of all that are gone, and to have opened the
eyes of all that are to come.' "*

* *Martyrs of Science*, 4th edit., 1858.

ISAAC NEWTON MAKES THE FIRST REFLECTING TELESCOPE.

ACCORDING to Newton's own confession, he was extremely inattentive to his studies while in the public school at Grantham; and Sir David Brewster, with the sympathy of a fond biographer, attributes this idleness to the occupation of the mind of the future philosopher with subjects in which he felt a deeper interest. He had not been long at school before he exhibited a taste for mechanical inventions. With the aid of little saws, hammers, hatchets, and tools of all sorts, he was occupied during his play-hours in constructing models of known machines and amusing contrivances. Thus he modeled a wind-mill from a mill, which he watched in course of construction, near Grantham. Dr. Stukeley describes this working model to have been "as clean and curious a piece of workmanship as the original." Newton next constructed a water-clock: it had a dial-plate at top, with figures of the hours; the index was turned by a piece of wood, which either fell or rose by water dropping. He also invented a mechanical carriage, a four-wheeled chair, which was moved by the handle or winch wrought by the person who sat in it. He also made for the amusement of his school-fellows paper kites and lanterns of crimpled paper. At home he drove wooden pegs into the walls and roofs of the buildings, as gnomons to mark by their shadows the hours and half hours of the day; and he carved two dials in stone upon the walls of 'us house at Woolsthorpe.*

Several years after, when Newton had entered Trinity College, Cambridge, we find in one of his commonplace books an entry, dated January, 1663–4, " on the 'grinding of spherical optic glasses'—on the errors of lenses, and the method of rectifying them," etc., to which Newton soon applied himself. Descartes had invented and described machines for grinding and polishing lenses

* See *School-days of Eminent Men,* 1858.

with accuracy, upon which the perfection of refracting telescopes and microscopes depended. Newton, however, by the first experiments which he made with a prism, found that the perfection of telescopes was limited not so much for want of glasses truly figured, as because *light* itself is a heterogeneous mixture of differently refrangible rays, so that an exactly figured glass could not collect all sorts of rays into one point.

This new branch of science now occupied much of Newton's attention; and among the articles which he purchased on his visit to London early in 1669 were lenses, two furnaces, and several chemicals. Toward the end of 1668, thinking it best to proceed by degrees, he first "made a small perspective, to try whether his conjecture would hold good or not." The telescope was six inches long. The aperture of the large speculum was something more than an inch; and as the eye-glass was a plano-convex lens, with a focal length of one sixth or one seventh of an inch, "it magnified about forty times in diameter," which he believed was more than any six-feet refracting telescope could do with distinctness. It did not, however, through the bad materials and the want of a good polish, represent objects so distinctly as a good six-feet refractor; yet Sir Isaac saw with it Jupiter, and also the horns or "moon-like phase of Venus." He therefore considered this small telescope as an "epitome" of what might be done by reflections; and he did not doubt that in time a six-feet reflector might be made which would perform as much as any 60 or 100-feet refractor.

Newton did not, however, resume the construction of reflectors till the autumn of 1671, when, finding that grinding and polishing the lenses produced very little change in the indistinctness of the image, he discovered that the defect arose from the different refrangibility of the rays of light. He took the glass prism which he had purchased at Stourbridge fair, and having made a hole in the window-shutter of his darkened room, he admitted through the prism a ray of the sun's light, which, after refraction, exhibited on the opposite wall the Solar or Prismatic Spectrum, and by a laborious investigation proved the different refrangibility of the rays of light to be the real cause of the imperfection of refracting tele-

scopes, which he proposed to remedy by a metallic speculum within the tube, by which the rays proceeding from
the object are reflected to the eye. Newton accordingly set about executing another reflecting telescope with
his own hands. This consisted of a concave metallic
speculum, the rays reflected by which were received
upon a plane metallic speculum inclined 45° to the axis
of the tube, so as to reflect them to the side of the tube,
in which there was an aperture to receive a small tube
with a plano-convex glass, by means of which the image
formed by the speculum was magnified thirty-eight
times; "whereas an ordinary telescope of about two
feet long only magnified thirteen or fourteen times."

At the request of some of the members of the Royal
Society, Newton sent this telescope for inspection, and
subsequently presented it to that distinguished body.
It was also shown to King Charles II. This instrument
is carefully preserved in the library of the Royal Society at Burlington House, Piccadilly, with the inscription

"THE FIRST REFLECTING TELESCOPE, INVENTED BY SIR ISAAC
NEWTON, AND MADE WITH HIS OWN HANDS."

Such was the first Reflecting Telescope that was successfully constructed and applied to the heavens; though Sir David Brewster describes it as a small and ill-made instrument, incapable of showing the beautiful celestial phenomena which had been long seen by refracting telescopes; and more than fifty years elapsed before telescopes of the Newtonian form became useful in astronomy.

Nevertheless, this " is the instrument which, under the hands of Herschel and Rosse, has grown to proportions so gigantic as to require the aid of vast machinery to elevate and depress the tube. Newton's first telescope is nine inches long, Lord Rosse's six-feet reflector is sixty feet in length !"*

At Grantham, in Lincolnshire, nigh to the hamlet wherein Newton was born, was reared in 1858 a statue of this " greatest genius of the human race." This was 131 years from the date of Newton's death, and verified the homely proverb that a prophet is honored every where save in his own country and among his own people. The statue is of bronze, and is nearly 13 feet high; and the sculptor, Mr. Theed, has copied the likeness of Sir Isaac from a mask of his face taken after death, and from the portrait-bust by Roubiliac. The statue was inaugurated September 21, when Lord Brougham (one of the editors of Newton's works) delivered the address.

In conclusion, his lordship said: Let it not be imagined that the feelings of wonder excited by contemplating the achievements of this great man are in any degree whatever the result of national partiality, and confined to the country which glories in having given him birth. The language which expresses her veneration is equaled, perhaps exceeded, by that in which other nations give utterance to theirs; not merely by the general voice, but by the well-considered and well-informed judgment of the masters of science. Leibnitz, when asked at the royal table in Berlin his opinion of Newton, said that, "taking mathematicians from the beginning of the world to the time when Newton lived, what he had done was much the better half." "The *Principia* will ever remain a monument of the profound genius which revealed to us the greatest law of the universe," are the words of Laplace. "That work stands pre-eminent above all the other productions of the human mind." "The discovery of that simple and general law, by the greatness and the variety of the objects which it em-

---

* Weld's *Hist. Royal Society*, vol. i.

Statue of Sir Isaac Newton, at Grantham.

braces, confers honor upon the intellect of man." Lagrange, we are told by D'Alembert, was wont to describe Newton as the greatest genius that ever existed; but to add how fortunate he was also, "because there can only once be found a system of the universe to establish." "Never," says the father of the Institute of France—one filling a high place among the most eminent of its members—"never," says M. Biot, "was the supremacy of intellect so justly established and so fully confessed: in mathematical and in experimental science without an equal and without an example, combining the genius for both in its highest degree." The *Principia* he terms the greatest work ever produced by the mind of man; adding, in the words of Halley, "that a nearer approach to the Divine nature has not been permitted to mortals." "In first giving to the world Newton's method of fluxions," says Fontenelle, "Leibnitz did like Prometheus—he stole fire from heaven to bestow it upon men." "Does Newton," L'Hôpital asked, "sleep and wake like other men?   I figure him to myself as a celestial genius, entirely disengaged from matter."*

---

* "The great discovery which characterizes the *Principia* is that of the principle of universal gravitation—*that every particle of matter in the universe is attracted by, or gravitates to, every other particle of matter, with a force inversely proportional* to the square of their distance. . . . The most complete and successful attempt to make the *Principia* accessible to those who are 'little skilled in mathematical science,' has been made by Lord Brougham in his admirable analysis of that work, which forms the greater part of the second volume of his edition of Paley's *Natural Theology.*"—SIR D. BREWSTER'S *Life of Newton.*

GUINAND'S GLASS FOR ACHROMATIC TELESCOPES.

THE refracting telescope, whose inventor we can not confidently name, was a small and useless toy till Galileo turned it to the heavens; and though, in the hands of Huyghens and Hevelius, it added new satellites to our system, and displayed new forms and structures in the primary planets, yet it was only when made achromatic, through the labors of Hall, Dollond, Frauenhofer, and others, that it became an essential instrument for the advancement of astronomy. The reflecting telescope presents to us the same peculiarity. We do not know the inventor. Even in Sir Isaac Newton's hands, and as constructed and applied by himself, it effected no discoveries in the heavens.

The difficulty of procuring flint-glass free from flaws and imperfections long checked the improvement of the Achromatic Telescope. All convex lenses of glass, with spherical surfaces, as the reader may be aware, form images of objects in their focus behind the lens; but, owing to the spherical and chromatic aberrations, a mass of images of different colors, and not coincident with each other, is the result. Sir Isaac Newton pronounced these imperfections to be incurable; but Mr. Chester More Hall, a gentleman of Essex, so early as 1733, in imitation of the organ of sight, combined media of different refractive powers, and constructed object-glasses of flint and crown glass, which corrected the chromatic and diminished the spherical aberration. The telescopes thus made (which Dr. Bliss named achromatic, *i. e.*, destitute of color) were neither exhibited nor sold, nor was any account of their construction made known to the world. In a trial at Westminster about the patent for making achromatic telescopes, Mr. Hall was allowed to be the inventor; but Lord Mansfield observed that "it was not the person who locked his invention in his escritoire that ought to profit from such invention, but he who brought it forth for the benefit of the public."

In 1758, however, John Dollond arrived at the same result: he reinvented the achromatic telescope, manufactured the instrument for sale, and for more than half a century supplied all Europe with this invaluable instrument.*

The monopoly of these telescopes, however, soon passed into foreign states. The manufacture of flint-glass had been so severely taxed by the British government—as though to put down the achromatic telescope by statute— that if a philosopher melted a pound of glass fifty times, he had to pay the duty upon fifty pounds. When the government understood their ignorance of British interests, a committee of the Royal Society was permitted to erect an experimental glass-house, and to enjoy the privilege of compounding a pot of glass without the presence and supervision of an exciseman! The experimental furnace was erected at Green and Pellatt's Falcon Glass-house, and subsequently a room and furnaces were built at the Royal Institution; Dr. Faraday superintended the chemical part of the inquiry, and by the year 1830 the committee had manufactured glass of a superior quality for optical purposes. Nevertheless, Dr. Faraday considered the results as negative, and the manufacture was laid aside.

The monopoly of the Achromatic Telescope was thus lost. What a conclave of English legislators and philosophers attempted in vain, was, however, accomplished by a humble peasant in the gorges of the Jura, where no patron encouraged and no exciseman disturbed him. M. Guinand, a maker of clock-cases in the village of Brenetz, in the canton of Neufchatel, had been obliged by defective vision to grind spectacle-glasses for his own use. Thus practically versed in the optics of lenses, he amused himself with making small refracting telescopes, which

* John Dollond was born at Spitalfields, in London, in the year 1706. He was descended from French ancestors, who were compelled to quit Normandy upon the revocation of the edict of Nantes by Louis XIV. In early life he worked at the loom, but in 1752 he joined his son as an optician. He died in 1761, having been struck with apoplexy while engaged in an intense study of Clairaut's Theory of the Moon.

he mounted in pasteboard tubes. Meanwhile an achromatic telescope of English manufacture had come into the possession of Guinand's master, Jacquet Droz. He was permitted to examine it, to separate its lenses, and to measure its curves ; and, after studying its properties, he resolved to attempt to imitate the wondrous combination. Flint-glass was only to be had in England ; and he and his friend, M. Riondon, who went to that country to take out a patent for his self-winding watches, purchased as much glass as enabled Guinand to supply several achromatic telescopes. The glass, however, was bad ; and the clock-case maker, seeing no way of getting it of a better quality, resolved upon making flint-glass for his own use. Studying the chemistry of fusion, he made daily experiments in his blast furnace, between 1784 and 1790, with meltings of three or four pounds each, and carefully noted down the circumstances and the results of each experiment. He succeeded, and abandoning his business for the more lucrative one of making bells for repeaters, he obtained more means and leisure. He purchased a piece of ground on the banks of the Doubs, where he constructed a furnace capable of fusing *two hundred weight of glass.* The failure of his crucibles, the bursting of his furnaces, and a thousand untoward accidents, which would have disconcerted less ardent minds, served only to invigorate the unlettered peasant. The threads, and specks, and globules, which destroyed the homogeneity of his glass, were the subjects of his constant study; and he at last succeeded in obtaining considerable pieces of uniform transparency and refractive powers, sometimes *twelve,* and in one case *eighteen* inches in diameter. He at last acquired the art of soldering two or more pieces of good glass ; and, though the line of junction was often marked with globules of air or particles of sand, yet by grinding out these imperfections on an emeried wheel, and by replacing the mass in a furnace, so that the vitreous matter might expand and fill up the excavations, he succeeded in effacing every trace of junction, and was consequently able to produce with certainty the finest disks of flint-glass.

Frauenhofer, the Bavarian optician, having heard of Guinand's success in the manufacture of flint-glass, re-

paired to Brenetz in 1804, and induced the Swiss artisan
to settle at Munich, where, from 1805 to 1814, he practiced his art, and taught it to his employer. Frauenhofer
was an apt and willing scholar, and, possessing a thorough
knowledge of chemistry and physics, he speedily learned
the processes of his teacher, and discovered the theory
of manipulation, of which Guinand knew only the results.
Thus supplied with the finest materials of his art, he
studied their refractive and dispersive powers; and by
his grand discovery of the fixed lines in the spectrum,
he arrived at methods of constructing achromatic telescopes which no other artist had possessed. In these
laborious researches he was patronized by Maximilian
Joseph, King of Bavaria; and, had he not been carried
off by an insidious disease in the prime of life, he would
have astonished Europe with the production of an achromatic object-glass eighteen inches in diameter.

Guinand remained at Munich until 1814, when he returned to his native village, where, in 1820, he was visited by M. Lerebours, a celebrated optician of Paris, who
had heard of the success of his processes. Lerebours
purchased all his glass, and left orders for more; and M.
Cauchoix, another skillful Parisian artist, procured from
him large disks of glass. In this manner refracting telescopes came to be constructed in France rivaling the
most finished productions of the Munich artist; and England, which was long the exclusive seat of the manufacture of achromatic telescopes, had the mortification of
finally seeing both Germany and France completely out-
strip her in this branch of practical optics. This she
owed to the shortsighted policy of the British government, which had placed an exorbitant duty on the manufacture of flint-glass. This vexatious fiscal interference
has, however, been repealed, and the enterprise with
which makers and opticians have taken up the construction of large object-glasses has led to important results.

Mr. Apsley Pellatt, in his *Curiosities of Glass-making*,
a work of sound practical value as well as popular interest, says: "The secret of Guinand's success is considered
not to have been in the novelty of the materials or proportions, but in agitating the liquid glass while at the
highest point of fusion; then cooling down the entire

contents of the pot in a mass, and, when annealed and cool, by cleavage separating unstriated portions, afterward softening into clay moulds. Guinand left two sons, one of whom subsequently operated in conjunction with M. Bontemps, a scientific French glass-maker, who succeeded in making good flint optical glass on the principle of mechanical agitation."*

In 1848, Bontemps, after attaining high eminence in his art, was induced to retire from France, and to co-operate with Messrs. Chance, Brothers, and Co., of Birmingham, in improving the quality of their manufactures. They conjointly succeeded in producing a disk in flint of 29 inches in diameter, weighing 2 cwt., and which, being submitted to the operation of grinding, finishing, and other processes, in order to prove its quality, received a Council Medal at the Great Exhibition of 1851. When we recollect that a glass, exceeding only the small diameter of six inches, undergoes the annealing process with difficulty, and is liable to cool at the surface *more* especially than in the interior, and that this tendency increases with the size, we must regard this production of a disk of 29 inches as a very remarkable work.

* The widow of Guinand, and her other son, set up works in Switzerland upon the father's principles; they were succeeded by M. Daguet, of Soleure, who sent to the Great Exhibition of 1851 some of his products, but only of moderate size.

SIR WILLIAM HERSCHEL AND HIS TELE-
SCOPES.

THE long interval of half a century seems to be the period of hybernation during which the telescopic mind rests from its labors in order to acquire strength for some great achievement. Fifty years elapsed between the dwarf Telescope of Newton and the large instrument of Hadley; other fifty years rolled on before Sir William Herschel constructed his magnificent Telescope; and fifty years more passed away before the Earl of Rosse produced that colossal instrument which has already achieved such brilliant discoveries.*

We have just described the construction of Newton's dwarf Telescope; fifty years after which, John Hadley, the inventor of the Reflecting Quadrant which bears his name, began his experiments, and, probably after many failures, completed a telescope in 1720. It was presented to the Royal Society (of which Hadley was a Fellow), as thus recorded in the *Journal* for January 12, 1721:

> Mr. Hadley was pleased to show the Royal Society his Reflecting Telescope, made according to our President's (Sir Isaac Newton) directions in his *Optics*, but curiously executed by his own hand, the force of which was to enlarge an object near *two hundred times*, though the length thereof exceeds *six* feet; and having shown it, he made a present thereof to the society, who ordered their hearty thanks to be recorded for so valuable a gift.

This instrument consisted of a metallic speculum about six inches in diameter, and its focal length was 5 feet $2\frac{1}{2}$ inches. Its plane speculum was made of the same metal, about the 15th of an inch thick; and it had six eye-pieces, three convex lenses, 1-3d, 3-10ths, and 11-40ths of an inch, magnifying 190, 208, and 220 times; and an erecting eye-piece of three convex lenses, magnifying about 125 times. It had also a small refracting telescope as a finder, which we believe was first suggested by Des-

* Sir David Brewster's *Life of Sir Isaac Newton*, vol. ii.

cartes, and the whole was mounted upon a stand ingeniously and elegantly constructed. The celebrated Mr. Bradley, and the Rev. Mr. Pound, of Wanstead, compared it with the great Huyghenian refractor, 123 feet long, and, though less brightly, they saw with the reflector whatever they had hitherto discovered with the Huyghenian, together with the belts of Saturn, and the first and second satellites of Jupiter as bright spots on the body of the planet.

After executing another Newtonian telescope of the same size, Mr. Hadley made great improvements in those of the Gregorian form. He was now Vice-president of the Royal Society, and set about enabling astronomers and opticians to manufacture these valuable instruments. Mr. Hawksbee first made them for public sale; others did the same. The opticians, with the aid of Molyneaux and Hadley, succeeded in the new art of grinding and polishing specula. Scottish makers followed; and, says Sir David Brewster, " in this way the Reflecting Telescope came into general use, and, principally in the Gregorian form, it has been an article of trade with every regular optician." Notwithstanding these great improvements, no discovery of importance had yet been achieved by the Reflecting Telescope, and nearly three quarters of a century had elapsed without any extension of our knowledge of the Solar and Sidereal Systems. " This, however," continues Sir David, " was only one of those stationary intervals during which human genius holds its breath, in order to take a new and loftier flight. It was reserved for Sir William Herschel and the Earl of Rosse to accomplish the great work, and, by the construction of telescopes of gigantic size, to extend the boundaries of the Solar System, to lay open the hitherto unexplored recesses of the sidereal world, and to bring within the grasp of reason those nebular regions to which imagination had not ventured to soar."

Sir William Herschel, one of the very greatest names in the modern history of astronomical discovery, was self-instructed in the science in which he earned his high reputation. He was born at Hanover in 1738, and was the son of a musician in humble circumstances. Brought up to his father's profession, he was placed, at the age of

fourteen, in the band of the Hanoverian Guards, a detach-
ment of which being ordered to England in 1757, young
Herschel accompanied it, and remained to try his fortune
in London.   Here he had to struggle with many difficulties.   He then passed several years principally in giving
lessons in music to private pupils in different towns in
the north of England.   In 1765 he obtained the situation
of organist at Halifax; and next year he was appointed
to the same office in the Octagon Chapel at Bath, where
he settled, with the certain prospect of deriving a good
income from his profession, if he had made that his only
or his chief object.   There is a mass of stories relating
to his musical occupations, none of which have any certain foundation : as, that he played in the Pump-room
band at Bath; and that, when a candidate for the situation of organist, he helped his performance by placing
upon holding notes little bits of lead, which he dexter-
ously removed in time.

  But long before this, while yet only an itinerant teacher of music in country towns, Herschel had assiduously
devoted his leisure to the acquiring of a knowledge of
the Italian, the Latin, and the Greek languages.   He then
applied himself to the study of Robert Smith's profound
Treatise on Harmonics, for which purpose it was necessary that Herschel should make himself a mathematician;
and to accomplish this, he laid aside all other pursuits of
his leisure.   At Bath he devoted still more time to mathematical studies.   In the course of time he obtained a
competent knowledge of geometry; he next studied the
different branches of science which depend upon the
mathematics, his attention being first attracted by the
kindred departments of astronomy and optics.   He now
became anxious to observe with his own eyes those wonders of the heavens of which he had read so much, and
for that purpose he borrowed from an acquaintance a
two-feet Gregorian telescope.   This instrument interested him so greatly that he commissioned a friend in London to purchase one for him of a somewhat larger size;
but, fortunately for science, he found the price beyond
what he could afford.   To make up for his disappointment, he resolved to attempt to construct with his own
hands a telescope for himself; and after encountering

innumerable difficulties in the progress of his task, he at last succeeded, and in the year 1774 he completed a five-feet Newtonian reflector, with which he distinctly saw the ring of Saturn and the satellites of Jupiter.

Herschel now, becoming dissatisfied with the performance of his first instrument, renewed his labors, and in no long time produced telescopes of seven, ten, and even twenty feet focal distance. In fashioning the mirrors for these instruments his perseverance was indefatigable. For his seven-feet reflector he actually finished and made trial of two hundred mirrors before he found one that satisfied him; one hundred and fifty for his ten-feet, and above eighty for his twenty-feet instrument. He usually worked at a mirror for twelve or fourteen hours, without quitting his occupation for a moment. He would not even take his hand from what he was about to help himself to his food, and the little that he ate when so employed was put into his mouth by his sister. He gave the mirror its proper shape more by a certain natural tact than by rule; and when his hand was once in, as the phrase is, he was afraid that the perfection of the finish might be impaired by the least intermission of his labors.

It was on the 13th of March, 1781, that Herschel made the discovery to which he owes, perhaps, most of his popular reputation. On the evening of the above day, having turned his telescope (an excellent seven-feet reflector of his own constructing) to a particular part of the sky, he observed among the other stars one which seemed to shine with a more steady radiance than those around it. He determined to observe it more narrowly; after some hours, it had perceptibly changed its place—a fact which the next day became still more indisputable. The Astronomer Royal, Dr. Maskelyne, concluded that the luminary could be nothing else than a new comet; but in a few days it became evident that it was, in reality, a hitherto undiscovered planet; this Herschel named the *Georgium Sidus*, or Georgian Star, in honor of the King of England; but it has been more generally called either *Herschel*, after its discoverer, or *Uranus*. The diameter of this new globe has been found to be nearly four and a half times larger than our own; its size altogether eighty times that of our earth; its year is as long as eighty-three of ours; its distance from the sun is nearly eighteen hundred millions of miles, or more than nineteen times that of the earth; its density, as compared with that of the earth, is nearly as 22 to 100, so that its entire weight is not far from eighteen times that of our planet. Herschel afterward discovered successively no fewer than six satellites or moons belonging to his new planet. Mr. De Morgan almost prophetically wrote: "Its name is appropriate, inasmuch as Uranus is the father of Saturn in mythology; but what will be done if a new planet should be discovered still more distant than Uranus?" A new planet has been discovered by two master minds, independently of each other, in a manner which renders the discovery of Uranus deeply interesting.

The merit of this discovery is in itself small. It is the method which gave rise to it, on which this part of Herschel's fame must rest. Perceiving how much depended on an exact knowledge of telescopic phenomena, and a perfect acquaintance with the effect produced by differences of instrumental construction, he commenced a regular examination of the heavens, taking the stars systematically in series, and using one telescope throughout. He was not a mere dilettante star-gazer, but a volunteer, carrying on, with no great pecuniary means, a laborious and useful train of investigation.

Herschel's name now became universally known. The Copley Medal was awarded to him by the Royal Society. The king attached him to his court as private astronomer, with a salary of £400 a year; and soon after this he came to reside first at Datchet, and then at Slough, near Windsor. He now devoted himself entirely to science. In this year, 1781, he began a thirty-feet aerial reflector, with a speculum three feet in diameter; but as it was cracked in the operation of annealing, and as another of the same size was lost in the fire from a failure in the furnace, his hopes were disappointed. This double accident, however, only acted as a stimulus to higher achievements, and no doubt suggested the idea of making a still larger instrument, and of obtaining pecuniary aid for its accomplishment. In 1785, at the request of

Sir William Herschel, and with the sanction of the Council of the Royal Society, the president, Sir Joseph Banks, laid before George III. the great astronomer's scheme for the construction of a Reflecting Telescope of colossal dimensions. The king approved of the plan, and offered to defray the whole expense of it; a noble act of liberality, which has never been imitated by any other British sovereign.

Herschel next conceived the happy idea of "Gauging the Heavens," by counting the number of stars which passed at different heights and in various directions, over the field of view of fifteen minutes in diameter of his twenty-feet reflecting telescope. The field of view each time embraced only 1-838,000th of the whole heavens; and it would therefore require, according to Struve, eighty-three years to gauge the whole sphere by a similar process.

Toward the close of this year Herschel began to construct his Reflecting Telescope, *forty feet in length,* and having a speculum *fully four feet* in diameter. It was completed August 27, 1789; and Sir William has left a very complete description of the operations:

I began (says Herschel) to construct the forty-feet telescope about the latter end of 1785. In the whole of the apparatus none but common workmen were employed; for I made drawings of every part of it, by which it was easy to execute the work, as I constantly inspected and directed every person's labor, though sometimes there were no less than forty different workmen employed at the same time. While the stand of the telescope was preparing, I also began the construction of the great mirror, of which I inspected the casting, grinding, and polishing; and the work was in this manner carried on with no other interruption than what was occasioned by the removal of all the apparatus from Clay Hall, where I then lived, to my present situation at Slough. Here, soon after my arrival, I began to lay the foundation upon which, by degrees, the whole structure was raised as it now stands; and the speculum being highly polished and put into the tube, I had the first view through it on February 9, 1787. I do not, however, date the completing of the instrument till much later; for the first speculum, by a mismanagement of the person who cast it, came out thinner on the centre of the back than was intended, and on account of its weakness would not permit a good figure to be given to it. A second mirror was cast Jan. 26, 1788, but it cracked in cooling. Feb. 16, we recast it with peculiar attention to the shape of the back, and it proved to be of a proper degree of strength. Oct. 24, it was brought to a pretty good figure and polish, and I observed the planet Saturn with it. But not being satisfied, I continued to work upon it till Aug. 27, 1789, when it was tried upon the fixed stars, and I found it to give a pretty sharp image. Large stars were a little affected with scattered light, owing to many remaining scratches in the mirror. Aug. 28, 1789, having brought the telescope to the parallel of Saturn, I discovered a sixth satellite of that planet; and I also saw the spots upon Saturn better than I had ever seen them before, so that I may date the finishing of the forty-feet telescope from that time.—*Phil. Trans. for* 1790.

The thickness of the speculum, which was uniform in every part, was 3½ inches, and its weight nearly 2118 pounds; the metal being composed of 32 copper, and 10·7 of tin. The speculum, when not in use, was preserved from damp by a tin cover, fitted upon a rim of close-grained cloth. The tube of the telescope was 39 feet 4 inches long, and its width 4 feet 10 inches; it was made of iron, and was 3000 pounds lighter than if it had been made of wood. The observer was seated in a suspended movable seat at the mouth of the tube, and viewed the image of the object with a magnificent lens or eye-piece. The focus of the speculum, or place of the image, was within 4 inches of the mouth of the lower side of the tube, and came forward into the air, so that there was a space for part of the head above the eye, to prevent it from interrupting many of the rays going from the object to the mirror. The eye-piece moved in a tube carried by a slider directed to the centre of the speculum, and fixed on an adjustible foundation at the mouth of the tube.

The very first moment this magnificent instrument was directed to the heavens, a new body was added to the Solar System, namely, Saturn and six satellites, and in less than a month after, the seventh satellite of Saturn; "an object," says Sir John Herschel, "of a far higher order of difficulty."

Herschel's Great Telescope stood on the lawn in the rear of his house at Slough, and some of our readers, like ourselves, may remember its extraordinary aspect when seen from the Bath coach-road and the road to Windsor. The difficulty of managing so large an instrument, requiring, as it did, two assistants in addition to the observer himself and the person employed to note the time, prevented its being much used; and in 1839, the wood-work of the telescope being decayed, Sir John Herschel had it cleared away: piers were erected on which the tube was placed; that was of iron, and so well preserved that, although not more than one twentieth of an inch thick, when in the horizontal position, it contained all Sir John's family, besides portions of the machinery and polishing apparatus to the weight of a great many tons. Sir John attributes this great strength and resistance to decay to its internal structure, very similar to that since patented as Corrugated Iron Roofing, the idea of which originated with Sir William Herschel at the time he constructed the Great Telescope. By the system of triangular arrangement or diagonal bracing adopted in the wood-work also, much strength was gained.

The entire expense of the Great Telescope, so munificently defrayed by George III., including, of course, the cost of the construction of tools and the apparatus for casting, grinding, and figuring the reflectors, of which two were constructed, amounted to £4000. His abode at Slough became, as Fourier remarks, one of the most remarkable spots of the civilized world. M. Arago says it may confidently be asserted that at the little house and garden at Slough more discoveries have been made than at any other spot on the surface of the globe. Herschel married a widow lady, Mrs. Mary Pitt. He soon rose to affluent circumstances, partly by the profits arising from the sale of his mirrors for reflecting telescopes; and he died wealthy on August 23, 1822. He left one son, Sir John Herschel, one of the most active and successful adherents of science that our day has produced, and who, for four years, at the Cape of Good Hope, was engaged in making a survey of the Southern Hemisphere similar to the surveys which his father made of the Northern.

Herschel must be remembered by the number of bodies which he added to the Solar System, making that number half as large again as he found it; and no one individual ever added so much to the facts on which our knowledge of the Solar System is grounded. Some idea may be formed of his wonderful diligence from the fact that there are no less than sixty-nine papers by him in the *Philosophical Transactions.* The earliest writing of Herschel is said to be the answer to the prize question in the *Ladies' Diary* for 1779.

Herschel, by the various means we have glanced at, acquired success such as the world had never seen before, and a reputation of two-fold splendor, appreciable in its different parts by men of the lowest as well as the highest order of cultivation. Admirable as were the immediate results of his telescopic observations, they would have failed to secure him the exalted place now univer-

sally assigned to him in the history of astronomical discovery if he had not at the same time been endowed with
a mind of rare originality and power, combined with a
strong turn for speculation (Grant's *Hist. Physical Astronomy*, p. 534). To him we owe the first proof that
there exist in the universe organized systems besides our
own; while his magnificent foreseeings of the Milky
Way, the constitution of nebulæ, etc., first opened the
road to the conception that what was called the universe
might be, and in all probability is, but a detached and
minute portion of that interminable series of similar formations which ought to bear the name. Imagination
roves with ease upon such subjects; but even that daring faculty would have rejected the ideas which, after
Herschel's observations, became sober philosophy.

THE EARL OF ROSSE'S REFLECTING TELESCOPES.

To Sir Humphrey Davy's remark that "the aristocracy may be searched in vain for philosophers," we find a brilliant exception in the genius, the talent, the patience, and the liberality with which an Irish nobleman has constructed telescopes far transcending in magnitude and power all previous instruments, whether they were the result of private wealth, or of royal or national munificence. That nobleman is Lord Oxmantown, who inaugurated his succession to the Earldom of Rosse by the construction of a colossal instrument which has already achieved brilliant discoveries. Dr. Robinson has eloquently expressed his delight "that so high a problem as the construction of a *six-feet* speculum should have been mastered by one of his countrymen—by one whose attainments are an honor to his rank, an example to his equals, and an instance of the perfect compatibility of the highest intellectual pursuits with the most perfect discharge of the duties of domestic and social life."

In the improvement of the Reflecting Telescope, the first object has always been to increase the magnifying power and light by the construction of as large a mirror as possible; and to this point Lord Rosse's attention was directed as early as 1828, the field of operation being at his lordship's seat, Birr Castle, Parsonstown, about fifty miles west of Dublin. For this high branch of scientific inquiry Lord Rosse was well fitted, by a rare combination of "talent to devise, patience to bear disappointment, perseverance, profound mathematical knowledge, mechanical skill, and uninterrupted leisure from other pursuits."* All these, however, would not have been sufficient, had not a command of money been added, the gigantic telescope we are about to describe having cost certainly not less than twelve thousand pounds.

* *Description of the Great Telescope*, by Thomas Wood, M.D.; 4th edit., 1851.

It is impossible here to detail the admirable contrivances and processes by which Lord Rosse prepared himself for the great work.   Like Herschel, he employed
common workmen.   Mr. Weld says, in his excellent account of the monster telescope, "All the workmen are
Irish; they were trained under the superintendence of
Lord Rosse, being taken from common hedge-schools,
and selected in consequence of their giving evidence of
mechanical skill.   The foreman, a man of great intelligence, is of similar origin; and Lord Rosse assured me
such was his skill, that during his lordship's absence he
felt confident that his foreman could construct a telescope with a six-foot speculum similar in all respects to
that now erected."

In order to grind and polish large specula, Lord Rosse
soon perceived that a *steam-engine* and appropriate machinery were necessary; for this purpose he constructed
and used an engine of two-horse power.

Lord Rosse ground and polished specula 15 inches, 2 feet, and 3 feet
in diameter, before he commenced the colossal instrument.   He first
ascertained the most useful combination of metals for specula, in
whiteness, porosity, and hardness, to be copper and tin.   Of this compound the reflector was cast in pieces, which were fixed on a bed of
zinc and copper—a species of brass which expanded in the same degree by heat as the pieces of the speculum themselves.   They were
ground as one body to a true surface, and then polished by machinery
moved by the steam-engine.   The peculiarities of this mechanism were
entirely Lord Rosse's invention, and the result of close calculation and
observation: they were chiefly, placing the speculum with the face
upward, regulating the temperature by having it immersed in water,
usually at 55° Fahr., and regulating the pressure and velocity.   This
was found to work a perfect spherical figure in large surfaces, with a
degree of precision unattainable by the hand; the polisher, by working above and upon the face of the speculum, being enabled to examine the operation as it proceeded without removing the speculum,
which, when a ton weight, is no easy matter.

The contrivance for doing this is very beautiful.   The machine is
placed in a room at the bottom of a high tower, in the successive floors
of which trap-doors can be opened.   A mast is elevated on the top of
the tower, so that its summit is about 90 feet *above* the speculum.   A
dial-plate is attached to the top of the mast; and a small plane speculum and eye-piece, with proper adjustments, are so placed that the
combination becomes a Newtonian telescope, and the dial-plate the
object.   The last and most important part of the process of working
the speculum is to give it a *true parabolic figure*, that is, such a figure
that each portion of it should reflect the incident ray to the same

focus.   Lord Rosse's operations for this purpose consist, 1st, of a
stroke of the first eccentric, which carries the polisher along *one third*
of the diameter of the speculum; 2d, a transverse stroke twenty-one
times slower, and equal to 0·27 of the same diameter, measured on
the edge of the tank, or 1·7 beyond the centre of the polisher; 3d, a
rotation of the speculum performed in the same time as thirty-seven
of the first strokes; and, 4th, a rotation of the polisher in the same
direction about sixteen times slower.   If these rules are attended to,
the machine will give the true parabolic figure to the speculum,
whether it be *six inches* or *three feet in diameter*.   In the three-feet
speculum, the figure is so true with the whole aperture that it is
thrown out of focus by a motion of less than the *thirtieth of an inch;*
"and even with a single lens of one eighth of an inch focus, giving
a power of 2592, the dots on a watch-dial are still in some degree defined."

Thus was executed the three-feet speculum for the
twenty-six-feet telescope placed upon the lawn at Parsonstown, which in 1840 showed with powers up to 1000
and even 1600, and which resolved nebulæ into stars, and
destroyed that symmetry of form in globular nebulæ
upon which was founded the hypothesis of the gradual
condensation of nebulous matter into suns and planets.

This instrument also discovered a multitude of new objects in the moon, as a mountainous tract near Ptolemy,
every ridge of which is dotted with extremely minute
craters, and two black parallel stripes in the bottom of
Aristarchus.   Dr. Robinson, in his address to the British
Association in 1843, stated that in this telescope a building the size of the Court House at Cork would be easily
visible on the lunar surface.

This instrument was scarcely out of Lord Rosse's
hands, before he resolved to attempt, by the same processes, to construct another reflector, which was completed early in 1845.   The speculum has six feet of clear
aperture, and therefore an area four times greater than
that of the three-feet speculum.   The focal length is fifty-
four feet.   It weighs four tons, and, with its supports, it
is seven times as heavy as the four-feet mirror of Sir
William Herschel.   The Rosse speculum is placed in one
of the sides of a cubical wooden box, about eight feet a
side, in which there is a door, through which two men
go in to remove or to replace the cover of the mirror.
To the opposite end is fastened the tube, which is made
of deal staves an inch thick, hooped with iron clamp-

rings like a huge cask. It carries at its upper end, and in the axis of the tube, a small oval speculum six inches in its lesser diameter. The tube is eight feet diameter in the middle, but tapering to seven at the extremities, and is furnished with internal diaphragms about six and a half feet in aperture. The late Dean of Ely (Dr. Peacock) walked through the tube with an umbrella up. The speculum was cast on the 13th of April, 1842, ground in 1843, polished in 1844, and in February, 1845, the telescope was ready to be tried. The speculum was polished in *six hours*, in the same time as a small speculum, and with the same facility, and no particular care was taken in preparing the polisher.

The casting of a speculum of nearly four tons was an object of great interest as well as of difficulty. In order to insure uniformity of metal, the blocks from the first melting, which was effected in three furnaces, were broken up, and the pieces from each of the furnaces were placed in three separate casks, A, B, and C; then, in charging the crucibles for the final melting of the speculum, successive portions from cask A were put into furnaces a, b, c; from B into b, c, a; and so on.

In order to prevent the metal from bending or changing its form, Lord Rosse made the specimen rest upon a surface of pieces of cast iron strongly framed, so as to be stiff and light, and carrying levers to give lateral support; it is attached to an immense joint, like that of a pair of compasses moving round a pin, in order to give the transverse motion for following the star in right ascension. This pin is fixed to the centre-piece between two trunnions, like those of an enormous mortar, lying east and west, and upon which the telescope has its motion in altitude. Two specula have been provided: one contains three and a half, and the other four tons of metal, the composition of which is one hundred and twenty-six parts in weight of copper to fifty-seven and a half of tin.

The enormous tube is established between two lofty castellated piers sixty feet high, and is raised to different altitudes by a strong chain-cable attached to the top of the tube. This cable passes over a pulley on a frame down to a windlass on the ground, which is wrought by two assistants. To the frame are attached chain-guys, fastened to the counterweight; and the telescope is balanced by these counterweights suspended by chains, which are fixed to the sides of the tube, and pass over large iron pulleys.

On the *eastern* pier is a strong cast-iron semicircle, with which the telescope is connected by a rack-bar attached to the tube by wheelwork; so that, by means of a handle near the eye-piece, the observer can move the telescope along the bar on either side of the meridian, to the distance of an hour for an equatorial star. On the *western* pier are stairs and galleries. The observing gallery is moved along a railway by means of wheels and a winch, and the galleries can be raised by ingenious mechanism to various altitudes. Sometimes the galleries, filled with observers, are suspended midway between the two piers, over a chasm sixty feet deep.

So exquisitely adjusted is the machinery connected with this gigantic instrument, that the tube is moved with all the ease and precision of that of a microscope.

In order to form an idea of the effective magnitude of this colossal telescope (says Sir David Brewster), we must compare it with other instruments, as in the following table, which contains the number of square inches in each speculum, on the supposition that they were square in place of round:

Names of makers.	Diameter of speculum.	Area of surface.
Newton	1 inch . . .	1 square inch.
"	2·37 inches. .	5·6 square inches.
Hadley	4·5 " . .	20 " "
"	5 " . .	25 " "
Hawksbee . . .	9 " . .	81 " "
Ramage	21 " . .	441 " "
Lassels	2 feet . . .	576 " "
Lord Rosse . . .	2 " . . .	576 " "
" . . .	3 "	1296 " "
Herschel	4 "	2304 " "
Lord Rosse . . .	6 "	5184 " "

This magnificent instrument, by far the most powerful which the genius of man has hitherto executed for the purpose of exploring the grand phenomena of the heavens, has already, in the hands of its noble owner, done valuable service to astronomy by the light which it has thrown upon the structure of the nebular part of the universe. Many nebulæ, which had hitherto resisted all attempts to resolve them with instruments of inferior power, have been found to consist wholly of stars. Others exhibit peculiarities of structure totally unexpected. Thus former observers suggested the probability of the nebula No. 51 in Messier's catalogue being a vast sidereal system, identical in structure with a smaller one in its immediate vicinity, and to which it offered a striking analogy. The telescope of Lord Rosse has, however, destroyed this interesting surmise, by showing the nebula to be of a totally different structure—to be, in fact, composed of a series of spiral convolutions, arranged with remarkable regularity: and a connection has also been traced by means of these spirals between the nebula and its companion.

By means of the telescope, the flat bottom of the crater in the moon called Albateginus is distinctly seen to be strewed with blocks, not visible with less powerful instruments; while the exterior of another (Aristillus) is intersected with deep gullies radiating from its centre.

" We have in the mornings" (says Sir David Brewster) " walked again and again, and ever with new delight, along the mystic tube; and at midnight, with its distinguished architect, pondered over the marvelous sights which it discloses: the satellites, and belts and rings of Saturn—the old and new ring, which is advancing with its crest of waters to the body of the planet—the rocks, and mountains, and valleys, and extinct volcanoes of the moon—the crescent of Venus, with its mountainous outline—the systems of double and triple stars—the nebulæ and starry clusters of every variety of shape—and those spiral nebular formations which baffle human comprehension, and constitute the greatest achievement in modern discovery."

The Astronomer Royal, Mr. Airy, alludes to the impression made by the enormous light of the telescope— partly by the modifications produced in the appearance of nebulæ already figured, partly by the great number of stars seen at a distance from the Milky Way, and partly from the prodigious brilliancy of Saturn. The account given by another astronomer of the appearance of Jupiter was that it resembled a coach-lamp in the telescope, and this well expresses the blaze of light which is seen in the instrument.

A new difficulty has, however, arisen from these vast successes in telescopic construction. To insure the best performance of a telescope, not only should there be a cloudless sky, but a perfectly quiescent state of the whole atmosphere—" a most serene and quiet air ;" and this is indispensable for high magnifying powers ; yet so rarely is this state of the air to be found at the sea-level, that Lord Rosse assures us that whole years have passed away without affording him, among an abundance of clear nights, one of such accurately defining quality as to enable him to use the higher magnifying powers of his great reflecting telescope to any advantage. And this is a difficulty which continually increases with the size and excellence of the telescopes employed. Hence was suggested the expediency of transporting powerful instruments to the southern hemisphere for the physical observation of the celestial bodies; and in 1856 the government consented to a summer expedition to the Peak of Teneriffe, when Mr. Piazzi Smyth, with a most valuable equatorial instrument, at elevations of 8903 and 10,702 feet, found the skies often freer from haze, the stars always decidedly brighter, and the definition very much better than near the level of the sea.

The Earl of Rosse's Great Reflecting Telescope.

THE INVENTION OF THE MICROSCOPE.

SIR DAVID BREWSTER has sagaciously observed that, previous to the introduction of glass, the microscopes of the present day could not have been constructed, even if their theory had been known; but it seems strange that a variety of facts, which must have presented themselves to the most careless observer, should not have led to the earlier construction of optical instruments. Through the spherical drops of water suspended before his eye, an attentive observer might have seen magnified some minute body placed accidentally in its anterior focus; and in the eyes of fishes and quadrupeds, which he uses for his food, he might have seen, and might have extracted, the beautiful lenses which they contain. Had he looked through these remarkable lenses and spheres, and had he placed the lens of the smallest minnow, or that of the bird, the sheep, or the ox, in or before a circular aperture, he would have possessed a microscope or microscopes of excellent quality, and of different magnifying powers. No such observations, however, seem to have been made; and even after the invention of glass, and its conversion into globular vessels, through which, when filled with any fluid, objects are magnified, the Microscope remained undiscovered.

The earliest magnifying lens of which we have any knowledge was one rudely made of rock-crystal, which Mr. Layard found among a number of glass bowls in the northwest palace of Nimroud; but no similar lens has been found and described to induce us to believe that the Microscope, either simple or compound, was invented and used as an instrument previous to the commencement of the seventeenth century. In the beginning of the first century, however, Seneca alludes to the magnifying power of a glass globe filled with water; but as he only states that it made small and indistinct letters appear larger and more distinct, we can not consider such a casual remark as the invention of the single microscope, though it might have led the observer to try the effect of smaller globes, and thus obtain magnifying powers sufficient to discover phenomena otherwise invisible.

Lenses of glass were undoubtedly in existence in the time of Pliny; but at that period, and for many centuries afterward, they appear to have been used only as burning, or as reading glasses, and no attempt seems to have been made to form them of so small a size as to entitle them to be regarded even as the precursors of the single microscope.

No person has claimed to be the inventor of the single microscope. According to Peter Borell, the Jansens, spectacle-makers at Middleburg, invented the compound microscope in 1590, and presented the first instrument to Charles Albert, Archduke of Austria. This microscope is stated to have been six feet long. The Dutch have claimed the invention for Cornelius Drebell, of Alkmaar, who resided in London as mathematician to James I. Fontana, an Italian, made the same claim for himself; Viviani asserts that Galileo, his master, was led to the discovery of the microscope from that of the telescope; and the author of the preface to the works of Galileo, published at Milan in 1808, states that Galileo invented the microscope and the telescope about the same time, and that he applied the former to examine objects otherwise invisible. The instrument consisted, like the telescope, of a convex and a concave lens, and also of one lens more convex, and exhibited the structure of insects, and made visible things of prodigious littleness. It has been conjectured that Galileo might have made the microscope in imitation of Jansen's, as he did the telescope, which is more probable than that he was the original inventor.

Neither of these assertions has, however, been proved; and, from these conflicting circumstances, it is obvious that no single individual can be considered as the inventor of the microscope. Huyghens is of opinion that the single microscope was invented not long after the telescope; and, says Sir David Brewster, "as soon as two lenses were combined to magnify distant objects, it was impossible to overlook their influence in the examination of objects that were near, and it is highly probable that

the different individuals whom we have mentioned may
have had the merit of inventing, constructing, and using
the microscope."

Dr. Hooke was the first person who made a microscope
from a single sphere of glass, from the twentieth to the
fiftieth of an inch in diameter, with which many interesting phenomena may be observed, and even important discoveries made. Having taken a clear piece of glass, Dr.
Hooke drew it out, by the heat of a lamp, into threads,
which he melted into a small round globule; and this
sphere being ground on a whetstone, and then polished
on a metal plate with tripoli, he placed it against a small
hole in a thin piece of metal, and fixed it with wax. Thus
filled up, Dr. Hooke says that "it will both magnify and
make some objects more distinct than any of the great
microscopes can do." There have been several improv-
ers of this single-glass sphere.

The celebrated Leuwenhoeck, who made so many important discoveries with the single miscroscope, was supposed to have used only glass globules formed by fusion;
but Mr. Baker, who had upon his table when he wrote
the twenty-six microscopes which Leuwenhoeck left as a
legacy to the Royal Society, informs us that a double
convex lens, and not a sphere or globule, was in each of
them. These small lenses are ground and polished by
the hand, like all other lenses; and when the radii of
their surfaces are as one to six, they make very good microscopes. Leuwenhoeck placed the lenses between two
plates of silver perforated with a small hole, and having
before it a movable pin, upon which to place the object,
and adjust it to distinct vision. With magnifying powers varying from forty to one hundred and sixty, Leuwenhoeck made such important discoveries, that the compound microscope was laid aside for a time, and superseded in England for many years by the ingenious pock-
et-microscope of J. Wilcox, which, for nearly three quarters of a century, was manufactured in England.

We select these details from a valuable contribution
by Sir David Brewster to the *North British Review*, No.
50, in which the author gives a popular account of the
various inventions by which the microscope has been
brought to its present state of perfection, and become

one of the most valuable instruments in extending almost every branch of science. At the commencement of the present century no attempt had been made to fit up the microscope as an instrument of discovery, and to accommodate it to that particular kind of preparation which is required for the preservation and scrutiny of minute objects. For a very long period the microscope of Drebell served but to astonish the young and amuse the curious; and without greatly detracting from the merits of Leuwenhoeck, and other naturalists who used it, we may safely assert that, till it became achromatic by the labors of Lister, Ross,* and others, it was not fitted for those noble researches in nartual history and physiology in which it has performed so important a part.

* This skillful optician died of heart disease, Sept. 5, 1859.

L 2

SIR DAVID BREWSTER'S KALEIDOSCOPE.

THIS optical instrument is named from three Greek words—*Kalon eidos*, a beautiful form, and *scopeo*, I see; and it has been extensively applied to the creation and exhibition of an infinite variety of perfectly symmetrical figures. The idea of the instrument first occurred to Sir David Brewster in 1814, when he was engaged in experiments on the polarization of light by reflections from plates of glass. Sir David observed that when two planes were inclined to one another, and the eye of the spectator was nearly in the produced line of the common section of their planes, the farther extremities of the plate were multiplied by successive reflections, so as to exhibit the appearance of a circle divided into sections; also, that the several images of a candle near those extremities were similarly disposed about a centre. In repeating, at a subsequent period, the experiments of M. Biot on the action of fluids upon light, Sir David Brewster placed the fluids in a trough formed by two plates of glass cemented together at an angle; and the eye being necessarily placed at one end, some of the cement, which had been passed through between the plates, appeared to be arranged into a regular figure. The remarkable symmetry which it presented led to the experimenter's investigation of the cause of this phenomenon, and in so doing he discovered the leading principles of the Kaleidoscope.

The first Kaleidoscopes constructed by Brewster consisted simply of two plane mirrors of glass, having their posterior surfaces blackened, in order to prevent any reflection of light from them, and fixed in a cylindrical tube. The objects were pieces of variously-colored glass, attached to the farther ends of the mirrors, and projecting on the sectional space between them; or the objects were placed between two very thin plates of glass, and held by the hand or fixed in a cell at the end of the tube: in some cases, these plates were moved across the field of view, and in others they were made to turn round upon the axis of the tube. The pieces of colored glass, or other objects which were situated in the section, were, by the different reflections, made to appear in all the other sections, and thus the field of view presented the appearance of an entire object or pattern, all the parts of which were disposed with the most perfect symmetry. By moving the glass plates between which the objects were contained, the pattern was made to vary in form; and pleasing varieties in the tints were produced by moving the instrument so that the light of the sky or of a lamp might fall on the objects in different directions.

The inventor subsequently found means to obtain multiplied images of such objects as flowers, trees, and even persons or things in motion, and thus the importance of the instrument was greatly increased. For this purpose he caused the two mirrors to be fixed in a tube as before, but this tube was contained in another, from which, like the eye-tube of a telescope, it could be drawn at pleasure toward the eye : at the opposite end of the exterior tube was fixed a glass lens of convenient focal length, by which were formed images of different objects in the upper section, and which, being multiplied by successive reflections from the mirrors, produced in the field of view symmetrical patterns of great beauty. The properties of the instrument have been greatly extended ; and when it is constructed so that there may be projected on a screen a magnified image of the whole pattern, and the tube is supported on a ball-and-socket joint, the figures in its field may be easily sketched by a skillful artist, and great assistance thus obtained in designing beautiful patterns.

Sir David Brewster obtained a patent for his Kaleidoscope, and opticians were duly authorized by him to execute and sell them. The public did not, however, adequately encourage the manufacture of instruments of a superior kind, which, moreover, were expensive; while, in violation of the patent, imitations of the Kaleidoscope, rudely and inaccurately constructed, were sold at low prices by unprincipled persons. It is calculated that not less than 200,000 Kaleidoscopes were sold in three months in London and Paris; though, out of this number, Sir

David Brewster says, not perhaps 1000 were constructed upon scientific principles, or were capable of giving any thing like a correct idea of the power of his Kaleidoscope; so that the inventor gained little beyond fame, though the large sale of the imperfect instrument must have produced considerable profit. The effects of the instrument have been rendered highly useful in the industrial arts, especially in suggesting patterns for carpets, and other products of the loom.

The writer well remembers, in 1814–15, in a large school, the avidity with which pseudo-Kaleidoscopes were formed of pasteboard cylinders, blackened planes of glass, and pieces of colored glass, when the fantastic variety of the results obtained by this rude means scarcely foreshadowed the symmetrical beauty of the forms subsequently obtained by more exact methods. To the school-boy of five-and-forty years since, the making of the Kaleidoscope was nearly as popular a recreation as is the photographic art to the tyro of the present day.

MAGIC MIRRORS AND BURNING LENSES.

THE famous mirror which Ptolemy Euergetes caused to be placed in the Pharos at Alexandria belongs to the first class. This mirror is stated by ancient authors to have represented accurately every thing which was transacted throughout all Egypt, both on water and on land. Some writers affirm that upon its surface an enemy's fleet could be seen at the distance of 600,000 paces; others say more than 100 leagues! Abulfeda, in his description of Egypt, states this mirror to have been of "Chinese iron," which Buffon considers to mean polished steel; but a writer in the *Philosophical Magazine*, 1805, supposes the metal to have been *tutenag*, a Chinese metallic compound capable of receiving the highest polish. The existence of Ptolemy's mirror has, however, been generally treated as a fiction; but Father Abbat, in his *Amusemens Philosophiques*, first published at Marseilles in 1763, considers that it may have been at the time the only mirror of its kind, and, being a great wonder, its effects may have been greatly exaggerated; making allowance for which, nothing remains "but that at some distance, provided nothing was interposed between the objects and the mirror, those objects were seen more distinctly than with the naked eye; and that with the mirror many objects were seen which, because of their distance, were imperceptible without it."

It is certain that, under some circumstances, objects may be seen at a much greater distance than is generally supposed. Thus it is stated that the Isle of Man is clearly visible from the summit of Ben Lomond, in Scotland, or 120 miles distant. Brydone states that from the summit of Etna mountains 200 miles off may be distinguished; and during his visit to Teneriffe in 1856, Mr. Piazzi Smyth saw objects at a much greater distance.

Burning Mirrors have been celebrated on account of their size and extraordinary effects. One of these optical machines was the work of Stettala, a canon of Milan; it was parabolic, and, acting as a burning-glass, inflamed wood at the distance of fifteen or sixteen paces. Leonard Digges, in his *Pantometria*, 1571, states that "with a glasse framed by a revolution of a section parabolicall, I have set fire to powder half a mile and more distant." In the prosecution of this subject, the celebrated Napier and Sir Isaac Newton experimented with parabolic reflectors before 1673. Vilette, an artist and optician of Lyons, constructed three mirrors about the year 1670: one of these, which was purchased by the King of France, was thirty inches in diameter, and of about three feet focus. The rays of the sun were collected by it into the space of about one inch. It immediately set fire to the greenest wood; it fused silver and copper in a few seconds; and in one minute vitrified brick and flint earth. A mirror, superior even to these, was constructed by Baron von Tchivnhausen about 1687: it consisted of a metal plate, twice as thick as the blade of a common knife; it was five feet three inches in breadth, and its focal distance was three feet six inches. It produced the following effects: wood, exposed to its focus, immediately took fire; copper and silver passed into fusion in a few minutes; and slate was transformed into a kind of black glass, which, when laid hold of with a pair of pincers, could be drawn out into filaments. Pumice-stone and fragments of crucibles, which had withstood the most violent furnaces, were also vitrified.

The burning lens constructed by Mr. Parker many years since, at an expense of upward of £7000, was of flint-glass, 3 feet in diameter, and weighed 212 pounds; the focal length being 6 feet 8 inches, and the diameter of the focus 1 inch. To concentrate the rays still farther, a second lens was used, and reduced the diameter of the focus to half an inch. Under this lens every kind of wood took fire in an instant, whether hard or green, or even soaked in water. Thin iron plates grew hot in an instant, and then melted. Tiles, slates, and all kinds of earth, were instantly vitrified. Sulphur, pitch, and all resinous bodies melted under water. Fir-wood, exposed to the focus under water, did not seem changed; but when broken, the inside was burnt to a coal. Any metal whatever, inclosed in charcoal, melted in a moment, the fire sparkling like that of a forge. When copper was melted, and thrown down quickly into cold water, it produced so violent a shock as to break the strongest earthen vessels, and the copper was entirely dissipated. Though the heat of the focus was so intense as to melt gold in a few seconds, yet there was so little heat at a short distance from the focus that the finger might be placed an inch from it without injury. Mr. Parker, having put his finger at the focus to try the sensation, found it not to resemble that produced by fire or a lighted candle, but like that of a sharp cut with a lancet.

DISCOVERY OF THE PLANET NEPTUNE.

"Nothing in the whole history of astronomy can be compared to this."—*The Astronomer Royal.*

SIR DAVID BREWSTER, in his admirable summary of the important discoveries in physical astronomy which illustrated the century that followed the publication of Newton's *Principia*, remarks that, "Brilliant as they are, and evincing as they do the highest genius, yet the century in which we live has been rendered remarkable by a discovery which, whether we view it in its theoretical relations or in its practical results, is the most remarkable in the history of physical astronomy. In the motions of the planet Uranus, discovered since the time of Newton, astronomers had been for a long time perplexed with certain irregularities, which could not be deduced from the action of the other planets. M. Bouvard, who constructed tables of this planet, seeing the impossibility of reconciling the ancient with the modern observations, threw out the idea that the irregularities from which this discrepancy arose might be owing to the action of another planet. Our countryman, the Rev. Dr. Hussey, conceived 'the possibility of some disturbing body beyond Uranus;' and Hanson, with whom Bouvard corresponded on the subject, was of opinion that there must be *two new planets beyond Uranus*, to account for the irregularities. In 1834 Dr. Hussey was anxious that the Astronomer Royal should assist him in detecting the invisible planet; and other astronomers expressed the same desire to have so important a question examined and settled. On his return to Berlin from the meeting of the British Association in 1846, the celebrated astronomer, M. Bessel, commenced the task of determining the actual position of the planet; but, in consequence of the death of M. Flemming, the young German astronomer to whom he had intrusted some of his preliminary calculations, and of his own death not long afterward, the inquiry was stopped.

" While the leading astronomers in Europe were thus thinking and talking about the possible existence of a new planet beyond the orbit of Uranus, two young astronomers (Mr. Adams, of St. John's College, Cambridge, and M. Leverrier, of Paris) were diligently engaged in attempting to deduce from the irregularities which it produced in the motions of Uranus the elements of the planet's orbit, and its actual position in the heavens. In October, 1845, Mr. Adams solved this intricate problem —*the inverse problem of perturbations*,* as it has been called—placing beyond a doubt the theoretical existence of the planet, and assigning to it a place in the heavens which was afterward found to be little more than a single degree from its exact place! Anxious for the discovery of the planet in the heavens, Mr. Adams communicated his results to the Astronomer Royal and Professor Challis, but more than nine months were allowed to pass away before a single telescope was directed in search of it to the heavens. On the 29th of July, Professor Challis began his observations; and on the 4th and 12th of August, when he directed his telescope to the theoretical place of the planet as given him by Mr. Adams, *he saw the planet, and obtained two positions of it.*

" While Mr. Adams was engaged in this important inquiry, M. Leverrier—who had distinguished himself by a series of valuable memoirs on the great inequality of Pallas, on the perturbations of Mercury, and on the rectifications of the orbits of comets—was busily occupied with the same problem. In the summer of 1845, M. Arago represented to Leverrier the importance of studying the perturbations of Uranus, which suggestion he followed; and on November 10, 1845, submitted to the Academy of Science his First Memoir on the Theory of Uranus, and in the following June his Second Memoir, in which,

* "The solution of the inverse problem of disturbing forces has led Leverrier and Adams to the discovery of a new planet merely by the deductions from the manner in which the motions of an old one are affected; and its orbit has been so calculated that observers could find it—nay, its disk, as measured by them, only varies 1-1200th of a degree from the amount given by the theory."—*Lord Brougham's Inaugural Address on the erection of a Statue of Sir Isaac Newton at Grantham,* 1858.

after examining the different hypotheses that had been adduced to explain the irregularities of that planet, he is driven to the conclusion that *they are due to the action of a planet situated in the ecliptic at a mean distance double that of Uranus.* He then proceeds to determine where this planet is actually situated, what is its mass, and what are the elements of the orbit which it describes. After giving a rigorous solution of this problem, and showing that there are not two quarters of the heavens in which we can place the planet at a given epoch, he computes its heliocentric place on the 1st of January, 1847, which he finds to be in the 325th degree of longitude; and he boldly asserts that in assigning to it this place, he does not commit an error of more than 10°. The position thus given to it is within a degree of that found by Mr. Adams. Anxious, like Mr. Adams, for the actual discovery of the planet, M. Leverrier naturally expected that practical astronomers would exert themselves in searching for it. The place which he assigned to it was published on the 1st of June, and yet no attempt seems to have been made to find it for nearly five months. The *exact position* of the planet was published on the 31st of August, and on the 13th of September was communicated to M. Galle, of the Royal Observatory of Berlin, who discovered it as a star of the eighth magnitude the very evening on which he received the request to look for it. Professor Challis had *secured* the discovery of this remarkable body six weeks before, but the honor of having actually found it belongs to the Prussian astronomer. With the universal concurrence of the astronomical world, the new planet received the name of Neptune. It revolves round the sun in 172 years, at a mean distance of thirty, that of Uranus being nineteen, and that of the Earth one; and by its discovery the Solar System has been extended *one thousand millions of miles* beyond its former limits.

"The honor of having made this discovery (continues Sir David Brewster most emphatically) belongs equally to Adams and Leverrier. It is the greatest intellectual achievement in the annals of astronomy, and the noblest triumph of the Newtonian Philosophy. To detect a planet by the eye, or to track it to its place by the mind, are acts as incommensurable as those of muscular and intellectual power. Recumbent on his easy-chair, the practical astronomer has but to look through the cleft in his revolving cupola in order to trace the pilgrim star in its course, or, by the application of magnifying power, to expand its tiny disk, and thus transfer it from its sidereal companions to the planetary dominions. The physical astronomer, on the contrary, has no such auxiliaries: he calculates at noon, when the stars disappear under a meridian sun; he computes at midnight, when clouds and darkness shroud the heavens; and from within that cerebral dome, which has no opening heavenward, and no instrument but the Eye of Reason, he sees in the disturbing agencies of an unseen planet, upon a planet by him equally unseen, the existence of the disturbing agent; and from the nature and amount of its action, he computes its magnitude and indicates its place. If man has ever been permitted to see otherwise than by the eye, it is when the clairvoyance of reason, piercing through screens of epidermis and walls of bone, grasps, amid the abstractions of number and of quality, those sublime realities which have eluded the keenest touch and evaded the sharpest eye."*

* We are indebted for the above excellent *précis* of this great discovery to Sir David Brewster's *Life of Sir Isaac Newton*, vol. i., p. 366-370.

PALISSY THE POTTER.

THE production of enameled Pottery from native materials in France is strikingly commemorated in the kind of ware which may be said to be peculiar to that country, and is known as *Palissy Ware.* There is a good deal of embellishment mixed up with the life of the inventor of this ware, "and his adventures, real or imaginary, have assisted in multiplying the numbers of those dangerous books which ascribe imaginary events to real characters."* There is, however, enough of truth in the life of Palissy to awaken our sympathies, and excite our admiration of his works, which represent the most interesting epoch in the history of his art, while his personal life is a romance.

Bernard Palissy was born at La Chapelle-Biron, a village in the old diocese of Agen, at the commencement of the sixteenth century. His parents were poor, but they had him taught reading and writing. A land-surveyor, who had come to Agen to lay down a plan of that part of the country, remarked the boy Bernard's quickness, and the attention with which he watched his operations, and by his parents' consent took him away with him to teach him his business. His progress in practical geometry was so rapid that he mapped out districts before he had ended his apprenticeship. In the intervals of employment he was much given to the study of the Italian masters: he was delighted to paint images and designs on glass, and, as his name became known, he was commissioned to adorn churches and the castles of the nobles. This enabled him to gratify his taste for traveling and for studying natural objects. Nature had implanted in him a love of the beautiful, which became his teacher. Meanwhile, he became acquainted with the chemistry and mineralogy of his day, such as it was. He did not, however, profit so largely as he might have done by the state of knowledge in his time. He had the failing, so common with practical men, of inveighing against theory;

* Charles Tomlinson; *Encyclopædia Britannica,* 8th edit.

yet, in the only work which he has left on the subject of his art, he is obscure in the few practical details which he gives, and has mixed them up with theories of his own, which only prove how much painful toil and how many abortive experiments he would have been spared had he consulted those who were qualified to inform him of the true principles of physical and chemical sciences applicable to his researches.*

In 1539 Palissy quitted his native village, and settled as an artist at Saintes, where he married. Here his modes of obtaining a livelihood became less profitable, and employment was often not to be had. He filled up his time with the indulgence of scientific theories, but felt within him the working of energies which had not yet been called into full action. While in this state of mind, a beautifully enameled cup, which had probably been made at Faenza, in Italy, fell into his hands. Struck with its beauty, he set about inquiring into its mode of manufacture and the secrets of its composition, especially the enamel. He undertook a course of experiments on the subject, but without success: he burnt the clay itself, mixed it with various ingredients, covered it with ever-varying preparations, and tried them, with renewed hopes, in the furnaces of glaziers and potters, but without success. He then built for himself a furnace, which he ultimately demolished and rebuilt; for this, he found, would be his main dependence. In those days, a man of genius, which placed him greatly in advance of his neighbors, was almost sure to be suspected of sorcery, and Palissy's friends began to look upon him with terror; others imagined him to be a coiner of false money, and others thought him to be insane.

The desire to master his object had now taken such possession of Palissy, that for several years he devoted nearly all his time and means to its pursuit, in spite of the claims of his wife and family, and the remonstrances of his friends. He has described with bitter feeling the conflict in his own breast at this time; yet he bore outwardly a cheerful countenance, and strove to inspire his family with the confidence he himself felt, that he should one day place them in affluence by his success, and thus

* Dr. Lardner on "The Potter's Art."

overpay them for all the privations they were enduring.
Fifteen years thus passed away. Palissy was still firm
in his conviction, yet had not succeeded; but nothing
short of producing enamel in all its perfection would satisfy him.   One day, when he thought himself on the point
of attaining the great object of his life, a workman, on
leaving him, demanded the wages that were due to him:
Palissy had no money, and paid him with the few clothes
he had left. He had now to work alone—to prepare his
colors, and to heat and watch the furnace which his own
hands had made.   Once more he was on the verge of
success: he placed in his oven a vase, on which his hopes
were centred, and ran for wood to feed the fire: it was
all consumed.   He stood for a moment overwhelmed
with despair; then rushing into his garden, he tore up
the trellis-work that supported his fruit-trees, broke it in
pieces, and heated his furnace. Up sprang the flame,
and then sank into the deep-red glow which promised
the realization of his hopes; again the fire was nearly exhausted, when he broke into pieces his chairs and tables,
then the door, next the window-frames, and at last the
very flooring of his house—to feed the furnace. This was
Palissy's final effort, and his triumph. He shouted with
joy as he showed his wife and children the vase he had
just taken out of the furnace; it was bright with the imperishable colors that till then he had only seen in dreams
since he had first beheld the cup of Faenza.

This was in the year 1550. He had now discovered
the composition of various enamels, and it was not long
before his beautiful works found their way into all parts
of France.   The king, Henri II., commissioned Palissy to
execute certain vases and figures to adorn his palace garden; he sent for the potter to Paris, gave him apartments in the Tuileries, with a patent, which set forth that
he was the inventor of a new kind of pottery; and, under the patronage of the king, the queen, Catharine de'
Medici, and the Constable Montmorency, Palissy was
known at Paris by no other name than that of Bernard
de Tuileries. He was employed by the Duke of Montmorency to decorate the Château d'Ecouen; and one of
the finest existing specimens of Palissy Ware is a flask
which bears the Montmorency arms.

BERNARD PALISSY, THE POTTER.

WEDGWOOD'S EARLY POTTERY AT BURSLEM.

- Palissy's *figulines*, or rustic pottery, became the fashion of the day, and his beautiful designs were every where admired. The general style of this ware is marked by quaintness and singularity. While the forms are in general correct and pure, there is no painting, properly so called. The figures are given in colored relief, and the enamel is hard and brilliant. The colors are usually bright, and mostly yellows, blues, and grays, sometimes extending to green, violet, and brown; but no fine white, nor any tint of red. He is considered "a great master of the power and effect of neutral tints." His *pièces rustiques*, intended to adorn the large sideboards, or *dressers*, of the dining-halls of the period, and the dishes and plateaux for the same purpose, and not for the table, are loaded with figures in relief. A favorite object with him was also a flat kind of basin, representing the bottom of the sea, with fishes, shells, sea-weeds, pebbles, snakes, etc.; and among his works are ewers and vases grotesquely ornamented, boars' heads, curious salt-cellars, figures of saints, wall and floor tiles.*

The natural objects represented on the pieces of Palissy are remarkable for truth of form and color, having been, with the exception of certain leaves, moulded from nature. He was more or less a naturalist: his shells are all tertiary fossil shells from the Paris basin; the fishes are those of the Seine; and the reptiles, a prevailing subject, those of the banks of the same river. He made use of no foreign natural production. He must be admired as well for the beauty as for the utility of his discovery. It was to him that France owed her high rank in the ceramic art. He formed the first cabinet of natural history collected in France; and he lectured on botany, chemistry, and agriculture before learned scholars. He wrote, though he knew neither Latin nor Greek, in a style which reminds one of Montaigne. In his *Traité*

* In the Bernal Collection, dispersed in 1855, was an extremely rare specimen of Palissy Ware—a circular dish on a foot, a lizard in the centre, and a very rich border, twelve and a half inches in diameter. This was purchased in a broken state at Paris for twelve francs, and after being admirably restored in England, was sold to Mr. Bernal for four pounds. At his sale, this very fine specimen brought £162. It is now in the collection of Baron Gustave de Rothschild. Very fine imitations of Palissy Ware are made in Staffordshire by the Mintons.

M

de l' Art de Terre, he tells the sad story of his twenty
years' anxiety, labor, and privation with touching truthfulness; the unparalleled difficulties he encountered, the
sacrifices he made, the sufferings he endured, and his
obstinate perseverance, amounted to a sort of heroism.

He tells us, in words of religious truth, the mainspring
of his hope throughout this long probation. " I have
found nothing better," he says, " than to observe the
counsel of God, His edicts, statutes, and ordinances; and
in regard to His will, I have seen that He has command-
ed His followers to eat bread by the labor of their bodies,
and to multiply the talents which he has committed to
them." The heroism which Palissy showed in the pursuit of his art he evinced in his religious faith; and on
Sunday mornings he would assemble four or five " simple and unlearned men" for religious worship, and exhort
them to good works. Such was " the beginning of the
Reformed Church of the town of Saintes." Some time
after, when the place was assailed by the fierce opponents
of the Reformers, the workshop of Palissy was broken
into by the mob, and the poor potter sought shelter in a
corner, but, being discovered, was dragged to a dungeon
at Bordeaux. Here he would have perished on the
gallows but that his country might thereby lose his valuable art.

The character of this great improver of Pottery was
strongly marked, not only by patience, perseverance, and
sagacity, but also by moral firmness and unshaken rectitude. He lived in troublous times, and, being a consci-
entious Protestant, he unhesitatingly avowed his religious
opinions, even in his discourses on art. He had warmly
embraced the principles of the Reformation; he was
arrested at the time of the first edict against Protestants,
framed at Ecouen by Henri II. in 1559; he recovered
his liberty through the intercession of the Constable of
Montmorency with the Queen, and through the same
powerful protection Palissy escaped from the massacre
of St. Bartholomew. He, however, thus escaped un-
scathed to endure greater sufferings. In his ninetieth
year he was again accused of heresy, and, refusing to
renounce his opinions, he was thrown into the Bastile.
There he was visited by Henri III. " My good man,"

said the king, "if you can not conform yourself on the matter of religion, I shall be compelled to leave you in the hands of my enemies." "Sire," replied the venerable old man, "I was already willing to surrender my life, and could any regret have accompanied the action, it must assuredly have vanished upon hearing the great King of France say 'I am compelled.' This, sire, is a condition to which those who force *you* to act contrary to your own good disposition can never reduce *me*, because I am prepared for death, and because your whole people have not the power to compel a single potter to bend his knees before images which he has made."

And so Palissy, to the eternal disgrace of the monarch and the priests, and of his country, whose art he had so signally ennobled, was detained in the Bastile, where he died, at little short of a hundred years of age.

The high moral firmness and unshaken rectitude of Bernard Palissy must ever command the admiration of mankind. No example can be found of one to whom the following lines of Horace (translated by Francis) are more truly applicable:

> "The man, in conscious virtue bold,
> Who dares his secret purpose hold,
> Unshaken hears the crowd's tumultuous cries,
> And th' impetuous tyrant's angry brow defies."

JOSIAH WEDGWOOD AND HIS WARES.

FEW men have labored so successfully to refine and elevate his art as Josiah Wedgwood, "the Father of the Potteries," and the first of a long succession of Staffordshire potters, who have applied the highest science and the purest art to the improvement of their commercial enterprise.

Wedgwood was born on the 12th of July, 1730, at Burslem, in Staffordshire, and was the son of a poor potter. His education was very limited; for "scarcely any person in Burslem learned more than mere reading and writing until about 1750, when some individuals endowed the free-school, for instructing youth to read the Bible, write a fair hand, and know the primary rules of arithmetic." Wedgwood had little time for self-improvement, since at the age of eleven years he worked in his elder brother's pottery as *thrower*, his father being then dead. The small-pox, which left an incurable lameness in his right leg, so as afterward to require amputation, compelled him to relinquish the potter's wheel. After a time he left Burslem for Stoke, where his talent for the production of ornamental pottery first developed itself. He next, in partnership with one Wheildon, manufactured knife-handles in imitation of agate and tortoise-shell, melon table-plates, green-pickle leaves, and similar articles. But Wheildon had little taste for the new branches of art-manufacture for which Wedgwood had so great a predilection; he therefore returned to Burslem in 1759, and set up for himself, in a small thatched manufactory, where he continued to make ornamental articles. He prospered, and soon took a second manufactory, where he made white stone-ware; and a third, at which he produced the improved cream-colored ware by which he gained so much celebrity. Of this new ware Wedgwood presented some articles to Queen Charlotte, who thereupon ordered a complete table-service, and was so pleased with its execution as to appoint

Wedgwood her potter, and to command that the ware should be called "Queen's Ware." It has a dense and durable substance, covered with a brilliant glaze, and is capable of bearing uninjured sudden alternations of heat and cold. It was from the first sold at a cheap rate, and the addition of embellishments very little enhanced the cost: first a colored edge or painted border was added to the Queen's Ware, and lastly printed patterns, which covered the whole surface. Nor was this beautiful ware confined to England; for M. Faujas de Saint Fond shows how widely the fame of Wedgwood's pottery had spread before 1792, when, "in traveling from Paris to Petersburg, from Amsterdam to the farthest part of Sweden, and from Dunkirk to the extremity of the south of France, one is served at every inn upon English ware. Spain, Portugal, and Italy are supplied with it; and vessels are loaded with it for the East Indies, the West Indies, and the continent of America." England is mainly indebted to Wedgwood for the extraordinary improvement and rapid extension of this branch of industry. Before his time our potteries produced only inferior fabrics, easily broken or injured, and totally devoid of taste as to form or ornament.

Wedgwood's success was not the result of any fortunate discovery, accidentally made, but was due to patient investigation and unremitting efforts. He called upon a higher class of men than was usually employed in this manufacture to assist in his labors, and in prosecuting his experiments he was guided by sound scientific principles. In partnership with Bentley (a descendant of the celebrated scholar, Richard Bentley), Wedgwood now devoted himself to the higher branches of his manufacture, and succeeded in obtaining from eminent patrons of art the loan of specimens of sculpture, vases, cameos, intaglios, medallions, and seals, suitable for imitation by some of the processes he had introduced. He obtained for this purpose valuable sets of Oriental porcelain; and Sir William Hamilton lent specimens of ancient art from Herculaneum, of which Wedgwood's ingenious workmen produced the most accurate and beautiful copies. Meanwhile the Portland or Barberini Vase was offered for sale, and Wedgwood, with the view of copying it, endeavored to purchase it, and for some time continued to offer an advance upon each bidding of the Duchess of Portland, until at length, his motives being ascertained, he was offered the loan of the vase on condition of withdrawing his opposition; consequently, the duchess became the purchaser, at the price of 1800 guineas. Wedgwood then made fifty copies of the vase, which he sold at 50 guineas each: he is said to have paid £400 for the model, and the entire cost of producing the copies is stated to have exceeded the amount of the sum received by him. Sir Joseph Banks and Sir Joshua Reynolds bore testimony to the excellent execution of these copies, which were chased by a steel rifle, after the bas-relief had been wholly or partially fired.

The Portland Vase is composed of two layers of vitrified paste or glass, one white, the other blue, so perfect an imitation of an onyx cameo that it was long regarded as a natural production. It was discovered about the middle of the tenth century, and said to be many centuries earlier, and of Greek workmanship. It has been deposited in the British Museum since 1810. It was exhibited in a small room of the old Museum buildings until Feb. 7, 1845, when it was wantonly dashed to pieces by a fanatic; but the fragments being gathered up, the vase has been restored by Mr. Doubleday so beautifully that a blemish can scarcely be detected. The vase is now kept in the Medal Room at the Museum. The mode in which it was manufactured was not known until it was broken, and it is now considered as satisfactory proof that the making of glass was carried on to a high state of perfection by the ancients. One of Wedgwood's copies of the Portland Vase was sold in 1859 for above 200 guineas.

Flaxman, the greatest English sculptor, was largely employed by Wedgwood in the preparation of models for the beautiful works of art which he was the first, in modern times, to execute in pottery. By numerous experiments upon various kinds of clay and coloring substances, he succeeded in producing the most delicate cameos, medallions, and miniature pieces of sculpture, in a substance so extremely hard that they appear likely to exceed even the bronzes of antiquity in durability. Another important discovery made by him was that of painting on vases without the glossy appearance of ordinary painting on porcelain or earthenware; an art which was practiced by the ancient Etruscans, but which appears to have been lost since the time of Pliny. The indestructibility of some of his wares rendered them extremély valuable in the formation of chemical vessels, particularly those opposed to the action of acids. The fame of Wedgwood's operations was such, that his works at Burslem, and subsequently at Etruria, a village built by him near Newcastle-under-Lyne, and to which he removed in 1771, became a point of attraction to visitors from all parts of Europe.

Wedgwood's more beautiful inventions were a *terra cotta*, which could be made to resemble porphyry, granite, Egyptian pebble, and other beautiful stones of the siliceous or crystalline kind; a black porcelainous biscuit called *basaltes;* a white and a cane-colored porcelain biscuit, smooth and wax-like; and another white porcelainous biscuit, which receives color from metallic oxides like glass on enamel in fusion. This property renders it applicable to the production of cameos, and all subjects required to be shown in bas-relief, as the ground can be made of any color, while the raised figures are of the purest white. Mr. Wedgwood likewise invented a porcelain biscuit nearly as hard as agate, which will resist the action of all corrosive substances, and is consequently well adapted for mortars in the chemist's laboratory.

Wedgwood's inventions greatly increased the number of persons employed in the Potteries, and improved them by mechanical contrivance and arrangement, his private manufactory having had, for thirty years and upward, all the efficacy of a public work of experiment. In 1785, Wedgwood stated in evidence before a committee of the House of Commons that from 15,000 to 20,000 persons were then employed in the Potteries, with much greater numbers in digging coals for them, and, in various parts of England and Ireland, in digging flints and clay for the earthenware manufacture, 50,000 or 60,000 tons of those materials being annually conveyed to Staffordshire by coasting and inland navigation.

In addition to the attention displayed by Wedgwood on the manufacture inseparably connected with his name, he displayed great public spirit in the encouragement of various useful schemes. By his exertions, and the engineering skill of Brindley, was completed the Trent and Mersey Canal, by which water communication was established between the Pottery district of Staffordshire and the coasts of Devonshire, Dorsetshire, and Kent, whence some of the materials of the manufacture are derived. Wedgwood also planned and carried into execution a turnpike road ten miles in length through the Potteries. He was a Fellow of the Royal Society and of the Society of Antiquities; he also invented a pyrometer, which, as a measure of expansion by heat, has not been surpassed. He made the most liberal use of his ample fortune. He died at Etruria in 1795, in his sixty-fifth year; and, although he had so largely contributed to the prosperity of his countrymen, it was not until more than sixty years after his decease that any fitting memorial of this eminent public benefactor was decided on. In 1859 it was resolved to erect at Stoke a statue of the great potter, holding in his hand the Portland Vase.

Wedgwood had many English imitators: he has even been imitated abroad, especially at Sèvres, Dresden, and Vienna.

JAMES WATT AND THE STEAM-ENGINE.

BEFORE we attempt an outline of the great discoveries of this scientific benefactor, it may be interesting to glance at the earliest employment of the mighty power of Steam, which carries us back to a remote classic age. It appears that the ascending vapor of fluids, as well as their downward tendency, was summoned by the ancients to the aid of superstition. Anthemius of Tralles, the architect of Justinian, being desirous to annoy the orator Zeno, his neighbor and his enemy, conducted steam in leather tubes from concealed boilers, and made them pass through the partition-wall to beneath the beams which supported the ceilings of Zeno's house. When the caldrons were made to boil, the ceilings shook as if they had been shaken by an earthquake.

Another example of the application of steam to the purposes of imposture is given by Tollius. History informs us that on the banks of the Weser, Busteric, the god of the ancient Teutons, sometimes exhibited his displeasure by a clap of thunder, which was succeeded by a cloud that filled the sacred precincts. The image of the god was made of metal, and the head, which was hollow, contained an amphora (nine English gallons) of water. Wedges of wood shut up the apertures at the mouth and eyes: while burning coals, artfully placed in a cavity of the head, gradually heated the liquid. In a short time the generated steam forced out the wedges with a loud noise, and then escaped in three jets, raising a thick cloud between the god and his astonished worshipers. In the Middle Ages the monks availed themselves of this invention, and the steam *bust* was put in requisition even before Christian worshipers.

The entry among the manuscripts of Leonardo da Vinci of the Architonnere of Archimedes, or the apparatus of a steam-gun, has been already noticed, in the sketch of the Discoveries of Leonardo, at page 131.

The *Æolopile*, or Ball of Æolus, was another ancient application of steam. It consisted of a hollow globe of metal, with a long neck, terminating with a very small orifice, which, being filled with water, and placed on a fire, exhibited the steam, as it was generated by the heat, rushing apparently with great force through the narrow opening. A common tea-kettle is, in fact, a sort of Æolopile. The ancients applied the current of steam, as it issued from the spout, to propel the vanes of a mill, or, by acting immediately upon the air, to generate a movement opposite to its own direction.

The Staffordshire Jack of Hilton, in 1680, was a small steam-boiler under the following guise: it was a little hollow image of brass, of about twelve inches high, kneeling upon the left knee, and holding the right hand upon the head, having a little hole in the place of the mouth about the bigness of a great pin's head, and another in the back about two thirds of an inch in diameter: at this last hole it was filled with water (about four pints and a quarter), which, when set to a strong fire, evaporated after the same manner as an Æolopile, and vented itself at the smaller hole in the mouth.

Father Verbiest, in his *Astronomia Europœa*, 1680, gives a curious account of some experiments that he made at Pekin. He placed an Æolopile upon a car, and directed the steam generated within it upon a wheel to which four wings were attached; the motion thus produced was communicated by gearing to the wheel of the car. The machine continued to move with great velocity as long as the steam lasted, and by means of a kind of helm it could be turned in various directions. An experiment was made with the same instrument applied to a small ship, and with no less success.

These facts belong to the *curiosities* of the subject. In tracing the practical history of the Steam-engine through some of its earlier modifications, we shall find that, although the present form of this stupendous machine almost deserves the title of an invention, yet many steps had been taken, and much labor and much ingenuity expended, before it was brought to that point from which the more modern improvements may be said to have begun.

The first apparatus of this description of which any authentic account has been preserved was suggested by Hero the elder, who lived at Alexandria about B.C. 100. It consisted of a vessel in which steam was generated by the application of external heat.   A ball was supplied with the elastic vapor thus procured by means of a bent pipe, a steam-tight joint being provided for that purpose. Two tubes, bent to a right angle, are the only parts open to the air, and as the steam rushes out from very minute apertures, a rotatory motion is produced.   A description of this apparatus is preserved in Hero's *Spiritalia*, published by the Jesuits in 1693; and an excellent account of Hero's inventions has been published by Mr. Bennet Woodcroft.

The next attempt was the experiment made in 1543 by Blasco de Garay, a sea-captain, to propel vessels by a machine having the appearance of a steam-engine.   This experiment was made before the Emperor Charles V. in the port of Barcelona, in the *Trinity*, 200 tons burden. All that could be discovered during the trial was that the machinery consisted of a large boiler containing water, and that wheels were attached to each side of the vessel, by the revolution of which it was propelled.   After the experiment Garay took away all the machinery, leaving only the framing of wood in the arsenals of Barcelona.   As a boiler was used, it is probable, though not certain, that steam was the agent.   It is most likely that the contrivance of Garay was identical with that of Hero. The experiment succeeded, Garay was rewarded, and the usefulness of the contrivance in towing ships out of port was admitted, yet it does not appear that a second experiment was ever made, much less that the machine was brought into practical use.   Mr. Macgregor impugns this report,* and states, as the result of his inquiries in Spain, that if Blasco de Garay used a steam-engine to propel a vessel, the evidence of the fact is not afforded by his two letters at Simancas, and is not produced, if it is known there or at Barcelona, by the public officers and others interested in supporting such a claim.

Seventeen years later, in 1615, Solomon de Caus, who had been engineer and architect to Louis XIII., King of

---

* In a Paper read to the Society of Arts, April 14, 1858.

France, published a work, in which he speaks of the great violence " when water exhales in air by means of fire, and the said air is inclosed; as, for example, take a ball of copper of one or two feet diameter, and one inch thick, which being filled with water by a small hole, which shall be strongly stopped with a peg, so that neither air nor water can escape, it is certain that, if we put the said ball upon a great fire, so that it will become very hot, it will cause a compression so violent that the ball will burst in pieces with a noise like a petard." This effect is due more to the high-pressure steam raised from the water than to the pressure of the heated air contained in the ball. It is, however, evident that De Caus ascribed the force entirely to the air, and not to the agency of steam, which he never mentions; wherefore he can not be considered to have had a share in the invention of the steam-engine.

Next is Brancas's Revolving Apparatus, which was still more simple than that contrived by Hero. A copper vessel, filled with water (in the original figure made in the form of an ornamental head), was furnished with a pipe, through which the steam was propelled; and striking against the vanes of a float, readily gave motion to pestles and mortars for pounding materials to make gunpowder, and rolling-stones for grinding the same; machines for raising water by buckets, for sawing timber, for driving piles, etc. No very considerable force could have been obtained from this simple apparatus, as the steam, passing through the atmosphere in its passage to the wheel, must, to a certain extent at least, be converted into water; and the method has no analogy to any application of steam in modern engines.

After the publication of the work by Brancas, more than thirty years elapsed ere the appearance of the Marquis of Worcester's *Century of Inventions* recalled the attention of the scientific world to this important subject. His Hydraulic Machine is described at p. 161–168. It " raised water more than forty geometrical feet by the power of one man only, and in a very short space of time drawing up four vessels of water through a tube or channel not more than a span in width."

This contrivance was a great advance upon that of De

Caus; for, allowing that he knew the physical agent by which the water was driven upward in his apparatus, still it was only a method of causing a vessel of boiling water to empty itself; and, before a repetition of the process could be made, the vessel should be refilled, and again boiled. In the machine of Lord Worcester, on the other hand, the agency of the steam was employed in the same manner as it is in the steam-engine of the present day, being generated in one vessel, and used for mechanical purposes in another. Upon this distinction depends the whole practicability of using steam as a mechanical agent. Had its action been confined to the vessel in which it was produced, it never could have been usefully employed.

Sir Samuel Morland's "Principles of the New Force of Fire" has been noticed at page 159, but he does not indicate the form of the machine by which he proposed to render the force of steam a useful mover. It is, however, remarkable, that at this early period, before experiments had been made on the expansion which water undergoes in evaporation, Morland should have given so near an approximation to the actual amount of that expansion. It can scarcely be supposed that such an estimate could be obtained by him otherwise than by experiments.

To Denis Papin, a native of Blois, is due the discovery, in 1688, of one of the qualities of steam, to the proper management of which is owing much of the efficacy of the modern steam-engine. He conceived the idea of producing a moving power by means of a piston working in a cylinder, as in the motion of pumps; and he first proposed to produce the vacuum under the piston by means of common air-pumps worked by a water-wheel. This, however, would but amount to a mere transfer of power; but he subsequently produced the vacuum in another way. He constructed a small model cylinder, in which was placed a solid piston; and in the bottom of the cylinder, under the piston, was contained a small quantity of water, which being heated by fire, steam was produced, the elastic force of which raised the piston to the top of the cylinder: the fire being then removed, and the cylinder being cooled by the surrounding air, the steam was condensed and reconverted into water, leaving a vacuum in the cylinder, into which the piston was pressed by the force of the atmosphere. The fire being applied and subsequently removed, another ascent and descent was accomplished, and in the same manner the alternate motion of the piston was continued.

Nevertheless, Arago gives the invention of the steam-engine to Papin, who certainly imagined the formation of a vacuum by cooling the steam; and also heated the steam, and when he wanted it to cool, *took away the fire*. Papin did not, however, make any machine at all, although Arago thus speaks of it:

The machine, in which our countryman was the first to combine the elastic force of steam with the property possessed by this vapor of annihilating itself by cooling, he never made on a large scale: his experiments were always made with simple models. The water intended to generate the steam was not even contained in a separate vessel; inclosed in the cylinder, it rested on the metal plate that closed the orifice at the bottom. It was this plate that Papin heated directly, to transform the water into steam; it was from the same plate that he took away the fire when he wished for condensation to be effected. Such a proceeding, barely allowable in an experiment intended to verify the correctness of a principle, would evidently be still less admissible if the piston were required to move with some celerity. Papin, while saying that success might be attained "by various constructions easy to imagine," does not indicate any of them. He leaves to his successors both the merit of applying his fruitful idea, and that of inventing the details which alone could insure the success of the machine.

None of the several inventions hitherto noticed appear to have advanced beyond experimental models. About the close of the seventeenth century, Captain Thomas Savery proposed to combine the machine described by the Marquis of Worcester with an apparatus for raising water by suction into a vacuum produced by the condensation of steam. Savery appears to have been unaware of Papin's invention, and states that his discovery of the condensing principle arose as follows. Having drunk a flask of Florence wine at a tavern, and flung the empty flask on the fire, he called for a basin of water to wash his hands. A small quantity which remained in the flask began to boil, and steam issued from its mouth. He then put on a thick glove, seized the flask, and plunged its mouth in the cold water, which immediately rushed up into the flask and filled it.

According to another version of the story, it was the accidental circumstance of Savery's immersing a heated tobacco-pipe in water, and perceiving the water imme-' diately rush up the tube on the concentration by the cold of the warm and thin air, that first suggested to the captain the important use that might be made of steam, or any other gas expanded by heat, as a means of creating a vacuum.

This circumstance immediately suggested to Savery the possibility of giving effect to the atmospheric press-ure by creating a vacuum, first by exhausting the barrel of a pump by filling it with steam, and then condensing the same steam, when the atmospheric pressure would force the water from the well into the pump-barrel, pro-vided it were not more than thirty-four feet above the water in the well. He perceived also that, having lifted the water to this height, he might use the elastic force of steam, in the manner described by the Marquis of Worcester, to raise the same water to a still greater ele-vation ; and that the same steam which accomplished this mechanical effort would serve, by its subsequent condensation, to reproduce the vacuum and draw up more water. "It was on this principle," says Lardner, "that Savery constructed the first engine in which steam was ever brought into practical operation." He " enter-tained" the Royal Society with showing them his engine, for the success of which they gave him a certificate. The engine is thus referred to in Koitzer's *System of Hydrostatics* in 1729: "The first time a steam-engine played was in a potter's house at Lambeth, where, though it was a small engine, yet it (the water) forced its way through the roof and struck up the tiles in a manner that surprised all the spectators."

Captain Thomas Savery was descended from an old family in South Devon, where he was born about the middle of the seventeenth cen-tury. Mechanics appear to have been his favorite study, and as he pursued them practically, he was able to form a body of workmen to execute his various plans. He had a patent for his steam-engine in 1698, and the exclusive privilege of constructing it was confirmed to him in 1699 by Act of Parliament. Desaguliers has unjustly accused him of having derived his plans from the Marquis of Worcester; but all writers have acknowledged that he was the first who ever con-structed an engine of this kind which possessed any great and prac-

tical utility; and it must be stated, that the experiments, in 1690, of
Papin (to whom it has been attempted to transfer the honor of the
invention) were not productive of any useful results till followed out
in England in the beginning of the following century.   It is of no
consequence whether Savery was or was not acquainted with these
experiments, for he worked on essentially different principles.   His
moving power was the elasticity of steam, to which our engineers
have again returned since Watt demonstrated the great advantage of
it; whereas Papin used the pressure of the atmosphere (which can
never exceed a few pounds on the square inch of the piston), and
steam was only a subordinate agent by which he procured a vacuum.
The arrangement also of the different parts of Savery's engine, and
particularly the means he used for condensing the steam, are all his
own, and mark him for a man of truly inventive genius.   It is said
that Savery joined in a patent with Newcomen and Cawley for the
atmospheric engine; but this appears to be a mistake, since no traces
of such an instrument have been found at the Rolls Office.   He took
out a patent, however, in 1686, for polishing plate glass and for rowing vessels with paddle-wheels, and, in 1706, for a double bellows to
produce a continuous blast.   He published in 1698 *Navigation Improved;* in 1702, *The Miner's Friend;* and in 1705, a translation, in
folio, of Cohorn's *Fortification.*   This last was dedicated to George,
Prince of Denmark, to whom he was indebted, that same year, for
the office of treasurer to the sick and wounded.   Savery is understood
to have accumulated a considerable fortune.   He died in 1715.—
PROF. RIGAUD, F.R.S.

About 1717, the Safety-valve, which had been invented
about 1681 by Papin for his Digester, was applied to
Savery's engines by Desaguliers.

Papin, while making experiments for Boyle, discovered that if vapor be prevented from rising, the water becomes hotter than the usual
boiling-point.   This led to the invention of his "Bone Digester,"
which he presented to the Royal Society, with a letter describing its
uses for softening bones, and for "cookery, voyages at sea, confection-
ery, making of drinks, chemistry, and dyeing."   Charles II. com-
manded Papin to make a Digester for his laboratory at Whitehall,
and the invention excited great interest.   It was exhibited in operation once a week, in Water Lane, Blackfriars, in a house "over against
the Blue Boot," where the people crowded in such numbers that only
those were admitted who brought with them recommendations from
Fellows of the Royal Society.   In 1684, when Papin was appointed
temporary curator by the Royal Society, he invited certain Fellows to
a supper prepared by his Digesters.   John Evelyn was a guest; and
he tells us how the hardest beef and mutton bones were made by the
Digester as soft as cheese, without water or other liquor, and with less
than eight ounces of coal, producing "an incredible quantity of
gravy," and delicious jelly from the beef-bones.   The guests also ate
pike and other fish bones "without impediment," and pigeons "stewed in their own juice;" in such case the natural juice reducing "the
hardest bones to tenderness."   Evelyn sent a glass of the jelly to his

wife, "to the reproach of all that the ladies ever made of the best hartshorn."

The enormous strength required for Papin's Digester, and the means to which he was obliged to resort for confining the covers, must have early shown him what a powerful agent he was using. Subsequently he adapted the piston of the common sucking-pump to a steam-machine, making it work in the cylinder, and applying steam as the agent to raise it. It is a curious fact, that although Papin invented the safety-valve, he did not apply it to his steam-machine.

About the year 1711, Thomas Newcomen, an ironmonger, and John Cawley, a glazier, both of Dartmouth, Devon, in visiting the tin mines of Cornwall, saw Savery's engine at work, and detected the causes which led to its inefficiency for drainage. This Newcomen proposed to remedy by his atmospheric engine, in which he intended to work the mining pumps by connecting the end of the pump-rod by a chain with the arch-head of a working beam playing on an axis, the other arch-head of the beam being connected by a chain with the rod of a solid piston moving air-tight in a cylinder. If a vacuum be created beneath the piston, the atmosphere will press it down with a force of fifteen feet per square inch, and the end of the beam being thus raised, the pump-rod will be drawn up. If an equivalent pressure be introduced below the piston, it will neither rise nor fall; and if, in this case, the pump-rod be made heavier than the piston and its rod, so as to overcome the friction, it will descend and elevate the piston again to the top of the cylinder, and so the process may be continued.

The power of such a machine would depend entirely on the magnitude of the piston, the vacuum and the counterpoise being effected by the alternate introduction and condensation of the steam. We have only space for this general description of Newcomen's engine. It was worked by the alternate opening and closing of two valves, the regulating and condensing. When the piston reached the top of the cylinder, the former was to be closed and the latter opened; and on reaching the bottom, the former was to be opened and the latter closed.

It has been said that we are indebted for the important invention in this engine termed "Hand-gear," by which its valves or cocks are worked by the machine itself, to an idle boy named Humphrey Potter, who, being employed to stop and open a valve, saw that he could save himself the trouble of attending and watching it by fixing a plug upon a part of the machine which came to the place at the proper times, in consequence of the general movement. If this anecdote be true, what does it prove? That Humphrey Potter might be very idle, but that he was, at the same time, very ingenious. It was a contrivance, not the result of accident, but of acute observation and successful experiment.

Although we find in Newcomen's engine no new principle, its mechanism and combinations were very important. The method of condensing the steam by the sudden injection of water, and of expelling the air and water from the cylinder by the injection of steam, are two processes which are still necessary to the operation of the improved Steam-engine, and appear to be wholly due to the inventors of the Atmospheric Engine. After Mr. Beighton had, about 1718, made this machine itself shut and open the cocks for regulating the supplies of steam and water, for half a century no farther important progress was made, until Mr. Watt applied his vast genius to the adaptation of steam-power to the uses of life. The earlier steam-engines may be regarded as steam-pumps, and that of Newcomen the connecting link between the steam-pump and the modern engine, of which it contained the germ.

We have now to hail the appearance of the great improver of the Steam-engine.

JAMES WATT was born at Greenock on the 19th of January, 1736. He was the fourth child in a family which for a hundred years had more or less professed mathematics and navigation. His constitution was delicate, and his mental powers were precocious. He was distinguished from an early age by his candor and truthfulness; and his father, to ascertain the cause of any of his boyish quarrels, used to say, "Let James speak; from him I always hear truth." James also showed his constructive tastes equally early, experimenting on his playthings with a set of small carpenter's tools which his father had given him. At six he was still at home. "Mr. Watt," said a friend to the father, "you ought to send that boy to school, and not let him trifle away his time at home." "Look what he is doing before you condemn him," was the reply. The visitor then observed the child had drawn mathematical lines and figures on the hearth, and was engaged in a process of calculation. On putting questions to him, he was astonished at his quickness and simplicity. "Forgive me," said he, "this child's education has not been neglected; this is no common child."

Watt's cousin, Mrs. Marian Campbell, describes his inventive capacity as a story-teller, and details an incident of his occupying himself with the steam of a tea-kettle, and by means of a cup and a spoon making an early experiment in the condensation of steam. To this incident she probably attached more importance than was its due from reverting to it when illustrated by her after-recollections. Out of this story, reliable or not in the sense ascribed to it, M. Arago obtained an oratorical point for an *éloge* which he delivered to the French Institute. Watt may or may not have been occupied as a boy with the study of the condensation of steam while he was playing with the kettle. The story suggests a possibility, nothing more; though it has been made the foundation of a grave announcement, the subject of a pretty picture, and will ever remain a basis for suggestive speculation.

Watt was sent to a commercial school, where he was provided with a fair outfit of Latin and with some elements of Greek; but mathematics he studied with greater zest, and with proportionate success. By the time he was fifteen he had read twice, with great attention, S. Gravesande's *Elements of Natural Philosophy;* and "while under his father's roof he went on with various chemical experiments, repeating them again and again, until satisfied of their accuracy from his own observations." He even made himself a small electrical machine about 1750–53; no mean performance at that date, since, according to Priestley's *History of Electricity,* the Leyden phial itself was not invented until the years 1745–6.

His pastime lay chiefly in his father's marine store, among the sails and ropes, the blocks and tackle, or by the old gray gateway of the Mansion House on the hill above Greenock, where he would loiter away hours by day, and at night lie down on his back and watch the stars through the trees.

At this early age Watt suffered from continual and violent headaches, which often affected his nervous system for many days, even weeks; and he was similarly afflicted throughout his long life. He seldom rose early, but accomplished more in a few hours' study than ordinary minds do in many days. He was never in a·hurry, and always had leisure to give to his friends, to poetry, romance, and the publications of the day: he read indiscriminately almost every new book he could procure. He assisted his father in his business, and soon learned to construct with his own hands several of the articles required in the way of his parent's trade; and by means of a small forge, set up for his own use, he repaired and made various kinds of instruments, and converted, by the way, a large silver coin into a punch-ladle, as a trophy of his early skill as a metal-smith. From this aptitude for ingenious handiwork, and in accordance with his own deliberate choice, it was decided that he should proceed to qualify himself for following the trade of a mathematical-instrument-maker. He accordingly went to Glasgow in June, 1754, his list of personal property including "silk stockings, ruffled shirts, cut-velvet waistcoats, *one working ditto*, one leather apron, a quadrant, a score of articles of carpentry, and a pair bibels." From Glasgow, after a year's stay, he proceeded for better instruction to London, on the 5th of June, 1755, in charge of his connection, John Marr. They traveled on horseback, riding the same horses throughout, and taking twelve days for the journey.

On Watt's arrival in the metropolis he sought a situation, but in vain; and he was beginning to despond, when he obtained work with one John Morgan, an instrument-maker in Finch Lane, Cornhill. Here he gradually became proficient in making quadrants, parallel rulers, compasses, theodolites, etc., until, at the end of a year's practice, he could make "a brass sector with a French joint, which is reckoned as nice a piece of framing work as is in the trade." During this interval he contrived to live upon 8s. a week, exclusive of his lodging. His fear of the press-gang and his bodily ailments, however, led to his quitting London in August, 1756, and returning to Scotland, after investing twenty guineas in additional tools.

At Glasgow, through the intervention of Dr. Dick, he was first employed in cleaning and repairing some of the instruments belonging to the college, and, after some difficulty, he received permission to open a shop within the precincts as "mathematical-instrument-maker to the University." Here Watt prospered, pursuing alike his course of manual labor and of mental study, and especially extending his acquaintance with physics; endeavoring, as he said, "to find out the weak side of Nature, and to vanquish her." About this time he contrived an ingenious machine for drawing in perspective; and from fifty to eighty of these instruments, manufactured by him, were sent to different parts of the world. He had now procured the friendship of Dr. Black and another University worthy, John Robison, who, in stating the circumstances of his first introduction to Watt, says, "I saw a workman, and expected no more, but was surprised to find a philosopher as young as myself, and always ready to instruct me."

It was some time in 1764 that the Professor of Natural Philosophy in the University desired Watt to repair a pretty model of Newcomen's steam-engine. Like every thing which came into Watt's hands, it soon became an object of most serious study. Now the great defect of this engine was that more than three fourths of the whole steam was condensed and wasted during the ascent of the piston, and to this defect Watt applied himself, and so approached his great achievement, of which Robison records these incidents:

At the breaking up of the College (I think in 1765) I went to the country. About a fortnight after this I came to town, and went to have a chat with Mr. Watt, and to communicate to him some observations I had made on Desaguliers' and Belidor's account of the steam-engine. I came into Mr. Watt's parlor without ceremony, and found him sitting before the fire, having lying on his knees a little tin cistern which he was looking at. I entered into conversation on what we had been speaking of at our last meeting—something about steam. All the while Mr. Watt kept looking at the fire, and laid down the cistern at the foot of his chair. At last he looked up at me, and said briskly, "You need not *fash* yourself any more about that, man; I have now made an engine that shall not waste a particle of steam. It shall all be boiling hot—ay, and hot water injected, if I please." So saying Mr. Watt looked with complacency at the little thing at his feet, and, seeing that I observed him, he shoved it away under a table with his foot. I put a question to him about the nature of his contrivance. He answered me rather dryly. I did not press him to a farther explanation at that time, knowing that I had offended him a few days before by blabbing a petty contrivance which he had hit on for turning the cocks of the engine. I had mentioned this in presence of an engine-builder, who was going to erect one for a friend of mine, and this having come to Mr. Watt's ears, he found fault with it.

At a later period Watt frankly told Robison all his contrivance; and long after, the latter found that the little apparatus which he saw on Watt's knee, and which he pushed under the. table with his foot, was the condenser of his first experiment. In the summer of 1767 the whole contrivance was perfect in Watt's mind; and so well defined there was the date of his invention, that, on being asked in 1817 whether he recollected how the first idea of his great discovery occurred to him, he replied, " Oh yes, perfectly. One Sunday afternoon, I had gone to take a walk in the Green of Glasgow, and *when about half way between the Herd's house and Arn's Well*, my thoughts having been naturally turned to the experiments I had been engaged in for saving heat in the cylinder, *at that part of the road* the idea occurred to me that, as steam was an elastic vapor, it would expand, and rush into a previously exhausted space; and that if I were to produce a vacuum in a separate vessel, and open a communication between the steam in the cylinder and the exhausted vessel, such would be the consequence."

As the result of his examination of the Newcomen engine, Watt soon found, notwithstanding all his efforts, that it would not give the amount of work represented by the fuel consumed ; and, on examining the structure of the machine more closely, he was led to ask why the steam should first do its work in the cylinder, and then be condensed there by a jet of cold water. Steam, like air, is an elastic fluid, and will rush into a vacuum communicating with a vessel in which it is contained. Let the cylinder of the engine be filled with steam ; establish a communication between it and another vessel, kept as free as possible of air, and in which a jet of cold water is playing ; the steam will then be condensed, and the temperature of the cylinder will not be affected. *This is the great discovery of Watt ;* he made others more ingenious, but none of greater utility : the best proof of its excellence is, that it still keeps its place in the condensing engine after nearly a century of progress in the art. . . . . The pigmy cistern (in which Watt made his first experiment) has been the parent of a progeny of giants, and has astonished the world by the magnitude of the results produced from a cause apparently so insignificant.— JAMES SIMF, M. A.

The interesting little model, as altered by the hand of Watt, was long placed beside the noble statue of the engineer in the Hunterian Museum at Glasgow. Watt himself, when he had got the bearings of the invention, could think of nothing else but his machine, and addressed himself to Dr. Roebuck, of the Carron Iron-works, with the view of its practical introduction to the world. A partnership ensued, but the connection did not prove satisfactory. Watt went on with his experiments, and in September, 1766, wrote to a friend, "I think I have laid up a stock of experience that will *soon* pay me for the trouble it has cost me." Yet it was between eight and nine years before that invaluable experience was made available, so as either to benefit the public or repay the inventor, and a much longer term elapsed before it was possible for that repayment to be reckoned in the form of substantial profit.

Watt now began to practice as a land-surveyor and civil engineer. His first engineering work was a survey for a canal to unite the Forth and Clyde, in furtherance of which he had to appear before the House of Commons. His consequent journey to London was still more important, for then it was that he saw for the first time the great manufactory which Boulton had established at Soho, and of which he was afterward himself to be the guiding intelligence. In the mean time, among his other performances, he invented a Micrometer for measuring distances; and, what is still more remarkable, he entertained the idea of moving canal-boats by the steam-engine through the instrumentality of a *Spiral Oar*, which as nearly as possible coincides with the screw-propeller of our day.

Watt's negotiations for partnership with Boulton were long and tedious. Dr. Roebuck's creditors concurred because, curiously enough, *none of them valued Watt's engine a farthing*. Watt himself now began to despair, and his health failed; yet in 1774, when he had removed to Birmingham, he wrote to his father: "The Fire-engine I have invented is now going, and answers much better than any other that has yet been made, and I expect that the invention will be very beneficial to me."

A long series of experimental trials was nevertheless

requisite before the engine could be brought to such perfection as to render it generally available to the public, and therefore profitable to its manufacturers. In January, 1775, six years of the patent had elapsed, and there seemed some probability of the remaining eight running out as fruitlessly. An application which was made for the extension of its term was unexpectedly opposed by the eloquence of Burke; but the orator and his associates failed, and the extension was accorded by Act of Parliament.

The first practical employment of Watt's engines to any considerable extent was in the mining districts of Cornwall, where he himself was, in consequence, compelled to spend much of his time subsequent to 1775. Here he had to contend not only with natural objects in the dark abysses of deeply-flooded mines, but with a rude and obstinate class of men as deeply flooded with inveterate prejudices. The result in the way of profit was not, however, satisfactory, notwithstanding the service to the mining interest was enormous. "It appears," says Watt in 1780, "by our books, that Cornwall has hitherto eat up all the profits we have drawn from it, and all we have got by other places, and a good sum of our own money to the bargain." Even in 1783 he writes, "We have altered all the engines in Cornwall but one, and many in other parts of England, but do not acquire riches so fast as might be imagined; the expenses of carrying on our business are necessarily very great, and have hitherto consumed almost all our profits; but we hope to do better by continuing our attention and exertions, and by multiplying the number of our works."

At this stage Watt himself was more fertile in mechanical inventions than in any other portion of his busy life. Taking his patents in their chronological order, the first (subsequent to that of 1769) was "for a new method of copying letters and other writings expeditiously" by means of copying *presses*. Of the same date was his invention of a machine "for drying linen and muslin by steam." On the 25th of October, 1781, he took out his third patent (the second of the steam-engine series) "for certain new methods of applying the vibrating or reciprocating motion of steam or fire engines, to produce a continued rotative motion round an axis or centre, and thereby to give motion to the wheels of mills or other machines. One of these methods was that commonly known as the *sun and planet wheels ;* they were five in all. A favorite employment of his in the workshops at Soho, in the latter month of 1783 and the earlier ones of 1784, was to teach his steam-engine, now become nearly as docile as it was powerful, to work a tilt-hammer for forging iron and making steel. "Three hundred blows per minute—a thing never done before," filled him, as his biographer* says, with feelings of excusable pride. Another patent in the steam-engine series, taken out in 1784, contained, beside other methods of converting a circular or angular motion into a perpendicular or recti-lineal motion, the well-known and much-admired *parallel motion,* and the application of the steam-engine to give motion to wheel-carriages for carrying persons and goods. To ascertain the exact number of strokes made by an engine during a given time, and thereby to check the cheats of the Cornish miners, Watt also invented the "Counter," with its several indexes. Among his leading improvements, introduced at various periods, were the *throttle-valve,* the application of the *governor,* the *barometer* or float, the *steam-gauge,* and the *indicator.* The term during which he seems to have thus combined the greatest maturity with the greatest activity of intellect, and the portion of his life which they comprehended, was from his fortieth to his fiftieth year. Yet it was a term of increased suffering from his acute sick headaches, and remarkable for the infirmities over which he triumphed; notwithstanding, he himself complained of his "stupidity and want of the inventive faculty."

Watt's chemical studies in 1783, and the calculations they involved from experiments made by foreign chemists, induced him to make a proposal for a philosophical *uniformity of weights and measures ;* and he discussed this proposal with Priestley and Magellan. While Watt was examining the constituent parts of water, he had op-

* Mr. James Patrick Muirhead, in his *Life of James Watt, with Selections from his Correspondence;* from which work, and an able review of the same in the *Times* journal, the leading data and characteristics in the present sketch have been in part derived.

N

portunities of familiar intercourse not only with Priestley, but with Withering, Keir, Edgeworth, Galton, Darwin, and his own partner Boulton—all men above the average for their common interest in scientific inquiries. Dr. Parr frequently attended their meetings, and they kept up a correspondence with Sir William Herschel, Sir Joseph Banks, Dr. Solander, and Afzelius. Mrs. Schimmelpenninck, who was greatly given to physiognomical studies, has left us this picture of Watt at this period :

Mr. Boulton was a man to rule society with dignity; Mr. Watt to lead the contemplative life of a deeply introverted and patiently observant philosopher. He was one of the most complete specimens of the melancholic temperament. His head was generally bent forward, or leaning on his hand in meditation; his shoulders stooping, and his chest falling in; his limbs lank and unmuscular, and his complexion sallow. His intellectual development was magnificent; comparison and causality immense, with large ideality and constructiveness, individuality, and enormous concentrativeness and caution.

He had a broad Scottish accent; gentle, modest, and unassuming manners; yet, when he "entered a room, men of letters, men of science, nay, military men, artists, ladies, even little children, thronged round him. Ladies would appeal to him on the best means of devising grates, curing smoky chimneys, warming their houses, and obtaining fast colors. I can speak from experience of his teaching me how to make a dulcimer and improve a Jew's harp."

In the year 1786 Watt and Boulton visited Paris, on the invitation of the French government, to superintend the erection of certain steam-engines, and especially to suggest improvements in the great hydraulic machine of Marly, which Watt himself designates a " venerable" work. In Paris Watt made many acquaintances, including Lavoisier, La Place, Fourcroy, and others scarcely less eminent; and while here he discovered with Berthollet a new method of *bleaching* by chlorates, an invention of the latter which Watt subsequently introduced into England.

Meanwhile Watt had vigilantly to defend his patents at home, which were assailed by unworthy and surreptitious rivals as soon as it was proved that they were pecuniarily valuable. Some of the competing engines, as Watt himself described them, were simply asthmatic. " Hornblower's, at Radstock, was obliged to stand still once every ten minutes to snore and snort." " Some were like Evans's Mill, *which was a gentlemanly mill;* it would go when it had nothing to do, but it refused to work." The legal proceedings both in equity and at common law which now became necessary were numerous. One bill of costs, from 1796 to 1800, amounted to between £5000 and £6000; and the mental and bodily labor, the anxiety and vexation which were superadded, involved a fearful tax on the province of Watt's discoveries.

With the year 1800 came the expiration of the privilege of the patent of 1769, as extended by the statute of 1775, and also the dissolution of the original copartnership of Messrs. Boulton and Watt, then of five-and-twenty years' duration. The contract was renewed by their sons, the business having become so profitable that Watt and his children were provided with a source of independent income; and at the age of sixty-four the great inventor had personally realized some of the benefits he contemplated.

Soho, to some extent, maintained its reputation as a steam foundry after Watt himself had ceased to manage it. By 1824, when a monument in Westminster Abbey was voted to him, it had created power in round numbers equal to that of 100,000 horses. By 1854 an addition of nearly two thirds of that amount had been made, giving a total amount equal to that of 170,000 horses; and this was the amount of power supplied from the forges of one manufactory only.

Henceforth Watt's ingenuity became discursive, discretionary, almost capricious, but in every phase and form it continued to be beneficent. In 1808 he founded a prize in Glasgow College, as an acknowledgment of "the many favors" which that learned body had conferred upon him. In 1816 he made a donation to the town of Greenock, "to form the beginning of a scientific library" for the instruction of its young men. Nor, amid such donations, were others wanting on his part, such as true religion prescribes—to console the poor and relieve the suffering.

While resting in his latter days from severer labors, Watt's mind still dwelt on their great development in the form of *Steam Navigation.* It was long since he had posed his significant question as to whether "a spiral oar" or "two wheels" were to be preferred for this purpose. But he lived to know that a steam-boat had been successfully used in America, that the British Channel had been crossed, and the Rhine navigated by another; both vessels, the American and the British, having been impelled by engines manufactured at Soho, constructed on the principles invented by himself, and not without the benefit of his own direct inspection and counsels.

In 1816, on a visit to Greenock, Watt made a voyage in a steam-boat to Rothsay and back again. In the course of this experimental trip he pointed out to the engineer of the boat the method of "backing" the engine. With a foot-rule he demonstrated to him what was meant. Not succeeding, however, he at last, under the impulse of the ruling passion (and we must remember he was then eighty), threw off his over-coat, and, putting his hand to the engine himself, showed the practical application of his lecture. Previously to this, the "back-stroke" of the steam-boat engine was either unknown or not generally known. The practice was to stop the engine entirely a considerable time before the vessel reached the point of mooring, in order to allow for the gradual and natural diminution of her speed.

With regard to the application of steam power to *locomotion on land*, it is remarkable enough that, when Watt's attention was first directed, by his friend Robison, to the steam-engine, "he (Robison) at that time threw out an idea of applying the power to the moving of wheel-carriages." "But the scheme," adds Watt, "was not matured, and was soon abandoned on his going abroad."

In 1769, however, when he heard that a linen-draper, one Moore, had taken out a patent for moving wheel-carriages by steam, he replied, "If linen-draper Moore does not use my engine to drive his chaises, he can't drive them by steam." In the specification of his patent of 1784 he even described the principles and construction of "steam-engines which are applied to give motion to wheel-carriages for removing persons or goods, or other matters, from place to place;" and in 1786, Watt himself had a steam-carriage "of some size under hand;" but his most developed plan was to move such carriages "on a hard, smooth plain;" and there is no evidence to show that he even anticipated the union of the rail and wheel.

Among Watt's mechanical recreations, soon after the date of the last of his steam-engine patents, were four plans of making lamps, which he describes in a letter to Argand; and for a long time lamps were made at Soho upon his principles, which gave a light surpassing, both in steadiness and brilliancy, any thing of the kind that had appeared. About a year after, in 1788, he made "a pretty instrument for determining the specific gravities of liquids," having, he says to Dr. Black, improved on a hint he had taken.

Watt also turned his "idle thoughts" toward the construction of an *Arithmetical Machine;* but he does not appear ever to have prosecuted this design farther than by mentally considering the manner in which he could make it perform the processes of multiplication and division.

Early in the present century Watt devised, for the Glasgow Water-works, to bring pure spring-water across the Clyde, an articulated suction-pipe, with joints formed on the principle of those in a lobster's tail, and so made capable of accommodating itself to all the actual and possible bendings at the bottom of the river. This pipe was, moreover, executed at Soho from his plans, and was found to succeed perfectly.

Watt describes, as his hobby-horse, a *machine to copy sculpture,* suggested to him by an implement he had seen and admired in Paris in 1802, where it was used for tracing and multiplying the dies of medals. He foresaw the possibility of enlarging its powers so as to make it capable of working even on wood and marble, to do for solid masses and in hard materials what his copying-machine of 1782 had already done for drawings and writings impressed upon flat surfaces of paper—to produce, in fact, a perfect fac-simile of the original model. He worked at this machine most assiduously; and his "likeness lathe," as he termed it, was set up in a garret, which, with all its mysterious contents, its tools and models included, have been carefully preserved as he left them.

It is gratifying to find that the charms of Watt's presence were not dimmed by age. "His friends," says Lord Jeffrey, speaking of a visit which he paid to Scotland when upward of eighty, "in that part of the country never saw him more full of intellectual vigor and collo-

quial animation, never more delightful or more instructive." It was then also that Sir Walter Scott, meeting
him " surrounded by a little band of northern literati,"
saw and heard what he felt he was never to see or hear
again—"the alert, kind, benevolent old man, his talents
and fancy overflowing on every subject, with his attention alive to every one's question, his information at every one's command." Campbell the poet, who saw him
later, in the beginning of 1819 (he was then eighty-three),
describes him as so full of anecdote that he spent one of
the most amusing days he had ever had with him. Lord
Brougham, later still, in the summer of the same year,
found his instructive conversation and his lively and even
playful manner unchanged. But in the autumn of this
year, on the 19th of August, he expired tranquilly at his
house at Heathfield. He was buried at Handsworth. A
tribute to his memory was but tardily rendered by the
nation. Five years subsequent to Watt's death, in 1824,
a meeting was held, at which the erection of a statue was
proposed by Baron Dupin : there were present the prime
minister, the Earl of Liverpool, and his colleagues, Mr.
(afterward Sir Robert) Peel, and Mr. Huskisson ; the
other principal speakers were Sir Humphrey Davy, Mr.
Wilberforce, Sir James Mackintosh, and Mr. (now Lord)
Brougham : yet of these illustrious men, two only, Peel
and Brougham, lived to see completed the memorial
which their eloquence so honorably advocated, for the
statue was not erected until eleven years after it had
been proposed—that is, in 1835.

In Westminster Abbey—in the chapel of St. Paul, on
the north side of the choir of the chapel of Edward the
Confessor—is placed a marble sitting statue of James
Watt, by Chantrey, which was voted at the above meeting. It is a fine work, badly located, as classic sculpture
in a Gothic edifice ever must be. The pedestal bears an
eloquent inscription from the pen of Lord Brougham, and
is remarkable for not containing a word of monumental
flattery.* It is as follows :

Not to perpetuate a name
Which must endure while the peaceful arts flourish,
But to show

---

* For this portrait-statue Chantrey received 6000 guineas.

That mankind have learned to honor those
Who best deserve their gratitude,
The king,
His ministers, and many of the nobles
And commoners of the realm,
Raised this monument to
JAMES WATT,
Who, directing the force of an original genius
Early exercised in philosophic research
To the improvement of
The steam-engine,
Enlarged the resources of his country,
Increased the power of man,
And rose to an eminent place
Among the most illustrious followers of science
And the real benefactors of the world.

Born at Greenock, MDCCXXXVI. Died at Heathfield, in Staffordshire, MDCCCXIX.

Jeffrey and Arago added more elaborate tributes to Watt's genius; and Wordsworth has declared that he looked upon him, considering its magnitude and universality, " as perhaps the most extraordinary man that this country has ever produced." His noblest monument is, however, his own work.

Wherever the Steam-engine is applied to manufactures or arts, to travel and transport by sea or land, to agriculture, even to war, there is Watt's instrumentality. The steam power of Great Britain alone is a stupendous item to contemplate in this sense. It is estimated in a recent number of the *Quarterly Review as equivalent to the manual labor of* 400,000,000 *of men, or more than double the number of males supposed to inhabit the globe.* Such *power* did Watt confer upon his nation, and in a still larger degree upon his species.

A century ago (says Dr. Arnott), no man had conceived it possible that human ingenuity would one day devise a machine like the modern Steam-engine, which, at small comparative cost and with perfect obedience to man's will, should be able to perform the work of millions of human beings, and of countless horses and oxen, and of water-mills and wind-mills; and which, in doing such complex and delicate labor as formerly was supposed to be obtainable only from human hands and skill, as of spinning, weaving, embroidering flower-patterns on cloth, etc., should work with speed and exactness far surpassing the execution of ordinary human hands.

Watt's patent for his first improvements in the Steam-engine was taken out in the same year as Arkwright's patent for Spinning with rollers, viz., 1769—one of the most brilliant eras in the annals of inventive genius—when Black and Priestley were making their great discoveries in chemistry; when Hargreaves, Arkwright, and Watt revolutionized the processes of manufacture; when Smeaton and Brindley executed prodigies of engineering science.

THE COTTON MANUFACTURE:

HARGREAVES AND HIS SPINNING-JENNY; ARKWRIGHT AND THE SPINNING-FRAME.

SCARCELY a century has elapsed since a native of Lancashire, of very humble origin, began to devote his attention to the application of machinery to the preparation and spinning of raw cotton for weft. In the year 1760, or soon after, a *Carding Engine*, not very different from that now in use, was contrived by James Hargreaves, an untaught weaver, living near Church, in Lancashire; and in 1767 the *Spinning-Jenny* was invented by the same person. This machine, as at first formed, contained eight spindles, which were made to revolve by means of bands from a horizontal wheel. Subsequent improvements increased the power of the Spinning-Jenny to eighty spindles; when the saving of labor which it thus occasioned produced considerable alarm among those persons who had employed the old mode of spinning, and a party of them broke into Hargreaves' house, and destroyed his machine. The great advantage of the invention was so apparent, however, that it was soon again brought into use, and nearly superseded the employment of the old spinning-wheel, when a second rising took place of the persons whose labor was thus superseded by it. They went through the country destroying, wherever they could find them, both Carding and Spinning Machines, by which means the manufacture was for a time driven away from Lancashire to Nottingham.

Hargreaves stated that he derived the idea of the Jenny from the following incident: Seeing a hand-wheel with a single spindle overturned, he remarked that the spindle, which was before horizontal, was then vertical; and as it continued to revolve, he drew the roving of wool toward him into a thread. It then seemed to Hargreaves plausible that, if something could be applied to hold the roving as the finger and thumb did, and that contrivance to travel backward on wheels, six or eight, or even twelve threads, from as many spindles, might be spun at once. This was done, and succeeded; but Hargreaves, driven by mobs, as we have described, to Nottingham, unable to bear up against such ill treatment, there died in obscurity and distress, having given the property of his Jenny to the Strutts, who thereon laid the foundation of their industrial success and opulence.

The cotton yarn produced by the common spinning-wheel and spinning-jenny could, however, not be made sufficiently strong to be used as warp, for which purpose linen yarn was employed; and it was not until another machine, invented by an individual of as humble origin as Hargreaves, was brought into successful operation, that the above disadvantage was overcome. This machine, which took up what Hargreaves had begun, was the *Spinning-Frame*, invented by Richard Arkwright, who was born at Preston in 1732, and, being the youngest of a poor family of thirteen children, he received but little education, if he ever was at school at all. He was bred to the business of a barber, which he carried on in the town of Bolton. "Two shops are mentioned as having been occupied by Arkwright when he lived in Bolton: one in the passage leading to the Old Millstone Inn, Deansgate; the other, a small shop in Churchgate. The lead cistern in which his customers washed after being shaved is still in existence, and is in the possession of Mr. Peter Skelton, of Bolton."*

About 1760 Arkwright became a dealer in hair, which he collected by traveling up and down the country, and, having dressed the hair, he sold it again to the wig-makers. He kept a better article than either of his competitors in the same trade, and he had a profitable secret method of dyeing hair.

Up to this time the English cotton cloths (called *calico* from Calicut in India, the place of their production) had only the weft of cotton, the warp, or longitudinal threads, being of linen; it being impossible, by any means then-known, to spin the cotton with a sufficiently hard twist to be used as a warp. The raw materials were then delivered by the master-manufacturers to cottagers.

* *Life and Times of Samuel Crompton.* By Gilbert J. French. 1859.

N 2

living in the villages of the district, who both carded and spun the cotton, and wove the cloth. The demand for these cottons soon became so great, that, although there were 50,000 spindles constantly at work in Lancashire alone, each occupying an individual spinner, they could not supply the quantity of thread required. To remedy this state of things, several ingenious individuals had thought of spinning by machinery instead of by the one-thread wheel. A Mr. Wyatt, of Lichfield, is stated to have invented a spinning apparatus as early as 1733, and had factories built with his machines both at Birmingham and Northampton: but these undertakings failed; the machines perished, and no model or description of them was preserved.* Wyatt's claim to the invention has, however, been disproved. A Mr. Laurence Earnshaw, of Mottram in Cheshire, in 1753, invented a machine to spin and reel cotton at one operation, which he showed to his neighbors, and then destroyed it, through the generous apprehension that he might deprive the poor of bread.†

Arkwright had also turned his attention to mechanics. His first effort was an attempt to discover the perpetual motion; and in seeking for a person to make him some wheels for a project of this kind, he got acquainted with one Kay, a clockmaker at Warrington, where they jointly devised a model of a machine for spinning cotton thread. Next year, 1768, they began to erect this machine at Preston, in the parlor of the dwelling-house attached to the Free Grammar School. Arkwright and Kay, however, soon left Preston, dreading the hostility of the Lancashire people to their attempt to introduce spinning by machinery. They next removed to Nottingham, where, wanting capital, Arkwright took his model to Messrs. Need and Strutt, stocking-weavers of that place; and Mr. Strutt, being a man of scientific acquirements, was satisfied of the great value of the proposed machine, and he and Mr. Need entered into partnership with Arkwright, who, in 1769, took out a patent for the machine as its inventor.‡ A spinning-mill, driven by horse-power,

* *Manchester Memoirs*, Second Series, vol. iii.
† Paines' *History of Lancashire*, vol. iii.
‡ It is related that when Arkwright applied to Mr. Strutt, his ma-

was at the same time erected, and filled with the frames, being (unless we include Wyatt's at Lichfield) the first work of the kind that had been known in this country. In 1771 Arkwright and his partners established another mill at Cromford, in Derbyshire, the machinery in which was set in motion by a water-wheel; and in 1775 he took out a second patent, with additions to his original apparatus.*

The most important of Arkwright's contrivances was a device for drawing out the cotton from a coarse to a finer and harder-twisted thread, and so rendering it fit to be used for warp as well as weft. This was most ingeniously managed by the application of a principle which had not yet been introduced in any other mechanical operation. The cotton was, in the first place, drawn off from the skewers on which it was placed by one pair of rollers, which were made to move at a comparatively slow rate, and which formed it into threads of a first or coarser quality; but at a little distance behind the first was placed a second pair of rollers, revolving three, four, or five times as fast, which took it up when it had passed through the others, the effect of which would be to reduce the thread to a degree of fineness so many times greater than that which it originally had. The first pair of rollers might be regarded as the feeders of the second, which could receive no more than the others sent to them; and that, again, could be no more than these others themselves took up from the skewers. As the second pair of rollers, therefore, revolved, we will say, five times for every revolution of the first pair—or, which is the same thing, required for their consumption in a given time five times the length of thread that the first did—

chines were much embarrassed by the fibres of the wool sticking to the roller. This circumstance greatly annoyed Mr. Arkwright; and it is said that Mr. Strutt engaged to remove the evil on condition of participating in the profits of the result. They repaired to the mill, when Mr. Strutt, taking a lump of chalk out of his pocket, and applying it to the roller, the sticking was instantly prevented.

* In Arkwright's apparatus, which was a combination of the carding and spinning machinery, this first part of the process was somewhat modified; but the principle of the two pairs of rollers, the one revolving faster than the other, which forms the peculiarity of the machine, was employed as here described.

they could obviously obtain so much length by drawing out the common portion of cotton into thread of five times the original fineness. Nothing could be more beautiful or more effective than this contrivance, which, with an additional provision for giving the proper twist to the thread, constitutes the water-frame, or throstle, so called from its being originally moved by water-power.

Spinning by rollers was an entirely original idea. Arkwright stated that he accidentally derived the first hint of his invention from seeing a red-hot iron bar elongated by being made to pass between rollers; and though there is no mechanical analogy between that operation and the process of spinning, it is not difficult to imagine that, by reflecting upon it, and placing the subject in different points of view, he might be led to this invention, which he particularly claimed as his own. Of other machines included in his patent he was rather the improver than the inventor; and the original spinning-machine for coarse thread, the Spinning-Jenny, Arkwright admitted to have been first conceived by Hargreaves.

Other parties disputed Arkwright's property in his inventions; his patents were invaded by the cotton-spinners, and he could only enforce his rights by long and costly litigation. Doubtless, to him alone belongs the merit both of having combined the different parts of the spinning machinery, of having first brought it into actual use on an extensive scale, and demonstrated its power and value. The great scene of his operations was at Cromford, in Derbyshire, about twenty miles from Manchester, where the work-people hailed him as a benefactor, and where water-power without limit was found to drive his machinery. It was not, however, until the lapse of five years from their erection that any profit was realized by the works at Cromford; but from that time Arkwright grew wealthy, notwithstanding his patent had been canceled by law. He built for himself a stately castellated mansion amid the scenes of industry where he had raised up his own fortune. He served as high-sheriff of Derbyshire in 1786, and received knighthood on presenting an address of congratulation to King George III.

Sir Richard Arkwright died at Cromford in 1792, in

his sixtieth year. A beautiful monument by Chantrey has been erected over his remains in Cromford Chapel.

To the close of his life, the management of his factories was his daily occupation, and even amusement. He scarcely took any out-door recreation, but employed his time either in superintending the daily concerns of these establishments, or in improving his machinery. His wealth increased to such an extent that, besides possessing, exclusive of his mill property, one of the largest landed estates in England, he presented on two occasions each of his ten children with the sum of ten thousand pounds. He left at his death half a million of money.

And thus it was that, from a poor barber, Arkwright raised himself not merely to rank and affluence, but to be one of the foremost founders of a new branch of national industry, and in a wonderfully short space of time to assume the very first place among the manufacturers of his country.

Cromford mills are delightfully placed on the Derwent, in one of the most picturesque dales in Derbyshire. Near the mills is Willersley Castle, where Arkwright lived in princely style. The mansion commands a fine prospect of the industrial valley.

Arkwright's Mills, from Cromford Heights.

SAMUEL CROMPTON AND THE SPINNING-MULE.

Hitherto our account of the Cotton manufacture has been chiefly illustrated by the inventions of Sir Richard Arkwright; contemporary with whom, though by the present generation only recognized as somewhat obscurely connected with the improvement of spinning machinery, was SAMUEL CROMPTON. "It is scarcely known that his discovery gave a wonderful impulse to the industry, and consequently to the wealth and population of South Lancashire, causing its insignificant villages to attain the importance of large and populous towns."*

Samuel Crompton was born December 3, 1753, of an ancient family, traceable to the time of Henry III. Crompton's parents resided at Firwood, near Bolton, occupying a farm, and, as was the custom of that time, employing their leisure hours in carding, spinning, and weaving. They removed when Samuel was five years old to a portion of the neighboring ancient mansion called Hall-in-the-Wood. The boy was well educated at Bolton; but it is probable that, owing to his mother's exigencies, "his little legs became accustomed to the loom almost as soon as they were long enough to touch the treadles." At the age of sixteen years he continued to reside with his mother, occupied at the loom, and attending an evening school at Bolton, where he advanced his knowledge of algebra, mathematics, and trigonometry. For six years previous to the above date, the increased demand for fine cottons led to a great scarcity of yarn for weft; and the invention of Kay's fly-shuttle, by doubling the speed of the weaver's operations, disturbed the natural balance between the quantity of yarn spun and the weavers' demand for it.

Such was the scarcity of yarn when, in 1767, Har-

* *Life and Times of Samuel Crompton*, by Gilbert J. French, 1859; whence, by permission, the leading data of this sketch are derived. This memoir is written in a bold and manly spirit, befitting the subject which it so eloquently rescues from neglect. It is the substance of two papers read to the members of the Bolton Mechanics' Institution by Mr. French, who has generously placed at the Society's disposal any profits that may arise from the publication of this edition of his work, which, we are happy to add, was sold within a few weeks.

THE HALL-IN-THE-WOOD, NEAR BOLTON.

greaves invented the Jenny (see page 296). "And two years afterward, when only sixteen years of age, Samuel Crompton spun on one of these machines, with eight spindles, the yarn which he afterward wove into quilting; and thus he was occupied for the five following years." At his solitary loom in the old Hall-in-the-Wood he became prematurely a thinker; and, debarred from company, he cultivated a taste for music, which led to the first trial of his mechanical skill in making a violin, which he commenced learning to play upon. He was master of Hargreaves' invention, the Jenny; and he was personally known to Arkwright, whose reputation as an inventor now rang through Lancashire. "This Bolton barber," says Mr. French, "without previous experience as a spinner, was now, in 1771 (Crompton being then eighteen years of age), building his famous mill at Cromford, in Derbyshire, and already obtaining the reputation of great wealth; while Samuel was passing half his working hours in piecing up the broken ends of the bad yarn, which prevented him from making satisfactory progress with his daily stint of weaving—for his mother insisted upon a certain amount of work being finished every day. A failure inevitably subjected him to her somewhat sharp vituperation; and if he succeeded in his allotted task, it was at the expense of so much time lost in mending the ever-breaking ends of his miserable yarn, that none remained for his darling fiddle, or for the few books he now desired to study."

The Hall-in-the-Wood is situated about a mile from Bolton, on elevated rocky ground, around which sweeps the Eagley brook or river; but few of the fine old trees remain to show that the name of the mansion was once entirely appropriate. The building, of post and plaster work, is mostly of the end of the fifteenth century; but the south front and porch are of stone, and the latter bears the date 1648. The dining-hall, and the room in which Crompton worked, now occupied as a bedroom, retain their original handsome windows in small leaden quarries. Here, in the year 1774, he commenced the construction of the Spinning Machine, which for many years was known as "the Hall-i'-th'-Wood Wheels." It took him five entire years to mature his improvement,

during which time he worked entirely alone, with no one
in his confidence to whom he could look for sympathy
or assistance; and he tells us that he succeeded at the
expense of every shilling he had in the world. All this
labor was in addition to his regular every-day work;
he toiled late and early. "Strange and unaccountable
sounds," says Mr. French, "were heard in the old Hall
at most untimely hours; lights were seen in unusual
places; and a rumor became current that the place was
haunted." Samuel was, however, soon discovered to be
himself the embodied *spirit* (of *invention*) which had
caused so much fear and trouble to the family. His difficulties were great, and the tools which he possessed
insufficient for the purpose; but, by devoting every shilling he could spare to the purchase of the requisite tools,
and aided by his clasp-knife, to which he is said to have
been greatly indebted, he at length triumphed. It is related that Crompton and his violin were frequently employed in the orchestra of the Bolton theatre at 1s. 6d.
each night; "but, small as it was, that payment greatly
assisted him in procuring the tools which he required for
his mechanical operations."

In our account of previous inventions for Cotton Spinning, we have already mentioned Kay's production of
the fly-shuttle in 1738; "and in the same year," says Mr.
French, "by a curious and interesting coincidence, and,
so far as can be learned, without any reference to the
recent improvement in weaving, a patent was obtained
by Louis Paul for spinning wool and cotton by passing
previously-prepared slivers between pairs of rollers turned with different degrees of velocity." Mr. Baines, however, whom we have already quoted, stated that Wyatt,
and not Paul, was the inventor of spinning by rollers.
This opinion remained undisturbed until September,
1858, when Mr. Robert Cole, F.S.A., read to the British
Association at the Leeds meeting a communication entitled "Some Account of Louis Paul and his invention of
the Machine for spinning Cotton and Wool by Rollers,
and his claim to such invention to the *exclusion* of John
Wyatt;" proving very satisfactorily that Louis Paul was
the original inventor of the method of spinning by rollers,
and that John Wyatt, whose family have claimed the

credit of the invention for him (he never appears to have made any such claim himself), had really little or nothing to do with the invention, though he certainly had a pecuniary interest in working it. The invention, though wonderfully ingenious, and supported by some of the distinguished men of the time, languished and died.

It next appears that Highs, or Hays, a reed-maker at Leigh, took up the plan of attempting to spin by rollers in 1767, and he was assisted in his experiment by Kay, the clockmaker, but with little success. Next appeared Arkwright, who is said to have adopted the plans of Highs and Kay, and the Spinning Jenny of Hargreaves.

Such was the position of Cotton Spinning, when, in 1774, Samuel Crompton commenced the experiments which eventuated in his *Hall-in-the-Wood Wheel*, or Muslin Wheel, because its capabilities rendered it available for yarn for making muslins; and, finally, it got the name of the *Mule*, from its partaking of the two leading features of Arkwright's machine and Hargreaves' Spinning Jenny. Crompton's first suggestion was to introduce a single pair of rollers, viz., a top and a bottom, which he expected would elongate the rove by pressure, like the process by which metals are drawn out, and which he observed in the wire-drawing for reeds used in the loom. In this he was disappointed, and afterward adopted a second pair of rollers, the latter pair revolving at a slower speed than the former, and thus producing a draught of one inch in three or four. This was neither more nor less than a modification of Mr. Arkwright's roller-beam. But Crompton assured Mr. Kennedy, his nearest friend, that when he constructed his machine he knew nothing of Arkwright's discovery;[*] and the rudeness of Crompton's machine, mostly of wood, shows that he was not acquainted with Arkwright's superior rollers and fixtures in iron, and their connection by clock-work. Mr. Kennedy says:

Crompton's first machine contained only about twenty or thirty spindles. He finally put dents of brass reed-wire into his under rollers, and thus obtained a fluted roller. But the great and important invention of Crompton was his spindle-carriage, and the principle of the thread having no strain upon it until it was completed. The carriage

---

[*] Paper read to the Literary and Philosophical Society of Manchester in 1830.

with the spindles could, by the movement of the hand and knee, recede just as the rollers delivered out the elongated thread in a soft state, so that it would allow of a considerable stretch before the thread had to encounter the stress of winding on the spindle. *This was the corner-stone of the merits of his invention.*

Just as Crompton had completed his first Mule in 1779, and was about to put it to actual work, the Blackburn spinners and weavers, who had previously driven poor Hargreaves from his home, renewed their tumults, and destroyed every jenny round Blackburn, except such as had less than twenty spindles. To save his new machine from destruction, Crompton took it to pieces, and concealed the various parts in a loft near the clock in the old Hall. There they remained hid for many weeks ere he dared to put them together again; but in the same year the wheel was completed, and the yarn spun upon it used for fine muslins; and one of the earliest results of this success was Crompton's purchase of a silver watch out of the wheel's earnings. In 1780 he married, and the young couple went to reside in a cottage attached to the old Hall; but Crompton continued to occupy one of the large rooms in the mansion, and there operated upon the Mule, " with a success which startled the manufacturing world by the production of yarn which, both in *fineness* and *firmness*, had hitherto been unattainable by any means or at any price." Assisted by his amiable young wife, he industriously spun at the Hall, with the greatest possible privacy, small quantities of the much-coveted yarn, producing week after week higher counts and an improved quality, for which he readily obtained his own price. The supply, however, could not satisfy one hundredth part of the demand: the old Hall was besieged by cunning persons, who came not only to purchase, but also to get at the mystery of the wonderful new wheel. Admission was denied; when many climbed up to the windows outside by the aid of harrows and ladders to look in at the new machine. Crompton blocked the intruders out with a screen; but one inquisitive seeker concealed himself for some days in a loft, and watched Samuel at work by means of a gimlet-hole pierced through the ceiling. Even Arkwright traveled sixty miles to endeavor to discover the secret of the new wheel, which all but eclipsed his water-frame.

Crompton now found it impossible to retain the secret of his machine: he had no patent, nor the means of purchasing one; when, rather than destroy the mule, he gave it to the public, upon condition of certain "manufacturing friends" paying him a sum of money, which did not exceed £60; yet the list of half-guinea subscribers of this paltry amount contains "the names of many Bolton firms now of great wealth and eminence as mule-spinners, whose colossal fortunes may be said to have been based upon this singularly small investment" (*French*). The money received merely sufficed to replace the machine which Crompton had given up; for his time, study, and toil he received not a shilling. After the secret had been made public, many persons who had promised subscriptions refused to pay, and even denounced Crompton as an impostor. This shameful treatment made him, to some extent, a moody and mistrustful man. In the five following years the Mule was generally employed for fine spinning throughout the manufacturing districts of England and Ireland, and particularly Scotland. Before 1785 Crompton removed to a farm-house near Bolton, and there besides farming, he worked secretly at his machine in the upper story of his house. Curious visitors still came; and among them was Mr. (afterward the first Sir Robert) Peel, who attempted to get at the Mule in Crompton's absence, but was defeated. He offered the inventor a lucrative situation, and even a partnership, in his establishment, both which Crompton declined to accept.

In 1800 a subscription was opened at Manchester to reward Crompton, but it did not exceed £500. With this sum he rented a factory story in Bolton, and there had two Mules, with the *power* to turn the machinery. Crompton now toiled onward: he submitted his invention to the Royal Society and the Society of Arts, but by neither was it entertained. He had started the stream of manufacturing prosperity, but no portion of it had reached the poor inventor. In the hope of some remuneration, he visited the manufacturing districts of England, Scotland, and Ireland, to ascertain the results of his invention, when he found the number of mule-spindles in use to be 4,600,000, spinning 40,000,000 of

pounds of cotton wool in a year. Armed with these
data, and a certificate signed by many manufacturing
and machine-making firms, Crompton petitioned Parliament for public remuneration, "and, after much delay,
the paltry sum of £5000 was granted him." In 1825 a
memorial was presented to Parliament for a second
grant, but without effect. On June 26, 1827, Crompton
died, in his seventy-fourth year, and was followed to the
grave by a host of Bolton worthies. Yet, in the next
page of his very interesting volume, Mr. French tells us,

From that day little has been said or thought of Samuel Crompton.
Men have been content to employ his great invention for their individual profit and for the benefit of the human race, but the memory of
the inventor has passed from the public mind almost like the shadow
of a summer cloud. The older manufacturers of the country have
been for the most part naturally willing to forget the man to whom
they were so greatly indebted, because they could not remember him
without taking shame to themselves for the injustice and ingratitude
with which he had been treated.

Without underrating the importance of other inventions, it may safely be asserted that Crompton's Mule is
the fulcrum which sustains that mighty lever, the Cotton
Trade, the most valuable and the most powerful of our
national resources. As the Jenny is now almost disused,
and all the finer yarns are spun exclusively upon the
Mule, its importance and value continue to increase.
During eighty years the principle of Crompton's invention has remained unchanged, while modifications, improvements, and auxiliaries have increased its productive
power a hundred-fold. In its infancy it was carefully
tended by the human hand; then it was nursed by water
power; next steam lifted the water back again to duplicate its work in turning the young machinery. But
steam was not long employed in this secondary office;
and as the powers and capabilities of the steam-engine
were developed, they were laid hold of by the cotton-
spinner, and riveted to his machinery, thus raising the
art to a stupendous power.

Meanwhile, the results of Crompton's genius have been
practically commemorated upon the site of his invention.
Near the Hall-in-the-Wood rises an octagonal chimney-
shaft 366 feet in height, in connection with steam-engines
and furnaces in a huge factory, where some thousands of

men and boys are employed in making mule-spinning machinery, and in the weekly production of thousands of mule-spindles. The old Hall has become the veritable centre of the existing cotton-manufacturing district. "Could we," says Mr. French, "tie a cord twenty miles in length to the top of the tall chimney that marks the spot, and sweep it round the country, the circle thus formed would embrace the populous towns and teeming villages engaged in spinning and weaving cotton : they radiate from that centre with compass-like regularity, Manchester, Preston, Oldham, and Blackburn being the cardinal points."

To this small spot of earth (remarkable only the other day for nothing beyond the sterility of its surface), and to its indefatigable inhabitants, Providence appears to have assigned the particular and special duty of clothing mankind. In furtherance of this work, they have dragged to the surface much of the mineral wealth which it contained, and have perforated it with thirty miles of subterranean canals, and countless miles of buried railways. They have crusted over its surface with factories and mills. Wealth, which can scarcely be reckoned, is represented by millions of spindles, which, with their auxiliary engines, are revolving day by day. The land on which they stand has been quadrupled in value. Railways spread over it like a close net-work of iron ; and it is covered with a conglomerated mass of towns and villages, so large and so closely set together that in many instances their longer streets meet each other, and populous places said to be seven miles asunder are really connected by continuous rows of gas-lights.

Many great and active minds have been at work to produce this unprecedented result ; but to *one*, more than all others collectively, it is due. It was the mind of Samuel Crompton which, under Providence, vivified this crowded area, and now fills it with a vitality not the less true that its action is unseen and unacknowledged.—FRENCH's *Life and Times of Samuel Crompton*.

DR. CARTWRIGHT AND THE POWER-LOOM.

This stupendous weaving-machine, a crowning achievement of the Cotton Manufacture, we owe to the genius of Edmund Cartwright, born, in 1743, at Marnham, in Nottinghamshire. He was educated for the Church in the University of Oxford, and published a volume of poems while yet a young man. He had reached his fortieth year before he had given any attention to mechanics.

Happening, in 1784, to be at Matlock, in the company of some gentlemen of Manchester, he maintained the practicability of inventing a machine to weave the vast additional quantity of cotton spun by Arkwright's machinery, and this Cartwright asserted was not a whit less practicable than the construction of the Automaton Chess-player then exhibiting in London. Soon afterward it occurred to Cartwright that, as in plain weaving, according to the conception he then had of the business, there could be only three movements to follow each other in succession, there could be little difficulty in producing and repeating them. He then employed a carpenter and smith to construct for him upon this principle a machine, and getting a weaver to put in the warp, to his great delight a piece of cloth was the produce. The warp was laid perpendicularly; the reed fell with a force of at least half a hundred-weight and the springs which threw the shuttle were strong enough to have thrown a Congreve rocket. Conceiving this to be a valuable invention, in 1785 Cartwright secured it by patent. He then condescended to see how other persons wove (for he had never before seen a loom), when he was astonished at their easy modes of operation compared with his powerful machine, which he did not patent till 1787.

Some time after, a manufacturer, on seeing Cartwright's first loom at work, observed that, wonderful as was the inventor's mechanical skill, he would be baffled in weaving patterns in checks, *i. e.*, combining in the same web a pattern or fancy figure with the crossing colors to form the check. Cartwright made no reply to the manufacturer's observation, but some weeks after showed him a piece of muslin beautifully woven in checks by machinery.

After this Dr. Cartwright made some valuable improvements in the combing of wool by machinery, in rope-making, and other departments of agriculture and manufactures. Even the steam-engine engaged his attention; and he used frequently to tell his son that, if he lived to be a man, he would see both ships and land-carriages impelled by steam. As early as 1793 he constructed a model of a steam-engine, attached to a barge, which he explained, in the presence of his family, to

Robert Fulton, whose zeal and activity afterward, as is well known, perfected the project of steam navigation in America. Even so late as 1823, Dr. Cartwright, then in his seventy-ninth year, contrived a plan of propelling land-carriages by steam.

Dr. Cartwright was defrauded of the pecuniary profits from his great invention of the power-loom by persons who devised contrivances for the same purpose slightly different from his. A manufactory containing 500 of Cartwright's machines was destroyed by fire almost immediately after it was built. On these and other accounts, the power-loom only began to be extensively introduced about 1801, the year in which Cartwright's patent expired. He was, however, in some degree subsequently compensated by a Parliamentary grant of £10,000.

Power-looms were not immediately introduced into factories. They remained an unprofitable speculation until it was discovered, in 1803, that the warp might be dressed before being put into the loom, and the service of the man employed for that purpose dispensed with. The construction of the machine, and the method of dressing, have been improved since that time, and cloth is now woven by the help of steam with a rapidity and to an extent formerly unknown.

A steam-engine of forty or sixty horse-power gives motion to thousands of rollers, spindles, and bobbins for spinning yarns, and works four or five hundred looms besides. This gigantic spinner and weaver needs very little assistance from man. It undertakes, and faithfully discharges, all the heavy work of putting shafts, wheels, and pulleys in motion, of throwing the shuttle, working the treadles, driving home the weft, and turning round the warp and cloth beams. One man may now do as much work as two or three hundred ninety years ago.

CALICO-PRINTING AND THE RISE OF THE PEELS.

The process of Calico-printing is not confined to cotton cloth, as the former term would lead us to suppose;

O

it is applied also to linen, silk, and woolen cloth. The art is supposed to have originated in India, and to have been known in that country for a very long period. From a passage in Pliny's Natural History, it is evident that Calico-printing was understood and practiced in Egypt in his time, but was unknown in Italy. "There exists," says Pliny, "in Egypt a wonderful method of dyeing. The white cloth is stained in various places, not with dye-stuffs, but with substances which have the property of absorbing (fixing) colors. These applications are not visible upon the cloth; but when the pieces are dipped in a hot caldron containing the dye, they are drawn out an instant after dyed. The remarkable circumstance is, that though there be only one dye in the vat, yet different colors appear on the cloth; nor can the colors be again removed." This description of Pliny evidently applies to Calico-printing. It is little more than a century and three quarters since the art was transferred from India to Europe, and little more than a century and a quarter since it was first understood in Great Britain, where, by the application of machinery and improved chemical processes, the rapidity of the execution, and the beauty, and variety, and fastness of the colors are unequaled. In this triumph of art stand pre-eminent the family of the Peels.

At Bamber Bridge, about the year 1763, the art and mystery of Calico-printing in Lancashire was first attempted by the Claytons. Near Knuydon Brook, about two miles east of Blackburn, there lived a tall, robust man, whose ordinary dress was a woolen apron, a calf-skin waistcoat, and wooden-soled clogs, and whose grisly hair was of a reddish color; he owned forty acres of poor grass-land, and three of his sons worked each at a loom in the dwelling-house. About 1765, one of these sons chanced to spoil in the weaving a piece of cloth made of linen and thread; it was therefore unsalable, and the father took the spoiled cloth to the Claytons at Bamber Bridge, requesting to have it printed of a pattern for kerchiefs, which was done, and the articles were worn by the family. The high price charged for printing this piece of cloth induced the owner to attempt the art himself, which he did in a secret apartment of his house at Peel Fold, the name of the above-mentioned forty acres of grass-land. The experimenter was Robert Peel, father of the first Sir Robert Peel, the great calico-printer of Bury in Lancashire, and of Fazely, in Stafford-shire.

The first successful experiment was a "Parsley-leaf," which Peel engraved upon a pewter plate and transfer-red in color to a piece of cloth; and, as this experiment was made in the absence of Peel's family, Mrs. Milton, a next-door neighbor, performed the calendering process with a flat smoothing-iron. It was requisite that, in ad-dition to a sharply-defined vivid impression of the pat-tern, the mordant should so bite-in the colors that they should resist the dissolving action of soap and water. In this, too, the experiment succeeded to admiration; and "Parsley Peel," as he was afterward called, exclaimed, with a shout of exclamation, that he was "a made man." The women of the family ironed the pieces of cloth in the secret room, to prevent prying neighbors seeing what they did. But this Robert Peel did more: he was the first person to supersede the hand-carding of cotton wool, and this he did by using the cards, one fixed in a block of wood, and the other slung from hooks fixed in a beam, where they remained in the kitchen beams at Peel Fold in 1850. Peel's carding-machines were broken by a mob of persons who came from Blackburn to Peel Fold for that purpose, and they afterward destroyed his works at Althain. Peel was at length driven out of the county by the violence of his neighbors, and took refuge at Bur-ton-on-Trent, in Staffordshire. The son of this humble inventor, the first Sir Robert Peel, established his print-works at Bury; and in the neighborhood was born his son, the great statesman, Sir Robert Peel, whose statue has been set up in the market-place of the town of Bury.

To detail fully the results of the Cotton Manufacture, and how largely it has contributed to the financial and national greatness of England, would fill a large volume. Its salient points have been thus glanced at by Mr. Hen-ry Ashworth, in a paper read to the Society of Arts in 1858.

The origin of the uses of Cotton is very remote. Its

production over many parts of the earth is spontaneous, and for 3000 years it has been wrought into garments by the people of India. This knowledge was also, at a very early period, possessed by the people of Egypt and other Eastern countries. The Egyptian looms (says Wilkinsón) were famed for their fine cotton fabrics, and many of these were worked with the needle in patterns in brilliant colors, but some were woven in the piece. Of these last were the cotton fabrics with blue borders, some of which are in the Louvre: though their date is uncertain, they suffice to show that the manufacture was Egyptian; and the many dresses painted on the monuments of the eighteenth dynasty prove that the most varied patterns were used by the Egyptians more than 3000 years ago, as they were at a later period by the Babylonians. In Spain, Cotton was known about the tenth century, and eventually it found its way to England. The Genoese were the first to supply this country with the raw material, probably from the Levant; and the Flemish emigrants are thought to have introduced the requisite skill to use it. Except, however, for candle-wicks, for which use it was imported during the Middle Ages, cotton wool was not employed as a material for manufactures very long before the year 1641, when Manchester purchased cotton wool from Cyprus and Smyrna, with which to make fustians and dimities for home consumption and exportation.

The arts of Spinning and Weaving appear among the earliest inventions of our race. They are mentioned in the Scriptures, in the Homeric poems, and by Herodotus, Strabo, Arrian, Pliny, and other early historians. Yet, strange as it may appear, in past ages we find that no mention is made of any improved process. It would appear to have been reserved to modern times, and to the people of Lancashire, to subvert the old rustic contrivances, and to substitute the mechanical inventions of Hargreaves, Arkwright, and Crompton as the basis of a manufacturing system. We owe it to the genius of these inventors, subsequently aided by Watt, and carried into practical operation by the enterprising efforts of other men, that the previously obscure and humble pretensions of cotton have been raised from insignificance, and invested with an importance truly national; that, along with the progress of this manufacture, our population has increased beyond any previously-conceived limits, the bounds of our industrial pursuits have been immensely enlarged, and articles of clothing have been rendered abundant and cheap. Mr. Porter, in his *Progress of the Nation*, says, "It is to the spinning-jenny and the steam-engine that we must look, as having been the true moving powers of our fleets and armies, and the chief support also of a long-continued agricultural prosperity."

Among the results of Cotton-Spinning Machinery, the diminution of price is as extraordinary as the fineness of the fabric. The raw material is now brought from India, and manufactured into cloths in England, which, after being returned to India, are actually sold there cheaper than the produce of the native looms.

In Cotton Spinning, such is the economy of labor introduced by the use of machinery, that one man and four children will spin as much yarn as was spun by six hundred men and fifty girls eighty years ago. And in the present day Cotton is carded, spun, and woven into cloth in the same factory; these different operations being performed by machinery, the several parts of which are all set in motion by a single steam-engine.

By these combined agencies, the actual value of the Cotton Manufacture, which in 1787 was estimated at £3,304,371, rose in 1833 to £31,338,693, according to Mr. Baines, and the capital employed in the manufacture was £34,000,000; while Mr. M'Culloch, in the *Commercial Dictionary* (1849), gives £36,000,000 as the value of the goods annually made, and £47,000,000 as the estimate of the capital employed. The reports of the cotton manufacture of the United Kingdom amounted in 1849 to £26,775,135, and in 1858 to £33,421,843. Mr. Baines in 1833 estimated the number of persons employed at 237,000, supporting 1,500,000 by upward of £6,000,000 of annual wages; whereas, in 1849, Mr. M'Culloch calculates that 542,000 spinners, weavers, bleachers, etc., and 80,000 engineers, machine-makers, smiths, masons, joiners, etc., were employed at annual wages amounting to £17,000,000 for 622,000 workmen. The development from 1849 to 1859 has proceeded at a rate at least as great as that which preceded.—SIR J. KAY SHUTTLEWORTH.

To these notices of British Cotton Manufacture should be added some account of the beautiful products of the Indian art. Dr. Royle pictures the native woman spinning thread for those wonderful fabrics to which the names of "dew of night," "running water," are figura-

tively applied. He describes her first carding her cotton
with the jawbone of a boalee fish; then separating the
seeds by a small iron roller, worked backward and forward on a flat board; then with a small bone reducing
it to the state of a downy fleece; and finally working it
into thread in the warm, moist atmosphere of a tropical
morning or evening, sometimes over a shallow vessel of
water, the evaporation from which helps to impart the
necessary moisture. Her spindle is delicately made of
iron, with a ball of clay attached, to give it the requisite
weight in turning; and it revolves on a piece of hard
shell, imbedded in another lump of clay to avoid friction.
In spite of her delicate fingers and all her Old World ingenuity, the ruthless Manchester manufacturer, with his
mules and Australian-grown cotton, hastens to supersede
her; and so, one after another, die out the arts of our
older civilization, leaving to the governed and the governors of the East the mighty task of founding a new
system, and new means of employment, upon the wreck
left by the conquests of machinery and steam.

The weaving art is similarly primitive in India; but
the very fine muslins are viewed as curiosities, and made
in small quantities, so that their use is limited almost
exclusively to the princes of the land.

Note.—The first operation to be performed in Cotton, after it is
carried from the field, is to cleanse it from the *seeds.* Cotton was
long cultivated in America under the serious disadvantage that the
whole crop was to be cleansed of its seeds by hand. In 1795 Eli
Whitney of Massachusetts invented the machine known as the Cotton-
gin, by which the seeds could be extracted at an infinite saving of labor and expense; and this invention gave an impetus to the cultivation in our Southern States which has brought the crop up from
189,316 pounds in 1791 to 2,000,000,000 in 1859. Gins are of two
kinds. The Roller-gin consists essentially of two small cylinders revolving in contact, or nearly so, with each other. The cotton is drawn
between these rollers, while the seeds, being too large to pass, are left
behind, and fall out on one side. The Saw-gin, invented by Mr.
Whitney, is intended for those sorts of cotton the seeds of which ad-
here too strongly to be separated by the former method. It consists
of a receiver, having one side covered with strong parallel wires, placed
like those of a cage and about an eighth of an inch apart. Between
these wires enter an equal number of circular saws, revolving on a
common axis. The teeth of these saws entangle the cotton and draw
it out through the grating of wires, while the seeds are prevented by
their size from passing. The cotton thus extricated is swept off from
the teeth of the saws by a revolving cylindrical brush, and the seeds
fall out at the bottom of the receiver.—*Am. Ed.*

JOHN LOMBE AND THE FIRST SILK-THROWING MILL IN ENGLAND.

To the Emperor Justinian we owe the introduction into Europe of the labors of the silk-worm, which, until his time, had been wholly confined to China. The means by which the secret of obtaining silk was conveyed to the emperor displayed furtive ingenuity, which bears some analogy to the stratagem by which the manufacture was conveyed to England. It appears that two Persian monks, employed as missionaries from India, having penetrated into China, " here, amid their pious occupations, viewed with a curious eye the common dress of the Chinese, the manufactures of silk, and the myriads of silk-worms, whose education, either on trees or in houses, had once been considered the labor of queens. They soon discovered that it was impracticable to transplant the short-lived insect; but that in the egg a numerous progeny might be preserved, and multiplied in a distant climate." On their return to the West, instead of communicating the knowledge they had acquired to their own countrymen, they proceeded on to Constantinople, and there imparted to Justinian the secret hitherto so well preserved by the Chinese, that silk was produced by a species of worm; and they added that the eggs might be successfully transported, and the insects propagated in his dominions. They likewise explained to the emperor the modes of preparing and manufacturing the slender filament—mysteries hitherto altogether unknown, or but imperfectly understood in Europe. By the promise of a great reward, the monks were induced to return to China; and there, with much difficulty, they succeeded in obtaining a quantity of silk-worms' eggs; these they concealed in a hollow cane, and at length, in the year 552, conveyed them in safety to Constantinople. The eggs were hatched in the proper season by the warmth of manure, and the worms were fed with the leaves of the wild mulberry-tree. These worms in due time spun their silk, and propagated, under the careful attendance of the monks, who also instructed the Romans in the whole process of manufacturing their production.

The insects thus produced were the progenitors of all the generations of silk-worms which have since been reared in Europe and the western parts of Asia—of the countless myriads whose constant and successive labors are engaged in supplying a great and still increasing demand. A caneful of eggs of an Oriental insect thus became the means of establishing a manufacture which fashion and luxury had already rendered important, and of saving vast sums annually to European nations, which, in this respect, had been so long dependent on, and compelled to submit to the exactions of, their Oriental neighbors. Justinian, however, took the infant manufacture into his own hands, made it an imperial monopoly, and raised the prices of silk higher than those which he had formerly prohibited as excessive, so that an ounce of the fabric could not be obtained under the price of six pieces of gold. Thus the emperor proved any thing but a free-trader when he had obtained the secret. However, the rearing and manufacture did not long remain merely an imperial prerogative, but were extended to Greece, and particularly in the Peloponnesus. The Venetians opened commercial relations with the Greek Empire, and continued for many centuries the channel for supplying the western parts of Europe with silks, which were now highly prized; for in the year 790 the Emperor Charlemagne sent two silken vests to Offa, King of Mercia. The Roman territories continued to supply most parts of Europe until Roger I., King of Sicily, upon his invasion of the territories of the Greek Empire, led into captivity a considerable number of silk-weavers, whom he compulsorily settled in Palermo, obliging them to teach his subjects their art; and in twenty years the silks of Sicily had become famous.

The knowledge of the several processes spread over Italy, and was carried into Spain, but it was not until the reign of Francis I. that the silk manufacture took root in France; and at this date, even our magnificent Henry VIII. could only obtain a pair of silk stockings for gala-days from Spain. His daughter Elizabeth was presented by her silk-woman with a pair of English-knit black silk stockings; but the manufacture in England did not make much progress in her reign until 1585, when many of the silk manufacturers of Antwerp fled to England from the persecutions of the Duke of Parma, then governor of the Spanish Netherlands. Near the close of his reign, Elizabeth's successor, James I., encouraged a London merchant to bring from the Continent of Europe some silk throwsters, silk dyers, and broad weavers; and a beginning was made in the manufacture of raw silk into broad silk fabrics, which increased so rapidly that, in 1629, the Silk Throwsters of London were incorporated, and the trade had its dye, called "London black." In 1661, the Company of Silk Throwsters in London employed above 40,000 men, women, and children. The revocation of the Edict of Nantes in 1685 compelled Protestant merchants, manufacturers, and artificers to emigrate from France in great numbers, when about 70,000 reached England and Ireland, and there established such seats of manufacture as that of Spitalfields, in silk of the highest styles of art and ingenuity of fabric then known. In 1713, the petition of the Weavers' Company to Parliament at the peace of Utrecht against the commercial treaty with France represents the silk manufacture as twenty times greater in amount than it had been in 1664, and that it had caused a great exportation of woolen and other manufactured goods to Turkey and Italy, whence the raw silk was imported.

Up to the year 1718, however, the whole of the silk used in England, for whatever purpose, was imported "thrown," i. e., formed into threads of various kinds and twists. In 1702 a Mr. Crotchet had attempted to establish the silk-throwing trade in a small mill which he built at Derby, but, from defects in his machinery and other difficulties, he was soon compelled to abandon his project. In 1715, John Lombe, whose name will always be remembered with veneration in connection with the Silk Trade, resolved upon visiting Italy, and acquiring, at any risk and any cost, a knowledge of the process adopted in that country, and of introducing it to England. Having well matured his plan, he started on his enterprise.

O 2

On reaching Italy, he found difficulties greater than he had anticipated; for the jealousy of the Italians guarded their secret with the most watchful care. At Piedmont, finding that an examination of the silk machinery and processes was strictly prohibited, and failing to gain open admission to the works, he bribed some of the work-people, and by their connivance, in the disguise of a common workman, he made several secret visits to the mills, and at each time carefully noted down every thing he saw, and made sketches of parts of the machinery, so as to perfect himself in the operation of throwing. His plot was before long discovered, and he was obliged to fly with the utmost precipitancy, bringing with him, however, his notes, sketches, and portions of the machinery, and, better still, a mind which had grasped and comprehended the whole process. He fled to avoid assassination, and took refuge on board ship, and returned to England with a full knowledge of the trade he had run such imminent risk to acquire.

Lombe was accompanied in his flight by two Italian workmen, whom he had bribed, and who risked their lives in his scheme. On arriving in England, he at once fixed on Derby as the scene of his operations, and in 1717 arranged with the Corporation for an island on the River Derwent, at the yearly rent of £8. On this island Lombe erected, at a cost of £30,000, the mill, yet standing, called "the Old Silk Mill." The ground being swampy, Lombe, before he began to build his mill, caused immense piles of oak, twenty feet in length, to be driven close together by means of an engine which he contrived for the purpose, and on these piles was laid a stone foundation, on which were turned the stone arches that support the walls.

During the four years occupied in the erection of the mill, Lombe, in order to save time and to raise money to carry on the works, hired rooms in various parts of Derby, and arranged with the corporation to use the town-hall, where he set up machines, which were for the time worked by hand. These engines more than fulfilled his expectations, and he was enabled to sell thrown silk at much lower prices than it could be obtained for from the Italians. By the time his large mill was completed and his machinery in active operation, he had permanently established the silk-throwing trade. In 1718 he obtained a patent for the sole and exclusive property in the mill for fourteen years, and, with the aid of his Italian workmen, carried on his new manufacture with great success.

John Lombe did not, however, long enjoy this prosperity; for soon afterward he died, at the early age of twenty-nine, from the effects of poison administered to him by the Italians through whom he had learned the art. William Hutton, the venerable historian, and a native of Derby, whose early days were spent toiling wearily in this very mill, says quaintly, among other interesting references:

But, alas! he had not pursued this lucrative commerce more than three or four years, when the Italians, who felt the effect of the theft from their want of trade, determined *his* destruction, and hoped that of his works would follow. An artful woman came over in the character of a friend, associated with the parties, and assisted in the business; she attempted to gain both the Italians, and succeeded with one. By these a slow poison was supposed, and perhaps justly, to have been administered to John Lombe, who lingered two or three years in agony. The Italian fled to his own country, and the woman was interrogated, but nothing transpired except what strengthened suspicion. Grand funerals were the fashion; and perhaps the most superb inhumation known in Derby was that of John Lombe. He was a man of quiet deportment, who had brought a beneficial manufactory into the place, employed the poor, and at advanced wages, and thus could not fail to meet with respect; and his melancholy end excited much sympathy.

Lombe was buried in All Saints' Church, Derby. Dying a bachelor, his property fell into the hands of his brother, William Lombe, who shortly afterward, being of a melancholy temperament, shot himself. About 1726 the mills passed to his cousin, Sir Thomas Lombe. In 1732 the patent expired, when Sir Thomas petitioned Parliament for a renewal, and pleaded "that the works had taken so long a time in perfecting, and the people in teaching, that there had been none to acquire emolument from the patent." "But he forgot," says Hutton, "to inform them that he had accumulated more than £120,000!" The government declined to renew the patent, but granted the sum of £14,000 to Sir Thomas as compensation, on condition that he would prepare, and deposit in the Tower of London, an exact and faithful model of his machinery, for the inspection and advantage of others who might purpose constructing and carrying on similar works.

The act authorizing the issue of the money mentions, among other causes which justified the grant, the great obstruction offered to Sir Thomas Lombe's undertaking by the King of Sardinia, in prohibiting the exportation of raw silk which the engines were intended to work.

The account of the machinery of this immense mill, five stories in height, and one eighth of a mile in length, has been much exaggerated. The grand machine is stated to have been constructed with 26,586 wheels and 96,746 movements, which worked 73,726 yards of organzine silk thread with every revolution of the water-wheel whereby the machinery was driven; and as this revolved three times in each minute, the almost inconceivable quantity of 318,504,960 yards of organzine could be produced daily! Hutton's authority is, however, to be preferred, for he served an apprenticeship of seven years in the mill, and he reduces the number of wheels to 13,384.

Soon after Lombe's patent had expired a mill was erected at Stockport, and this was followed by others in Derby and in various places, until now there are about 400 silk-throwing factories in England, employing, it is computed, considerably more than 100,000 operatives.

The chest in which John Lombe brought over to England his spindles, and various matters connected with the trade, we here engrave. It is one of the most richly carved and painted chests of its kind which is extant. Since Lombe's time, it has, until within the last few years, been preserved in the mill which he built, but is now the property of Mr. Llewellynn Jewitt, F.S.A., of Derby. The chest is, of course, much older than Lombe's time, and, apart from its association with his name and career, is a remarkably fine example of art. The mill is picturesquely situated on the Derwent: since Lombe's time it has received many additions; but the old mill, as built by him, still remains, and is likely to last through many generations. The accompanying view has been sketched from St. Michael's Mill.

Various attempts have been made to rear silk-worms in England. James I., to obtain the requisite food for the silk-worms, in 1608 sent

LOMBE'S SILK MILL, DERBY.

THE CHEST IN WHICH JOHN LOMBE BROUGHT FROM PIEDMONT THE FIRST
SILK MACHINERY INTO ENGLAND.

circular letters to all the counties of England, strongly recommending the inhabitants to plant mulberry-trees; and he directed to be distributed 10,000 mulberry-plants, which were to be procured in London at three farthings per plant. In 1609 James expended £935 in the planting of mulberry-trees upon the site of the present Buckingham Palace and Gardens, St. James's Park. It was at this time that Shakspeare planted his mulberry-tree. King James's garden did not succeed; but Charles I., by letters-patent, in the fourth year of his reign, granted to Walter Lord Aston the custody and keeping of the garden, and of the mulberries and silk-worms there, and of all the houses and buildings to the same garden belonging, for his own and his son's life. In the next two reigns "the Mulberry Garden" became a place of public refreshment: it is a favorite locality in the gay comedies of Charles the Second's time. The Silk-Garden scheme was revived in 1718, when part of the estate of Sir Thomas More (Chelsea Park) was leased to a company, and 2000 mulberry-trees were planted. Thoresby, in his Diary, 1723, tells us that he saw "a sample of the satin lately made at Chelsea of English silk-worms for the Princess of Wales, which was very rich and beautiful." This scheme also failed; but the Clock-house in Lower Chelsea was long after famous for the sale of mulberries from the trees planted for silk-rearing.

In 1790 the Society of Arts awarded a premium for silk grown in the neighborhood of London. No similar success is recorded until 1839, when Mr. Felkin produced at Nottingham some fine cocoons from eggs from Italy. Mrs. Whitby, at Newlands, near Lymington, Hants, has plantations of mulberry-trees, and has for many years reared silk with success from eggs of the large Italian sort, of four changes, from which she obtains as great a proportion and as good a quality of silk as they do in Italy or France. Mrs. Whitby has presented to the queen twenty yards of rich and brilliant damask manufactured from silk raised at Newlands. The obtaining a sufficient quantity of food for the worms at the right time had hitherto been the great difficulty of growing silk in England. This has been surmounted by Mrs. Whitby, whose silk is worth as much in the market as the best foreign silks; and, making allowance for unfavorable seasons, labor, machinery, outlay of money, etc., Mrs. Whitby states that land laid out for the silk-worm's food will afford a large profit. Some of the silk grown by her has been pronounced superior to the best Italian raw silk.

In 1846 scarfs were manufactured in Spitalfields from the produce of between 700 and 800 worms kept in an attic room in Truro. In size and weight the worms surpassed those in Italy; the cocoons were larger; the quality of silk, when reeled, was fully equal to the best imported, and the quantity exceeded the Italian average, and this in a season not remarkably propitious.

The home culture of silk is an important object, since the value of silk brought to England is above £2,000,000 annually; and the silk manufacture engages perhaps fifty millions of our capital, and employs one million of our population.

WILLIAM LEE AND THE STOCKING-FRAME.

KNIT Silk Stockings *made in England* were first worn by Queen Elizabeth, who refused to wear any cloth hose afterward. An apprentice, soon after, borrowed a pair of knit worsted stockings, made at Mantua, and then made a pair like them, which he presented to the Earl of Pembroke; and these are the first worsted stockings known to be knit in England. This humble process of knitting *seems* to have been superseded by the stocking-frame almost immediately after the introduction of knit stockings; for the invention of the stocking-frame dates from 1589, the thirty-first year of Elizabeth's reign.

A singular confusion pervades the early history of the stocking-frame; there is a strange jumble of persons, places, and dates in the accounts given of the invention and the inventor, which it is difficult to reconcile, unless we implicitly believe the evidence of a painting which long hung in Stocking-Weavers' Hall, in Redcross Street, London. This picture contained the portrait of a man in collegiate costume, in the act of pointing to an iron stocking-frame, and addressing a woman who is knitting with needles by hand. The picture bore the following inscription: "In the year 1589, the ingenious . William Lee, A.M., of St. John's College, Cambridge, devised this profitable art for stockings (but, being despised, went to France), yet of iron to himself, but to us and to others of gold: in memory of whom this is here painted."

From .Deering's *Account of Nottingham*, it appears that William Lee (whose name is sometimes written Lea) was a native of Woodborough, a village about seven miles from Nottingham. He was heir to a considerable freehold estate, and a graduate of St. John's College, Cambridge. It is reported that, being enamored of a young country-girl, who, during his visits, paid more attention to her work, which was knitting, than to her lover and his proposals, he endeavored to find out a machine which might facilitate and forward the operation of knitting, and by this means afford more leisure to the object of his affection to converse with him. Beckmann says, " Love indeed is fertile in inventions, and gave rise, it is said, to the art of painting; but a machine so complex in its parts and so wonderful in its effects would seem to require longer and greater reflection, more judgment, and more time and patience than could be expected in a lover. But, even if the case should appear problematical, there can be no doubt in regard to the inventor, whom most of the English writers positively assert to have been William Lee." Deering expressly states that Lee made the first loom in the year 1589, the date named on the painting.

Another version of the story states that Lee was expelled from the University for marrying contrary to the statutes. Having no fortune, the wife was obliged to contribute to their joint support by knitting; and Lee, while watching the motion of his wife's fingers, conceived the idea of imitating those movements by a machine. According to another version, Lee, while yet unmarried, excited the contempt of his mistress by contriving a machine to imitate the primitive process of knitting, and was rejected by her. But both accounts agree that the Stocking-frame was invented by Lee, and that about the date assigned. A writer in the *Quarterly Review*, 1816, however, observes, " This painting might give rise to the story of Lee's having invented the machine to facilitate the labor of knitting, in consequence of falling in love with a young country-girl, who, during his visits, was more attentive to her knitting than his proposals; or the story may, perhaps, have suggested the picture."

But there is another claimant. Aaron Hill ascribes the invention to a young *Oxonian*, who, having contracted an imprudent marriage, and having nothing to support his family but the produce of his wife's knitting, invented the stocking-frame, and thereby accumulated a large fortune. Evelyn, in his *Diary*, records having seen this machine as follows: " 3 May, 1661. I went to see the wonderful engine for weaving silk stockings, said to have been the invention of an Oxford scholar forty years since;" thus placing the invention many years later than the date of the picture in Stocking-Weavers' Hall.

The story of Lee's after-life, however, corroborates his being the inventor; his name is mentioned as such in the petition of the Stocking-Weavers of London to allow them to establish a guild. It is related that Lee, having taught the use of the machine to his brother and the rest of his relations, established himself at Culverton, near Nottingham, as a stocking-weaver. After remaining there five years, he applied to Queen Elizabeth for countenance and support; but, finding himself neglected both by the queen and her successor, James I., he transferred himself and his machines to France, where Henri IV. and his sagacious minister Sully gave the inventor a welcome reception. Lee is said to have carried over nine journeymen and several looms to Rouen, in Normandy. Nevertheless, after the assassination of Henri, Lee shared in the persecutions suffered by the Protestants, and is said to have died in great distress, of grief and disappointment, in Paris. Some of his workmen made their escape to England, and, under one Aston, who had been Lee's apprentice, established the stocking-manufacture permanently in England. Of Aston we find the following account in Thornton's *Nottinghamshire*, 1677, fol., p. 297:

At Culverton was born William Lee, Master of Arts in Cambridge, and heir to a pretty freehold here; who, seeing a woman knit, invented a loom to knit, and which he or his brother James performed and exercised before Queen Elizabeth; and leaving it to . . . . Aston, his apprentice, went beyond the seas, and was thereby esteemed the author of that ingenious engine wherewith they now weave silk and other stockings. This . . . . Aston added something to his master's invention; he was some time a miller at Thoroton, nigh which place he was born.

Lee's invention was important, as it not only enabled our ancestors to discard their former inelegant hose, but it likewise caused the English manufactures to excel all of foreign production, and to be sought for accordingly. Our makers soon exported vast quantities of silk stockings to Italy: these maintained their superiority for so long a period, that Keyslar, in his *Travels through Europe* as late as the year 1730, remarks, "At Naples, when a tradesman would highly recommend his silk stockings, he protests they are right English." In 1663 Charles II. granted to the Framework-Knitters' Society of London a charter, which Oliver Cromwell had refused them.

The painting of Lee and his wife, however, was parted with by the Company at a period of pecuniary embarrassment. Mr. Bennet Woodcroft has collected some particulars of the disposal of the picture, in the hope that they may lead to its restoration. In a list, dated 1687, of plate, paintings, etc., belonging to the Company, is an item—"Mr. Lee's picture, by Balderston:" it is also described in Hatton's *London*, 1708. From 1732, the Company's books show no more meetings at their Hall, or any farther entry of the picture. The Company subsequently let their Hall, and met at various taverns. The head of the Court Summons, dated 1777, is engraved from Lee's picture; and from this plate is copied an engraving in the Gallery of Portraits of Inventors in the Great Seal Patent Office. The picture is thought to have passed, about 1773, into the hands of an influential member of the Court of Framework Knitters, who from time to time lent the Company money, as their books testify. The Hall in Redcross Street has long been taken down.

JACQUARD AND HIS LOOM.

THE several looms employed in weaving appear to have been alike eclipsed by the exquisite apparatus of M. Jacquard, which is very properly named after the inventor. Like too many other inventors, he was treated with coldness and ingratitude by the community which he has so largely benefited.

Joseph-Marie Jacquard was born at Lyons in 1752, of humble parents, both of whom were weavers. He is said to have been left even to teach himself to read and write; but at a very early period he displayed a taste for mechanics by constructing neat models of buildings, furniture, etc. At the age of twelve his father placed him with a book-binder; he was subsequently engaged in type-founding and the manufacture of cutlery, in both which occupations he gave evidence of skill. Upon the death of his father, young Jacquard, with the small property left him, attempted to establish a business in weaving figured fabrics, but failed, and he was compelled to sell his looms to pay his debts. He subsequently married, and, disappointed of a portion with his wife, he was forced to sell his paternal residence. After occupying himself with ingenious schemes for improvements in weaving, cutlery, and type-founding, which produced nothing for the support of his family, Jacquard was driven into the service of a lime-burner at Bresse, while his wife had a small straw-hat business at Lyons, whither, in 1793, Jacquard returned, and assisted in the defense of that place against the army of the Convention, his only son, then a youth of fifteen, fighting by his side. They were compelled to fly, and, joining the army of the Rhine, his son was killed in battle, and Jacquard returned to Lyons, where he assisted his wife in her business of straw-hat making. Lyons at length began to rise from its ruins, and its artisans returned from Switzerland, Germany, and England, where they had taken refuge. Jacquard now applied himself with renewed energy to the completion of a machine for figure-weaving, of which he had conceived the idea as early as 1790. He succeeded, though imperfectly; and in 1801 he received from the National Exposition a bronze medal for his invention, which he patented. He set up a loom on this new principle, which was visited by Carnot, the celebrated mathematician.

About this time Jacquard's attention was directed by an English newspaper to a reward offered by a society for the invention of a machine for weaving nets for fishing and maritime purposes. Jacquard made the apparatus, but threw it aside; and his machine-made net falling into the hands of the prefect at Lyons, he and his machine were placed under arrest and conveyed to Paris, where the invention was submitted to inspectors, upon whose report a gold medal was awarded to Jacquard in February, 1804. He was now introduced to Napoleon and Carnot, when the latter, not understanding his mechanism, roughly asked him if he were the man who pretended to do that impossibility—to tie a knot in a stretched string. Jacquard, not disconcerted, explained the action of his machinery with simplicity, and convinced Carnot that the supposed impossibility was accomplished by it. He was then employed to repair and put in order the models and machines in the Conservatoire des Arts et Métiers, and while there he made some ingenious advances in weaving machinery, one of which was for producing ribbons with a velvet face on each side. He also contrived some improvements upon a loom invented by Vaucanson, which improvements have been stated to be the origin of the Jacquard machine. According to another account, Vaucanson's loom is in no way connected with Jacquard's; and, as its mechanism is very complex, its application limited to very small patterns, its action slow, and its cost very great, it belongs rather to the class of curious than of useful machines.

In 1804 Jacquard returned to Lyons to superintend his inventions for figure-weaving and for making nets, and in 1806 the municipal administration of Lyons purchased the loom for the use of the public. For some years, however, Jacquard had to struggle against the prejudice of the Lyonnese weavers, who conspired to discourage his machinery; and eventually it was publicly broken up and sold as old materials, while the inventor's personal safety was at times endangered. At length, under the effect of foreign competition, the value of Jacquard's loom was acknowledged, and it was brought very extensively into use, not only in France, but in Switzerland, Germany, Italy, and America, and it has even been introduced into the empire of China.

Jacquard was solicited by the manufacturers of Rouen and St. Quentin to organize their factories of cotton and batiste, and he received a similar offer from England; but he preferred remaining at Lyons, and continued to promote the use of his great invention until he retired to the neighboring village of Oudlins, where he died in 1834, at the age of eighty-two. During his life he received the cross of the Legion of Honor, and in 1840 a public statue was raised to his memory at Lyons.

The introduction of Jacquard's cheap and simple machine, coming within the reach of the humble weaver, forms a memorable epoch in the textile art. By its agency the richest and most complex designs are produced with facility at the most moderate price; and so far from diminishing employment, as some feared on its first introduction, it is stated to have increased the number of workmen in the manufacture in which it is used tenfold. Many ingenious applications of the Jacquard loom have been made, either to produce novel combinations or to work with more than usual rapidity.

Jacquard's invention is not, strictly speaking, a loom, but an appendage to the loom, intended to elevate or depress, by bars, the warp-threads for the reception of the shuttle; the patterns being produced by means of bands of punched cards acting on needles, with loops or eyes, which regulate the figure. The apparatus was first applied to silk-weaving only, but it has been extended to bobbin-net and other fancy manufactures, carpet-weaving, etc. Formerly the most elaborate brocades could only be produced by the most skillful weavers and the most painful labor; now, by aid of the Jacquard loom, the most beautiful products may be accomplished by men possessing only the ordinary amount of skill, while the labor attendant upon the actual weaving is little more than that required for making the plainest goods. The name of Jacquard has become, so to speak, technical in both the Old and New World, and his loom will prove a lasting record of his mechanical talent, though it has not uniformly secured him the respect of his own countrymen.

In 1853 a strange instance of ingratitude was added to the history of Jacquard and his Loom. Two of the inventor's nieces were compelled by poverty to offer for sale the gold medal bestowed by Louis XVIII. on their uncle, the sum asked being the intrinsic value of the gold, £20. The Chamber of Commerce of Lyons being acquainted with the circumstance, agreed to purchase the medal for £24! Such was the gratitude of the manufacturing interest of Lyons to the memory of a man to whom it owes so large a portion of its splendor.

DR. FRANKLIN PROVES THE IDENTITY OF LIGHTNING AND ELECTRICITY.

THE Abbé Nollet and other investigators had already made some ingenious suggestions respecting the analogies between Electricity and Lightning, when, in 1752, their truth was amply proved by Franklin, who, like his predecessors, meditating upon the similarity of their effects, traced out farther resemblances, and at length hit upon the happy expedient of sending up a common kite to an electric cloud, and thus experimentally demonstrating their identity. The following are the particulars of this great discovery:

Franklin begins his account of the similarity of the Electric Fluid and Lightning by cautioning his readers not to be staggered at the great difference of effects in point of degree, since from that no fair argument could be drawn of the actual disparity of their natures. It is, he says, no wonder that the effects of the one should so far exceed those of the other; for if two gun-barrels electrified will strike at two inches distance, and make a report, at how great a distance 10,000 acres of electric cloud must strike and give its fire, and how loud must be the crack! He then adds that flashes of lightning are generally crooked and waving, and so is a long electric spark; that lightning, like common electricity, strikes the highest and most pointed objects in its way in preference to others, such as hills, trees, towers, spires, masts of ships, points of spears, etc.; that it takes the readiest and best conductor; that it sets fire to inflammable bodies, rends others to pieces, and melts the metals. Lightning, he adds, has often been known to strike people blind, and the same happened to a pigeon which had received a violent shock of electricity; in other cases it killed animals, and they have also been killed by electricity.

Reasoning on these effects, and having observed that pointed conductors appear to attract electricity, he conceived that pointed rods of iron, fixed in the air, might draw from clouds their electric matter without noise or danger, and dissipate it at their termination in the earth. The following is his memorandum on this subject: "The electric fluid is attracted by points; we do not know whether this property be in lightning; but since they agree in all particulars in which we can already compare them, it is not improbable that they agree likewise in this. *Let the experiment be made.*"

In the year 1752, while waiting for the erection of a spire in the city of Philadelphia,* not imagining that a pointed rod of any moderate height would answer the purpose, it occurred to Franklin that by means of a common kite he might have ready access to the higher regions of the atmosphere.   Preparing, therefore, a large silk handkerchief, and two cross-sticks to extend it on, he took the opportunity of the first approaching thunderstorm, and went into a field, where there was a shed proper for the purpose; but, dreading the ridicule which he feared might attend an unsuccessful attempt, he communicated his intention to no one but his son, who assisted him in flying his kite. A considerable time elapsed without any appearance of success, and a promising cloud passed over the kite with no effect; when, just as he was beginning to despair, he observed some loose threads upon the string of the kite begin to diverge and stand erect: on this, he fastened a key to the string, and on presenting his knuckle to it was gratified by the first electric spark which had thus been drawn from the clouds: others succeeded; and when the string had become wet by the falling rain, a copious stream of electric fire passed from the conductor to his hand.   What were Franklin's emotions upon this interesting occasion it is not difficult to conceive: we are told that, when he saw the fibres of the string diverge and the spark pass, "he uttered a deep sigh, and wished that the moment were his last;" he felt that his name would be immortalized by the discovery.

Dr. Franklin pursued these experiments with much assiduity and success. He erected an insulated rod to draw the lightning from the clouds into his house, and performed, with the electricity thus derived, nearly all

* Proud as are the people of Philadelphia of their illustrious townsman, they pay little respect to his remains.   These lie within a very short distance of Arch Street, in the northeast corner of Christ Church grave-yard, at Fifth and Arch Streets.   The spot is marked by a large marble slab, laid flat on the ground, with nothing carved upon it but these words:

BENJAMIN }
and   } FRANKLIN,
DEBORAH }   1790.

Franklin, it will be recollected, wrote a humorous epitaph for himself: but his good taste and good sense showed him how unsuitable to his living character it would have been to jest in such a place.   After all, his literary works, scientific fame, and his undoubted patriotism, form his best epitaph.   Still, it may be thought, he might have been distinguished in his own land by a more honorable resting-place than the obscure corner of an obscure burying-ground, where his bones lie indiscriminately along with those of ordinary mortals; and his tomb, already well-nigh hid in the rubbish, may soon be altogether lost. We doubt much if one in a hundred of the present generation of Philadelphia have ever seen Franklin's grave.   Thousands pass daily within a few feet of the spot where his ashes and those of his wife repose, without being conscious of the fact, or, if aware of it, they are unable to obtain a glimpse of the grave.

P

the experiments for which he had before employed the common machine; and, that no opportunity might be lost of making such experiments, he attached a chime of bells to the electric rod, which gave him notice by their ringing of the electric state of his apparatus.

It should, however, be stated that two French gentlemen, Messrs. Dalibard and Deloz, were probably the first who experimentally verified Franklin's hypothesis, although the doctor was unacquainted with their proceedings. The former prepared his apparatus at Marly, near Paris; the latter at his house, which stood upon high ground in that city. M. Dalibard's apparatus consisted of an iron rod forty feet long, the lower end of which was brought into a sentry-box, where the rain could not enter, while on the outside it was fastened to three wooden posts by silken strings defended from the rain. This machine was the first that happened to be visited by the ethereal fire. M. Dalibard himself was from home; but in his absence he had intrusted the care of his apparatus to one Coisier, who was directed to call some of his neighbors, particularly the curate of the parish, whenever there should be any appearance of a thunder-storm. At length, on May 10, 1752, between two and three in the afternoon, Coisier heard a loud clap of thunder; he immediately ran to the sentry-box, and, in the presence of the curate and several neighbors, drew sparks from the conductor. A few days afterward a successful repetition of the experiment was made by M. Deloz at Paris.

These important and interesting experiments were repeated in almost every civilized country with varied success. In France a grand result was obtained by M. de Romas. He constructed a kite seven feet high and three feet wide, which was raised to the height of 550 feet by a string with a fine wire interwoven through its whole length, to render it a better conductor. On the 26th of August, 1756, sparks, or rather streams of light, were darted from the string of this kite of an inch in diameter and ten feet long.

Considering the facility, and, at the same time, the danger of these experiments, it is curious that they have only in one instance been attended by a fatal result, namely, in the case of Professor Richman of St. Petersburg. He had constructed an apparatus for experiments on atmospherical electricity which was entirely insulated, and had no contrivance for discharging it when too strongly electrified. On the 6th of August, 1753, he was exhibiting the electricity of his apparatus in company with a friend; while attending to an experiment, his head accidentally approached the insulated rod, and a flash of lightning immediately passed from it through his body, and deprived him of life. A red spot was produced upon his forehead, his shoe was burst open, and a part of his waistcoat singed; his companion was for some time rendered senseless; the door of the room was split and torn off its hinges.

Franklin's discovery of the identity of lightning and electricity has not been without its important practical results, among which is the application of conductors to buildings and ships, by which their safety during a thunder-storm is almost insured. The discovery has been most extensively applied by Sir William Snow Harris in his lightning-rods, which, by insuring the security of ships and buildings, have saved many lives and much valuable property.

CHEMISTRY OF THE GASES: DISCOVERY OF CHOKE-DAMP AND FIRE-DAMP.

In the time of Van Helmont, early in the seventeenth century, the workmen in certain German mines were molested, just as our colliers still are, by poisonous choke-damp and explosive fire-damp; that is to say (for the words were German, though only too easily domesticated in England), by suffocating and by fiery vapors, the former of which put out life silently but summarily, while the latter might blow its unfortunate victims to pieces. In sarcastic playfulness with the popular superstition as to these guardians of the mineral treasures of the old earth, Van Helmont imposed upon them the name of *Ghosts* or *Gases;* but he knew little or nothing positively about them. Boyle was probably the first to suspect that some solid bodies do in certain circumstances—when they are heated, for instance—throw off artificial airs, resembling the common atmospheric gases in thinness and in elasticity, as well as in dryness and permanency, but differing from them he could not tell how.

It was young Black, the greatest chemist Scotland has produced, and the discoverer of that fact of latent heat which Watt has embodied in the steam-engine, who took the first positively chemical step in the progress. He discovered that limestone (or chalk, or marble, or oyster-shell), when burned in the kiln, and thereby rendered quick, parts with a kind of air in which no animal can breathe or live; and also that it is owing to its setting free this air or gas that the change from inactive limestone to caustic quicklime is due. He called it fixed air, imprisoned in the rock till the furnace, or oil of vitriol, or the spirit of salt, extricated it from its fixture. He perceived and proved that this fixed air was neither more nor less than of the nature of an acid, but existing, alone of all acids, in the airy or gaseous state; and it was then conceived that there may exist many different kinds of airy matter, just as there are many kinds of solid and liquid substances.

This magnificent discovery was made at Edinburg almost within the memory of its present inhabitants, and it is the greatest discovery in natural science that has ever been made there. Dr. Chalmers said of this chemistry of the gases, "Think of Black catching fixed air, and discerning it to be an acid, at a time when nobody thought of such things; that was the great stroke; it was a very great thing to do."

Soon after this initiative had been· taken by Joseph Black, Priestley invented an easy way of collecting and handling gaseous bodies (the pneumatic trough, with its jars), and actually came upon some nine kinds of gas (all differing from ordinary air, and one from another) in a few years. Scheele had, meanwhile, been making conquests of the same sort in an obscure Swedish town, with no apparatus but phials and bladders, and had added two or three more to the list of new gases. All Europe followed these sagacious leaders—Cavendish, the discoverer of hydrogen; Watt, who first suggested that water is composed of two gases; Rutherford, the discoverer of nitrogen; Lavoisier, the interpreter, though not the first discoverer of oxygen, and the rest—until every body has at length become aware that gases are just the steams of liquids which boil at immensely low points of temperature, these liquids being the liquefactions of solid bodies which melt at temperatures lower still; and that, therefore, there may be no end to the number of the kinds of gaseous matter, precisely as there is no known limit to the vast variety of liquids and solids.—*North British Review*, No. 35.

Of Joseph Black it has been said he lived as fine a life of science as was ever lived, and died with a cup of milk unspilled in his hand.

The gas called by miners Fire-damp, or simply *damp*, is only met with in mining certain kinds of coal. It is especially abundant in the Newcastle coal-field. Elsewhere what is called *Choke-damp* prevails, this being carbonic acid gas; and it is not unlikely that other gases are mixed from time to time with these. When it is remembered that a large number of men, and often many horses, are employed underground, and that frequently there are miles of underground passages, and hundreds of miners, without more than two or three shafts communicating with the upper air, and these only chimneys many hundred feet long, and of small area, no one will be surprised that the air becomes vitiated, and that a small addition of foul gas renders it unfit for the support of life. When, however, gas of whatever kind comes off regularly, the mechanical means of ventilation commonly adopted are sufficient. It is only when there are sudden, unexpected, large jets of gas instantaneously poured forth, and when this gas, mixed with common air, becomes highly explosive, that the real danger arises.

SIR HUMPHREY DAVY AND THE SAFETY-LAMP.

THE origin of this great "invention for the preservation of human life" greatly partakes of that interest which is always concentrated on the struggle of life. Its principle was doubtless experimented on by Davy when a young man at Penzance, and writing his *Essays on Heat and Light*, even before he had commenced the study of chemistry. It is true that he shone early in the eye of the world, and was by nature much more than equal to the kind of researches he undertook; yet his great achievement of the Safety-lamp was the result of many years' patient and enlightened research, and may be traced from the commencement of his career of original research in the most remote town of Cornwall, to his construction of the Lamp itself in the theatre of the Royal Institution in London; where, in like manner, he developed heat by *rubbing two pieces of ice together*, which he had many years before rehearsed with Tom Harvey, one winter's day, beside Larigan River.

The boyhood of Davy has been sketched in some of the most fascinating pieces of biography ever written;* the annals of science do not present us with any record that equals the school-days and self-education of the boy Humphrey in popular interest; and, unlike many bright mornings, this commencement in a few years led to a brilliant meridian, and by a succession of discoveries, accomplished more, in relation to change of theory and extension of science, than in the most ardent and ambitious moments of youth he could either hope to effect or imagine possible.

Humphrey Davy was born at Penzance in 1778; was a healthy, strong, and active child, and could speak fluently before he was two years old; copied engravings

* Among these interesting records, entitled to foremost mention is the eloquent article in No. 3 of the *North British Review*, on Dr. Davy's edition of the works of his illustrious brother.

before he learned to write, and could recite part of the *Pilgrim's Progress* before he could well read it. At the age of five years he could gain a good account of the contents of a book while turning over the leaves; and he retained this remarkable faculty through life. He excelled in telling stories to his playmates; loved fishing, and collecting and painting birds and fishes; he had his own little garden, and recorded his impressions of romantic scenery in verse of no ordinary merit. To his self-education, however, he owed almost every thing. He studied with intensity mathematics, and metaphysics, and physiology; before he was nineteen he began to study chemistry, and in four months proposed a new hypothesis on heat and light, to which he won over the experienced Dr. Beddoes. With his associate Gregory Watt (son of the celebrated James Watt), he collected specimens of rocks and minerals. He made considerable progress in medicine; he experimented zealously, especially on the effects of the gases in respiration: at the age of twenty-one he had breathed nitrous oxide, and nearly lost his life from breathing carbureted hydrogen. Next year he commenced the galvanic experiments which led to some of his greatest discoveries. In 1802 he began his brilliant scientific career at the Royal Institution, where he remained till 1812; here he constructed his great voltaic battery of 2000 double plates of copper and zinc, and commenced the mineralogical collection now in the Museum. His lectures were often attended by 1000 persons: his youth, his simplicity, his natural eloquence, his chemical knowledge, his happy illustrations and well-conducted experiments, and the auspicious state of science, insured Davy great and instant success.

The enthusiastic admiration with which he was hailed can hardly be imagined now. Not only men of the highest rank—men of science, men of letters, and men of trade—but women of fashion and blue-stockings, old and young, pressed into the theatre of the institution to cover him with applause. His greatest labors were his discovery of the decomposition of the fixed alkalies, and the re-establishment of the simple nature of chlorine: his other researches were the investigation of astringent vegetables in connection with the art of tanning; the analysis of rocks and minerals in connection with geology; the comprehensive subject of agricultural chemistry; and galvanism and electro-chemical science. He was also an early but unsuccessful experimenter in the photographic art.

Of the lazy conservative spirit and ludicrous indolence in science which at this time attempted to hoodwink the public, a quaint instance is recorded of a worthy professor of chemistry at Aberdeen. He had allowed some years to pass over Davy's brilliant discovery of potassium and its congeneric metals without a word about them in his lectures. At length the learned doctor was concussed by his colleagues on the subject, and he condescended to notice it. "Both potash and soda are now said to be metallic oxydes," said he; "the oxydes, in fact, of two metals, called potassium and sodium by the discoverer of them, one Davy, in London, a verra troublesome person in chemistry."[*]

Turn we, however, to the brightest event in our chemical philosopher's career. By his unrivaled series of practical discoveries, Davy acquired such a reputation for success among his countrymen that his aid was invoked on every great occasion. The properties of fire-damp, or carbureted hydrogen in coal mines, had already been ascertained by Dr. Henry. When this gas is mingled in certain proportions with atmospheric air, it forms a mixture which kindles upon the contact of a lighted candle, and often explodes with tremendous violence, killing the men and horses, and projecting much of the contents of the mine through the shafts or apertures like an enormous piece of artillery. Soon after, a detonation of fire-damp occurred within a coal mine in the north of England, so dreadful that it destroyed more than a hundred miners. A committee of the proprietors besought our chemist to provide a method of preparing for such tremendous visitations, and he did it. He tells us that he first turned his attention particularly to the subject in 1815; but he must have been prepared for it by the researches of his early years. Still, there appeared little hope of finding an efficacious remedy. The resources of modern mechanical science had been fully applied in

* *North British Review*, No. 25.

ventilation. The comparative lightness of fire-damp was well understood; every precaution was taken to preserve the communications open; and the currents of air were promoted or occasioned, not only by furnaces, but likewise by air-pumps and steam apparatus. We may here mention that, for giving light to the coal-miner or pitman, where the fire-damp was apprehended, the primitive contrivance was a steel-mill, the light of which was produced by contact of a flint with the edge of a wheel kept in rapid motion. A "safety-lamp" had already, in 1813, been constructed by Dr. Clanny, the principle of which was forcing in air through water by bellows; but the machine was ponderous and complicated, and required a boy to work it. M. Humboldt had previously, in 1796, executed a lamp for mines upon the same principle as that of Dr. Clanny.

Davy, having conceived that flame and explosion may be regulated and arrested, began a minute chemical examination of fire-damp. He found that carbureted hydrogen gas, even when mixed with fourteen times its bulk of atmospheric air, was still explosive. He ascertained that explosions of inflammable gases were incapable of being passed through long narrow metallic tubes, and that this principle of security was still obtained by diminishing their length and diameter at the same time, and likewise diminishing their length and increasing their number, so that a great number of small apertures would not pass explosion when their depth was equal to their diameter. This fact led to trials upon sieves of wire-gauze; he found that if a piece of wire-gauze was held over the flame of a lamp, or coal-gas, it prevented the flame from passing; and he ascertained that a flame confined in a cylinder of very fine wire-gauze did not explode even in a mixture of oxygen and hydrogen, but that the gases burnt in it with great vivacity.

These experiments served as the basis of the Safety-lamp. The apertures in the gauze, Davy tells us, in his work on the subject, should not be more than 1-22d of an inch square. The lamp is screwed on to the bottom of the wire-gauze cylinder, and fitted by a tight ring. When it is lighted, and gradually introduced into an atmosphere mixed with fire-damp, the size and length of the flame are first increased. When the inflammable gas forms as much as 1-12th of the volume of air, the cylinder becomes filled with a feeble blue flame, within which the flame of the wick burns brightly; its light continues till the fire-damp increases to 1-6th or 1-5th, when it is lost in the flame of the fire-damp, which now fills the cylinder with a pretty strong light; but when the foul air constitutes 1-3d of the atmosphere, it is no longer fit for respiration, and this ought to be a signal to the miner to leave that part of the workings.

Sir Humphrey Davy presented his first communication respecting his discovery of the Safety-lamp to the Royal Society in 1815. This was followed by a series of papers, crowned by that read on the 11th of January, 1816, when the principle of the Safety-lamp was announced, and Sir Humphrey presented to the Society a model made by his own hands, which is to this day preserved in the collection of the Royal Society at Burlington House. From this interesting memorial the accompanying vignette has been sketched.

Model of the Safety-lamp, made by Sir Humphrey Davy's own hands; in the possession of the Royal Society.

There have been several modifications of the Safety-lamp, and the merit of the discovery has been claimed by others, among whom was Mr. George Stephenson; but the question was set at rest in 1817 by an examination, attested by Sir Joseph Banks, P.R.S., Mr. Brande, Mr. Hatchett, and Dr. Wollaston, and awarding the independent merit to Davy.

It should be explained that Stephenson's lamp was formed on the principle of admitting the fire-damp by narrow tubes, and "in such small detached portions that it would be consumed by combustion." The two lamps were doubtless distinct inventions; though Davy, in all justice, appears to be entitled to precedence, not only in point of date, but as regards the long chain of inductive reasoning concerning the nature of flame by which his result was arrived at.

Meanwhile, the report by the Parliamentary Committee "can not admit that the experiments (made with the lamp) have any tendency to detract from the character of Sir Humphrey Davy, or to disparage the fair value placed by himself upon his invention. The improvements are probably those which longer life and additional facts would have induced him to contemplate as desirable, and of which, had he not been the inventor, he might have become the patron."

"I value it," Davy used to say with the kindliest exultation, "more than any thing I ever did: it was the result of a great deal of investigation and labor; but, if my directions be attended to, it will save the lives of thousands of poor men."

The principle of the invention may be thus summed up. In the Safety-lamp, the mixture of the fire-damp and atmospheric air within the cage of wire-gauze explodes upon coming in contact with the flame, but the combustion can not pass through the wire-gauze, and, being there imprisoned, can not impart to the explosive atmosphere of the mine any of its force. This effect has been attributed to the cooling influence of the metal; but, since the wires may be brought to a degree of heat but little below redness without igniting the fire-damp, this does not appear to be the cause.

Professor Playfair has elegantly characterized the Safety-lamp of Davy as a present from Philosophy to the Arts; a discovery in no degree the effect of accident or chance, but the result of patient and enlightened research, and strongly exemplifying the great use of an immediate and constant appeal to experiment. After characterizing the invention as the *shutting up in a net of the most slender texture* of a most violent and irresistible force, and a power that in its tremendous effects seems to emulate the lightning and the earthquake, Professor Playfair thus concludes: "When to this we add the beneficial consequences, and the saving of the lives of men, and consider that the effects are to remain as long as coal continues to be dug from the bowels of the earth, it may be fairly said that there is hardly in the whole compass of art or science a single invention of which one would rather wish to be the author. . . . This," says Professor Playfair, "is exactly such a case as we should choose to place before Bacon were he to revisit the earth, in order to give him, in a small compass, an idea of the advancement which philosophy has made since the time when he had pointed out to her the route which she ought to pursue."

Honors were showered upon Davy. He received from the Royal Society the Copley, Royal, and Rumford Medals, and several times delivered the Bakerian Lecture. He also received Napoleon's prize for the advancement of galvanic researches from the French Institute. The invention of the Safety-lamp brought him the public gratitude of the united colliers of Whitehaven, of the coal proprietors of the north of England, of the grand jury of Durham, of the Chamber of Commerce at Mons, of the coal-miners of Flanders, and, above all, of the coal-owners of the Wear and the Tyne, who presented him (it was his own choice) with a dinner-service of silver worth £2500. On the same occasion, Alexander, the Emperor of all the Russias, sent him a vase, with a letter of commendation. In 1817 he was elected to the dignity of an Associate of the Institute of France; next year, at the age of forty, he was created a baronet.

Davy's discoveries form a remarkable epoch in the history of the Royal Society during the early part of this century, and from 1821 to 1829 almost every volume of the *Transactions* contains a communication by him. He was President of the Royal Society from 1820 to 1827. His administration was not altogether satisfactory; he was too sensitive. "Above all, he was disappointed in his life-long foolish hope of one day moving the government of Britain to patronize the cause of science"—as great an improbability in the present day as it was in poor Davy's time.

Fond of travel, geology, and sport, Davy visited, for the purpose of mineralogy and the angle, almost every county of England and Wales. He was provided with a portable laboratory, that he might experiment when he chose, as well as fish and shoot. In 1827, upon resigning the presidency of the Royal Society, he retired to the Continent; in 1829, at Geneva, his palsy-stricken body returned to the dust. They buried him at Geneva, where a simple monument stands at the head of the hospitable grave. There is a tablet to his memory in Westminster Abbey; there is a monument at Penzance; and his widow founded a memorial chemical prize in the University of Geneva. "His public services of plate, his imperial vases, his foreign prizes, his royal medals, shall be handed down with triumph to his collateral posterity as trophies won from the depths of nescience; but his WORK, designed by his own genius, executed by his own hand, tracery and all, and every single stone signalized by his own private mark, indelible, characteristic, and inimitable—his WORK is the only record of his name. How deeply are its foundations rooted in space, and how lasting its materials for time!" (*North British Review*, No. 3.)

One of the most pleasing episodes in the life of Davy is the account of his first reception of Michael Faraday, described by the latter in a note to Dr. Paris:

"When I was a bookseller's apprentice," says Faraday, "I was very fond of experiment, and very averse to trade. It happened that a gentleman, a member of the Royal Institution, took me to hear some of Sir H. Davy's last lectures in Albemarle Street. I took notes, and afterward wrote them out more fairly in a quarto volume.

"My desire to escape from trade, which I thought vicious and selfish, and to enter into the service of science, which I imagined made its pursuers amiable and liberal, induced me at last to take the bold step of writing to Sir H. Davy, expressing my wishes, and a hope that, if an opportunity came in his way, he would favor my views; and at the same time I sent the notes I had taken of his lectures."

To this application Sir H. Davy replied as follows:

To MR. FARADAY.

"December 24, 1812.

"SIR,—I am far from displeased with the proof you have given me of your confidence, and which displays great zeal, power of memory, and attention. I am obliged to go out of town till the end of January: I will then see you at any time you wish.

"It would gratify me to be of any service to you. I wish it may be in my power.

"I am, sir, your obedient humble servant,   H. DAVY."

Early in 1813 Davy requested to see Faraday, and told him of the situation of assistant in the Laboratory of the Royal Institution, to which, through Sir Humphrey's good efforts, Faraday was appointed. In the same year he went abroad with Davy as his assistant in experiments and in writing. Faraday returned in 1815 to the Royal Institution, and has ever since remained there.

There can not be a better testimony than the above circumstance to Davy's goodness of heart.

CARCEL AND HIS LAMP.

To Carcel, the clockmaker of Paris, we owe the solution of an important difficulty in lamp-making—the avoidance of the projection of the shade from the reservoir. In a lamp which he constructed, Carcel made the reservoir for oil at the lower part of the lamp, and placed close to it a clock-work which moved a little force-pump, the piston of which raised the oil as far as the wick. The spring was reached by means of a key. The mechanical means employed by Carcel for raising the oil to the burner were as ingenious as elegant; therefore have we changed nothing of the principle of the inventor's lamp. The wheel-work that he adopted has always been retained, the improvements being secondary points in the mechanism.

Carcel drew but a small profit from his important discovery. Like many originators of useful inventions, to whom we are indebted for the luxury and ease of actual life, he left to others the profits and benefit of his works. He died in 1812, full of infirmities. Life had been to him but a long and painful struggle. When he wished to patent and secure to himself the property of his discovery, and to commence the use of it, he was obliged to have recourse to a partner to find the necessary funds. It was the apothecary Carreau who joined him: thus the patent which was delivered the 24th of October, 1800, to the inventor of the Mechanical Lamp, bore the two names of Carcel and Carreau. But the latter had nothing to do with the discovery, though his intervention in the enterprise was not without its advantages. Carcel, greatly discouraged, would not have followed up the work he had proposed for himself had it not been for the entreaties and encouragement of his friend. However, the term of the patent expired without having brought any important profit to the two partners. In the Rue de l'Arbre Sec at Paris may still be seen the old shop of Carcel, occupied to this day by a member of his family, bearing this sign—"*Carcel, Inventeur.*" In the door-way of this simple shop may be seen the first model of the lamp which Carcel constructed. The hot air which passes from the glass chimney of the lamp serves to put in motion the mechanism by which the oil is raised to the burner. On other lamps is clock-work, constructed as by Carcel, the needles of which are put in action by the same mechanism which raises the combustible liquid. —From the *Engineer* journal, 1857.

THE production of hydrogen gas in a tobacco-pipe by filling the bowl with powdered coal, then luting it over and placing it in a fire, is well known; but even more familiar are the alternate bursting out and extinction of those burning jets of pitchy vapor, which contribute to render a common fire an object so lively, and of such agreeable contemplation in the winter evenings. We may pursue the subject in tracing the brilliant lights by which our streets are illuminated from the obscure recesses of nature, and showing by what steps that which was once thought simply an object of curiosity has been applied to a practical purpose of the most useful and agreeable kind; which an able writer, in showing what had been done with the gases, felicitously illustrated: "One species, or rather a variable mixture of two or three, composed of carbon and hydrogen, is made in the outskirts of nearly every town now-a-days, in enormous quantities, and then sent away from a huge trough or jar, or from a heart, to circulate through a system of metallic arteries, for the purpose of lighting streets and houses."

The existence and inflammability of coal-gas have been known in England for two centuries. In the year 1659 Thomas Shirley correctly attributed the exhalations from "the burning well" at Wigan, in Lancashire, to the coal-beds which lie under that part of the county; and soon after, Dr. Clayton, influenced by the reasoning of Shirley, actually made coal-gas, and detailed the results of his labors in a letter to the Hon. Robert Boyle, who died in 1691. He says he distilled coal in a retort, and that the contents were phlegm, black oil, and a spirit which he was unable to condense, but which he confined in a bladder. These are precisely what we now find, but under different names: the phlegm is water, the black oil is coal-tar, and the spirit is gas. Dr. Clayton several times repeated the experiment, and frequently amused his friends with burning the gas as it came from the bladder through holes made in it with a pin. "This is a hint which, in an age more alive to economic improvement, might have brought Gas-lighting into operation a century earlier, though the mechanical difficulties might have been too great to overcome at that period; a circumstance which has retarded the introduction of so many valuable discoveries, as it did that of the steamboat and printing-machine."*

About a century later (1753) Sir James Lowther communicated to the Royal Society a notice of a spontaneous evolution of gas at a colliery belonging to him near Whitehaven. While his men were at work, they were surprised by a rush of air, which caught fire at the approach of a candle, and burned with a flame two yards high and one yard in diameter; they were much frightened, but put the flame out by flapping it with their hats, and then all ran away. The steward of the works, hearing this, went down himself, lighted the air again, which had now increased, and had some difficulty in extinguishing it. It was found to annoy the workmen so much that a tube was made to carry it off. The tube projected four yards above the pit, and at the extremity of it the gas rushed out with much force. "The gas being fired," says the account, "it has now been burning two years and nine months without any sign of decrease." Large bladders were filled in a few seconds from the end of the tube, and carried away by persons, who fitted little pipes to them, and burned the gas at their own convenience. We do not learn what became of this copious supply; it probably diminished as the coal-bed was exhausted.

Soon after the middle of the last century Bishop Watson made many experiments on coal-gas, which he details in his *Chemical Essays:* he distilled the coal, passed the gas through water, conveyed it through pipes from one place to another, and did so much that we are only surprised he did not introduce it into general use.

Meanwhile the use of Gas had long been known in a distant part of the world. "Whether, or to what extent," says Mr. R. C. Taylor, on the coal-fields of China, "the Chinese artificially produce illuminating gas from

* *Penny Cyclopædia,* art, "Gas-lighting."

bitumen coal, we are uncertain. But it is a fact that spontaneous jets of gas, derived from boring into coalbeds, have for centuries been burning, and turned to that and other economical purposes. If the Chinese are not manufacturers, they are nevertheless gas consumers and employers on a large scale, and have evidently been so ages before the knowledge of its application was acquired by Europeans. Beds of coal are frequently pierced by the borers of salt water, and the inflammable gas is forced up in jets twenty or thirty feet in height. From these fountains the vapor has been conveyed to the salt-works in pipes, and there used for the boiling and evaporating of the salt; and other tubes convey the gas intended for lighting the streets and the larger apartments and kitchens."*

To return to England. Although the properties of coal-gas were known here so long ago, no one thought of applying it permanently to a useful object until the year 1792, when Mr. Murdoch, an engineer at Redruth, in Cornwall, erected a little gasometer and apparatus, which produced gas enough to light his own house and offices. Murdoch appears to have had no imitators, but he was not discouraged; and in 1797 he erected a similar apparatus in Ayrshire, where he then resided. In the following year he was engaged to put up a gas-work at the manufactory of Boulton and Watt at Soho. This was the first application of gas in a large way; but, excepting in manufactories or among scientific men, it excited little attention until the year 1802, when the front of the great Soho manufactory was brilliantly illuminated with gas on the occasion of the public rejoicings at the Peace. All Birmingham poured forth to view the spectacle, and strangers carried to every part of the country an account of what they had seen. It was spread about every where by the newspapers; easy modes of making gas were described; and coal was experimentally distilled in tobacco-pipes at the fireside all over the kingdom.

---

* Mr. Taylor notices the singular counterpart to this employment of natural gas in the valley of Kanawha in Virginia. The geological origin, the means of supply, the application to all the purposes of manufacturing salt, and of the surplus to illumination, are remarkably alike at such distant points as China and the United States.

Soon after this, several manufacturers adopted the use of gas: a button manufactory at Birmingham used it largely for soldering; Mr. Samuel Clegg first began to construct gas apparatus, and about 1806 exhibited gas-lights in the front of his manufactory. Halifax, Manchester, and other towns followed.

A single cotton-mill at Manchester used about 900 burners, and had several miles of pipe laid down to supply them; and Mr. Murdoch, who erected the apparatus used in this mill, sent a detailed account of his operations to the Royal Society in 1808, for which he received their gold medal. The success of Gas-lighting in the cotton factory was striking: it was very soon adopted for the softness, clearness, and unvarying intensity of the light; and it was free from the inconvenience and danger resulting from the sparks and frequent snuffing of candles, which tended to diminish the hazard of fire, and lessen the high insurance premium on cotton-mills.

Previous to the public display of Gas at Soho, it had, however, been applied to similar purposes by a M. Le Bon at Paris, who in 1801 lighted up his house and gardens with the gas obtained from wood and coal, and had it in contemplation to light up the city of Paris; but we find nothing farther recorded of M. Le Bon's results.

Thus we see that, although the Chinese have for ages employed natural coal-gas for lighting their streets and houses, only within the present century has gas superseded in London the dim oil-lights and crystal-glass lamps of the preceding century. Dr. Johnson is said to have had a prevision of this change when, one evening, from the window of his house in Bolt Court, he observed the parish lamplighter ascend a ladder to light one of the small oil-lamps. He had scarcely descended the ladder half way when the flame expired. Quickly returning, he lifted the cover of the lamp partially, and thrusting the end of his torch beneath it, the flame was instantly communicated to the wick by the thick vapor which issued from it. "Ah!" exclaimed the doctor, "one of these days the streets of London will be *lighted by smoke*" (*Notes and Queries*, No. 127).

The use of gas, however, made but slow progress in the metropolis: it was dirty and disagreeable, and no means had yet been found for purifying the gas, though lectures were delivered and experiments made upon the subject by a German named Frederick Albert Winsor. In 1803 and 1804 he lighted the old Lyceum theatre. He took out a patent in 1804, and issued a prospectus of a National Light and Heat Company, promising sub-scribers of £5 at least £570 per cent. per annum, with a prospect of ten times as much. A subscription was raised, it is said, of £50,000, which was expended in ex-periments, without profit to the subscribers, although Winsor gained experience, and the important process of purifying gas by lime. In 1807 he lighted one side of Pall Mall; on the king's birthday, June 4, he brilliantly illuminated the wall between Pall Mall and St. James's Park; and on August 16 exhibited gas-light in Golden Lane. In 1809 the National Light and Heat Company applied to Parliament for a charter, but they were op-

Frederick Albert Winsor, Projector of Street Gas-lighting.

posed by Mr. Murdoch on the score of prior discovery, and the charter was refused. It was, however, subse-quently granted, and in 1810 was established the Gas-light and Coke Company, in Cannon Row, Westminster; removed to Peter Street, or Horse-ferry Road, previous-ly the site of a market-garden, poplars, and a tea-garden. Soon after an extensive explosion took place on the premises, when a committee of the Royal Society was, at the request of the government, appointed to investigate the matter. They met several times at the gasworks to examine the apparatus, and made a very elaborate report, in which they stated as their opinion that, if Gas-lighting was to become prevalent, the works ought to be placed at a considerable distance from all buildings, and that the reservoirs should be small and numerous, and always separated from each other by mounds of earth, or strong party-walls. This committee consisted of Sir Joseph Banks, Sir C. Blagden, Col. Congreve, Mr. Lawson, Mr. Rennie, and Dr. Young. In the company's application to Parliament, one of their witnesses, Mr. Accum, the chemist, was bitterly ridiculed by Mr. Brougham, F.R.S.; and Sir Humphrey Davy asked if it were intended to take the dome of St. Paul's for a gasometer! In short, as Dr. Arnott remarks, "Davy, Wollaston, and Watt at first gave an opinion that coal-gas could never be safely applied to the purpose of street-lighting." However, the invention progressed, and in 1822 St. James's Park was first lighted with gas. Its safety was not, however, yet established; for in 1825, on the part of government, a committee of the most eminent scientific men minutely inspected the gas-works, and reported that the occasional superintendence of all the works was necessary.

Of the general process of making Gas we need only state that it is obtained from coal inclosed in red-hot cast-iron or clay cylinders or retorts, when hydro-carbon gases are evolved, and coke left behind; the gas, being carried away by wide tubes, is next cooled and washed with water, and then exposed to lime in close purifiers. It is then stored in sheet-iron gas-holders, miscalled gasometers, some of which hold 700,000 cubic feet of gas; and the several London companies have storage for ten million cubic feet of gas. Thence it is driven by the weight of the gas-holders through cast-iron mains or pipes under the streets, and from them by wrought-iron service-pipes to the lamps and burners: of the gas-mains there are 2000 miles.

The London Gas Company's works at Vauxhall are the most powerful and complete in the world: from this point their mains pass across Vauxhall Bridge to western London, and by Westminster and Waterloo Bridges to Hampstead and Highgate, seven miles distant, where they supply gas with the same precision and abundance as at Vauxhall. Their pipes extend 150 miles.

Gas-lighting has been extended from London throughout Great Britain, so that there is now scarcely a small town not lighted by gas. The Continental cities slowly followed our example; and it has reached our antipodes.

Gas has been made from oil and resin, but is too costly for street-lighting. Wood and peat are also used. In Ireland a village has been lighted with gas made from bog-turf. Gas-lights are also used in coal-mines, greatly facilitating the operations of the colliers. The greater cheapness of coal, in those places where it can be procured, will probably always place it above any other material that could be proposed for the manufacture of gas.

The Lime-ball, the Bude, and the Electric Lights are too expensive for street-lighting. Some of the processes of artificial illumination have been costly failures: upon the Patent Air-light (from hydrocarbons mixed with atmospheric air), proposed in 1838, upward of £30,000 were expended unsuccessfully. The Atmospheric Bude Light is the result of numerous experiments made by Mr. Goldsworthy Gurney, of Bude, in Cornwall, and is now extensively employed in lighting churches and other large buildings. Originally it was obtained from an oil lamp, the flame from which was acted upon by a current of oxygen: subsequently oil-gas was substituted for the liquid oil; but now the gas which is made for lighting the streets of towns is employed to produce the flame, and the brilliancy is increased by a current of atmospheric air ingeniously introduced. The Bude Light was first used for lighting the House of Commons in the year 1842: its cost is about one third the expense of common oil, and about one ninth that of composition candles.

JAMES BRINDLEY AND CANAL NAVIGA-
TION.

THE Canal, an artificial channel filled with water, is used for the transit of goods, for irrigation, and for supplying towns with water. The New River, by which London is in great part provided with water from Hertfordshire, is a canal. The canals by which ancient Egypt was intersected were used both for navigation and irrigation. Canals are known to have existed in China before the Christian era. The first canal made in Europe, as far as we know, was cut by Xerxes across the low isthmus of Athos. Canals were made by the Romans in Italy, and in the Low Countries about the outlets of the Rhine; and we have reason to think that they also made canals in Britain. But canal-making in modern Europe was first practiced by the inhabitants of North Italy and Holland. Works of this kind, which are still admired by engineers, were executed in Lombardy between the eleventh and thirteenth centuries: the canal from Milan to the Ticino was made navigable in 1271. The formation of canals was begun in the Netherlands in the twelfth century, when Flanders became the commercial *entrepôt* of Europe. Holland is intersected with canals, which have been compared to the public roads in other countries.

The origin of the present system of English Canals dates from the year 1755, when an Act of Parliament was passed for constructing one eleven miles long, from the mouth of Sankey Brook, in the River Mersey, to Gerard's Bridge and St. Helen's. It should, however, be mentioned that canals had been previously known for centuries in this country. The canal from the Trent to the Witham, which is the oldest in England, is said to have been dug in the year 1134.

James Brindley, who rose from a childhood of poverty and neglect to be a celebrated engineer, was born in Derbyshire in 1716. Through his father's dissipated

Q

habits, the boy was employed in farm labor, and allowed to grow up almost totally uneducated; to the end of his life, he was barely able to read and write. He is supposed, however, to have shown some bias toward mechanical invention; for, at the age of seventeen, he bound himself apprentice to a millwright at Macclesfield. Here he was left frequently by himself for whole weeks together, to execute works concerning which his master had given him no previous instruction; these he finished in his own way. On one occasion his master was employed to construct the machinery of a new kind of paper-mill, and, although he had inspected a mill in which similar machinery was in operation, it was reported that he would be unable to finish his contract. Brindley was informed of this rumor; and, as soon as he had finished his week's work, he set out for the mill, took a complete survey of the machinery, and after a walk of fifty miles, reached home in time to commence work on Monday morning. Having thus made himself perfectly master of the construction of the mill, he completed the machinery, with several improvements of his own contrivance.

Brindley, on the expiration of his apprenticeship, started in business on his own account, but did not confine himself to the making of mill-machinery. In 1752 he contrived an improved engine for draining some coal-pits at Clifton, Lancashire; it was set in motion by a wheel 30 feet below the surface, and the water for turning it was supplied from the Irwell by a subterraneous tunnel 600 yards long. In 1755 he executed a portion of the complex machinery for a silk-mill at Congleton; and in the following year he erected a steam-engine at Newcastle-under-Lyne, which effected a saving of one half in fuel.

Brindley's genius was constantly displaying itself by the invention of the most beautiful and economical simplifications. One of these was a method which he contrived for cutting all his tooth and pinion wheels by machinery, instead of having them done by hand as hitherto. This invention enabled him to finish as much of that sort of work in one day as had formerly been accomplished in fourteen.

But the character of Brindley's mind was comprehensiveness and grandeur of conception; and there speedily arose an adequate field for the display of his vast ideas, and almost inexhaustible powers of execution. In 1755 was begun the first modern canal actually executed in England—the Sankey Brook Navigation, eleven miles long. In 1758 he commenced, for the Duke of Bridgewater, the celebrated Bridgewater Canal, which as now completed, commences at Manchester and terminates at Runcorn, and has a branch to Worsley and Leigh. One of his earliest great works was an aqueduct carrying the canal across the Irwell; so that from the aqueduct may often be seen seven or eight men slowly dragging a boat up the Irwell against the stream, while, about 40 feet immediately over the river, a horse or a couple of men are enabled to draw with much greater rapidity five or six barges fastened one to the other. The canal from Worsley to Manchester, with the underground course and tunnels, cost £168,000, and is eighteen miles in length. With the exception of the part between Worsley and Leigh, this canal was executed by Brindley in five years.

While the Bridgewater Canal was yet in progress, Brindley commenced another canal passing through Staffordshire, and uniting the Trent and the Mersey. This canal is ninety-three miles in length, has ninety-six locks, and passes over many aqueducts: it has five tunnels, one of which, 2880 yards in length, is cut through Harecastle Hill, at more than 200 feet below the surface of the earth. The canal was not completed at Brindley's death; but his brother-in-law, Mr. Henshall, successfully finished it. Brindley also designed a canal, forty-six miles long, called the Staffordshire and Worcestershire Canal, for the purpose of connecting the Grand Trunk with the Severn. He also planned the Coventry Canal, and superintended the execution of the Oxford Canal. These undertakings opened an internal water-communication between the Thames, the Humber, the Severn, and the Mersey, and united the great ports of London, Liverpool, Bristol, and Hull by canals which passed through the richest and most industrious districts of England.

The canal from the Trent at Stockwith to Chesterfield,

forty-six miles long, was Brindley's last public undertaking. Phillips, in his *History of Inland Navigation*, says that Brindley pointed out the method of building walls against the sea without mortar; and that he invented a mode of drawing water out of mines by a losing and gaining bucket.

Brindley's designs were the resources of his own mind alone. When he was beset with any difficulty, he secluded himself, and worked out unaided the means of accomplishing his schemes. Sometimes he lay in bed two or three days; but when he arose, he proceeded at once to carry his plans into effect, without the help of drawings or models. He knew something of figures, but did not much avail himself of their assistance in his calculations: his habit was, to work the question chiefly in his head, only setting down the results at particular stages; yet his conclusions were generally correct. He died in 1772, in his fifty-sixth year.

Brindley was an enthusiast in canal navigation. When giving his professional evidence before a committee of the House of Commons, he expressed himself with so much contempt of rivers as means of internal navigation that a member was tempted to ask him for what object rivers were created; when Brindley replied, "to feed navigable canals." This is characteristic, and probably authentic; but it was made public by an anonymous correspondent to a journal, whose communications respecting Brindley were stated by some of his friends to contain many inaccuracies.

JOHN SMEATON: LIGHT-HOUSES AND HARBORS.

Of John Smeaton, the Civil Engineer, it may well be said that he was one of the earliest of "a self-created set of men, whose profession owes its origin, not to favor or influence, but to the best of all protection, the encouragement of a great and powerful nation"—in the construction of light-houses and harbors, and the undertaking of other great public works.

Smeaton was born in 1724, at Austhorpe, near Leeds, in a house built by his grandfather. His father was an attorney, and brought him up with a view to the legal profession.

He exhibited at a very early age great strength of understanding and originality of genius. His playthings were not the toys of children, but the tools with which men work; and he appeared to take greater pleasure in seeing the men in the neighborhood work, and asking them questions, than in any thing else. One day he was seen, to the no small alarm of his family, on the top of his father's barn, fixing up something resembling a windmill. On another occasion, he watched some men who were sinking a pump in a neighboring village, and observing them cut off a piece of bored pipe, he procured it, and actually made with it a pump that raised water. All this was done while he was in petticoats, and before he had reached his sixth year. About his fourteenth or fifteenth year he had made himself an engine to turn rose-work; he also made a lathe, by which he turned a perpetual screw in brass, a machine but little known at that time. In this manner he had, by the strength of his genius and indefatigable industry, acquired at the age of eighteen an extensive set of tools, and the art of working at most of the mechanical trades without the assistance of a master.

In 1742, in pursuance of his father's design, young Smeaton came to London, and attended the courts of law at Westminster Hall; but, finding the bent of his mind averse to the law, his father yielded to his wishes, and allowed him to devote his energies to more congenial pursuits. About the year 1750 he took up the business of a mathematical-instrument maker; next year he experimented with a machine that he had invented for measuring a ship's way at sea; and in 1752 and 1753 was engaged in a course of experiments "concerning the natural powers of water and wind to turn mills and other machines depending on circular motion." From thence resulted the most valuable improvements in hydraulic machinery, increasing the power one third. For these experiments Smeaton received the Copley Gold Medal of the Royal Society, of which he had become a Fellow. In 1754 he visited Holland and the Netherlands, and the acquaintance he thus obtained with the construction of embankments, artificial navigations, and similar works, probably formed an important part of his engineering education.

In 1759 Smeaton communicated to the Royal Society an experimental investigation, by which he reduced the art of designing wind-mills to general principles. The details may be seen in Professor Rankine's *Manual of the Steam-engine and other Prime Movers*, 1859.

In 1766 Smeaton commenced the great work which, more than any other, may be looked upon as a lasting monument of his skill—the erection of the Eddystone Light-house, built on the Eddystone rock, about fourteen miles south of Plymouth. Two light-houses had before been erected on the rock: the first was swept away by a storm; and the second, which was built of timber, was destroyed by fire in December, 1755. The immediate re-erection of the beacon being highly important, application was made to the Earl of Macclesfield, then President of the Royal Society, for advice as to the person who should be intrusted with the difficult task. The previous light-houses had been designed by non-professional men, and it was felt now that to erect another "would not so much require a person who had merely been bred, or had rendered himself eminent, in this or that profession, but rather one who, from a natural genius, had a turn for contrivances in the mechanical branches of science." Lord Macclesfield immediately perceived that Smeaton was the man required, and therefore recommended him. He commenced the work, in the spring of 1756, by accurately measuring the very irregular surface of the rock, and making a model of it. The cutting of the rock for the foundation was commenced on August 5th of the same year; the first stone was landed on the rock June 20, 1757; the building was finished October 9, 1759, and the lantern lighted for the first time on the 16th, the whole being completed in considerably less than four years, the time originally proposed, during which there were 421 days' work done upon the rock.

The Eddystone Light-house is a circular tower of stone sweeping up with a gentle curve from the base, and gradually diminishing at the top, somewhat similar to the swelling of the trunk of a tree, the upper extremity being surmounted with a lantern and gallery. The materials of the tower are moorstone, a hard granite, and Portland stone. The granite rock was partially worked to form the foundations; and as the rock-joint would be more subject to the action of the sea than any other, it was found necessary not only that the bed of every stone should have a level bearing, but that every outside piece should be grafted into the rock, so as to be guarded by a border thereof at least three inches in height above it, which would in reality be equivalent to the founding of the building in a socket three inches deep in the shallowest part. On Aug. 3, 1756, Smeaton fixed the centre point of the building, and traced out part of the plan on the rock; and on the 6th nearly the whole of the work was set out. On Sept. 4, two new steps at the bottom of the rock, and the dovetails, were roughed out, and some of the beds brought to a level and finished, after very great labor. The stones for the several courses were rough-worked at the quarries according to the engineer's draughts.

A part of the upper surface of the rock having been taken carefully off, but without the use of gunpowder, lest it should loosen the rock, six foundation-courses, dovetailed together, were raised on the lower part of the rock, which brought the whole to a solid level mass.

These courses, with eight others raised above them, are the solid bed of the work.   The courses of masonry are skillfully dovetailed together, and each layer of masonry is very strongly cemented, and connected by oak trenails or plugs, the whole being strongly cramped.   The general weight of the stones employed is a ton, and some few are two tons.   In the solid work the centre stones were fixed first, and all the courses were fitted on a platform and accurately adjusted before they were removed to the rock.   The base of the tower is about 26 feet 9 inches in diameter, taken at the highest part of the rock; the height of the solid masonry to the top of the stone staircase, from the centre of the base, is 28 feet 4 inches. The whole height of the tower and lantern is 85 feet 7 inches, or rather more than two fifths the height of the London Monument.   The upper part of the light-house, originally constructed of wood, was burnt in 1770, and renewed in 1774.   The Eddystone Light-house was Smeaton's first work, and also his greatest; probably, the time and all things considered, it was the most arduous undertaking that has fallen to any engineer, and none was ever more successfully executed.   And now, having withstood the storms of a hundred years, the Eddystone remains, unmoved as the rock it is built on, a proud monument to its great architect.

Next to the Eddystone Light-house, among the many useful works executed by Smeaton, ranks Ramsgate Harbor.   To his skill the preservation of the old London Bridge for many years was attributable: in 1761, one of the piers being undermined, the bridge was considered to be in such danger that no one would pass over it; the engineers were perplexed, when an express was sent to Yorkshire for Smeaton, who immediately sunk a great number of stones about the endangered pier, and thereby preserved it.   The great canal from the Forth to the Clyde, the Spurn Light-house, the Calder navigation, and some important bridges in Scotland, are also prominent among Smeaton's works.   On the 16th of September, 1792, while walking in his garden at Austhorpe, Smeaton was attacked with paralysis, and on October 28 he died.

Smeaton left many valuable records of his professional career.   In 1771, under his auspices, was established "the

Smeatonian Society of Civil Engineers," who subsequently published his reports on public works. His deliberation and caution were very great; and so highly was his judgment appreciated, that he was called " the Standing Counsel" of his profession, and he was constantly appealed to by Parliament on difficult engineering questions. He greatly improved the atmospheric steam-engine of Newcomen; he introduced many improvements in mathematical apparatus; his ardent love of astronomy led him to build an observatory at Austhorpe.

Smeaton uniformly evinced a high feeling of independence in respect of pecuniary matters, and would never allow motives of emolument to interfere with plans laid on other considerations. The Empress Catharine of Russia was exceedingly anxious to have his services in some great engineering works in her dominions, and she commissioned the Princess Daschkaw to offer him his own terms. But his plans and his heart were bent upon the exercise of his skill in his own country, and he steadily refused all the offers made to him. It is reported that when the princess found her attempts unavailing, she said to him, "Sir, you are a great man, and I honor you. You may have an equal in abilities, perhaps, but in character you stand alone. The English minister, Sir Robert Walpole, was mistaken; and my sovereign, to her loss, finds in you a man who has no price."

After Smeaton had retired from his profession, he was often pressed to superintend engineering works: when these entreaties were backed by personal offers of emolument, he used to send for an old woman who took care of his chambers in Gray's Inn, and say, "Her attendance suffices for all my wants;" a reply which intimated that a man whose personal wants were so simple, was not likely to break through a prearranged line of conduct for mere pecuniary consideration.

INVENTIONS OF JOSEPH BRAMAH.

THIS ingenious mechanician was born at Stainsborough, in Yorkshire, in 1749, and was intended for his father's occupation of a farmer; but he very early evinced a taste for mechanical pursuits, and at the age of sixteen was apprenticed to a joiner. He subsequently removed to London, where he worked as a journeyman cabinet-maker, and next set up in the same business for himself. His adoption of the profession of engineer or machinist appears to have arisen from his contriving improvements in water-closets. He next invented, and patented in 1784, the celebrated Bramah Lock; when he pronounced it "not to be within the range of art to produce a key, or other instrument, by which a lock on this principle can be opened."

Bramah is an early example of a man of genius devising and carrying out large and extensive schemes for the application of machinery to manufactures. Thus, when he obtained the patent for his admirable lock, he immediately set about the construction of a series of machine-tools for shaping with the required precision the barrels, keys, and other parts of the contrivance, which, indeed, would have utterly failed unless they had been formed with the accuracy which machinery alone can give. In Bramah's workshop was educated the celebrated Henry Maudslay, who worked with him from 1789 to 1796, and was employed in making the principal tools for his lock. Its peculiarity consisted in a novel application of tumblers, or movable obstacles, and the abandonment of the use of wards. This lock was greatly improved by Bramah's sons: its security depends on the doctrine of combinations, or the multiplication of numbers into each other, which is known to increase in the most rapid proportion. Bramah's lock was, however, picked in 1817, when it was improved by the introduction of false notches; it was again picked in 1851; nevertheless, it is still one of the most inviolable locks ever contrived.

Among the numerous other inventions of Bramah were improvements in water-cocks, pumps, and fire-engines; but his greatest work is the Hydraulic Press, a machine acting on the principle of the philosophical toy called the hydrostatic paradox, and of very great power in compressing bodies or lifting weights, in drawing up trees by the roots, or piles from beds of rivers: woolen and cotton goods are compressed by it into the most portable dimensions; and even hay, for military service, is reduced to such a state of coercion as to be easily packed on board transports.

Pascal demonstrated this principle and its advantages by fixing to the upper end of a cask set upright a very long and narrow cylinder. In filling the barrel, and afterward the cylinder, the simple addition of a pint or two of water, which the latter was capable of containing, produced the same effect as if the cask, preserving its diameter throughout, had had its height increased by the whole length of the cylinder. Thus the increase of weight of a pint or two of water was sufficient to burst the bottom of the hogshead by the immense augmentation of pressure it occasioned. Now, if we suppose the water removed from the cylinder of narrow dimensions, and replaced by a solid of equivalent weight, such as a piston, it is evident that the pressure must remain every where the same. Again, if we suppose the weight of the piston to be multiplied by the power of a lever acting on its shaft, the pressure will be proportionally augmented, so as to produce on the bottom of the cask a pressure equivalent to an enormous weight with the exertion of very little primitive force on the piston.

In the Museum of the Commissioners of Patents at South Kensington is "the first Hydraulic Press ever made," inscribed " Bramah, Invt. et Fect., 1796."

Mr. Bramah next patented the elegant and convenient . beer-machine for drawing liquors in a tavern-bar from barrels in the cellar by means of a force-pump. He also improved steam-engine boilers and paper-making machinery, and invented a machine for making pens by a mechanical process, by which several nibs, resembling steel pens, are cut out of one quill, and fixed in a holder. In 1806 he contrived a mode of printing, which, being applied to the numbering of bank-notes during the issue of one-pound notes by the Bank of England, saved the labor of 100 clerks out of 120.   This machine consists of disks or wheels, with the numbers from 1 to 9 and 0 cut on the periphery of each, the whole being mounted upon one axle, but to be turning independently of each other.   By the action of mechanism which is incapable of error, the position of one wheel of the series is moved between each operation of printing, so that when the machine is properly adjusted, it will print a series of numbers in regular progression, without the possibility of twice producing the same number.

In 1812 Bramah patented a scheme for laying water-mains, with force-pumps to throw water for extinguishing fires, and to supply a lifting power for raising great weights.   This ingenious inventor died in consequence of cold contracted while superintending the uprooting of trees in Holt Forest by his Hydraulic Press, in his sixty-eighth year, in 1814.

THOMAS TELFORD AND THE MENAI SUSPENSION BRIDGE.

In the life of this eminent engineer "another striking instance is added to those on record of men who have, by the force of natural talent, unaided save by uprightness and persevering industry, raised themselves from the low estate in which they were born to take their stand among the master spirits of the age."* Telford's father was a shepherd in the pastoral district of Eskdaile, in Dumfriesshire, where, in the parish of Waterwick, Thomas, his only son, was born in 1757. He received the rudiments of education at the parish school; and while engaged during the summer season as a shepherd-boy in assisting his uncle, he diligently made use of his leisure in studying the books lent to him by his village friends. At the age of fourteen he was apprenticed to a stone-mason at Langholm: he was for several years employed chiefly in his native district; and in the construction of plain bridges and farm-buildings, small village churches and manses, he passed a valuable training, such as is of singular advantage to the future architect or engineer. In 1780, being then about twenty-three, he visited Edinburgh for employment, and there, for about two years, he paid much attention both to architecture and drawing. He then removed to London, and there worked upon the quadrangle of Somerset House, under Sir William Chambers, the architect. Telford was next engaged in Portsmouth Dock-yard upon various buildings for about three years, during which he became well acquainted with the construction of graving-docks, wharf-walls, and similar engineering works. In 1787 he removed to Shrewsbury: subsequently, in Shropshire, he built a stone bridge over the Severn; and, next, the iron bridge at Buildwas, consisting of a very flat arch, 130 feet span; these being followed by forty other bridges in the same county.

* *Transactions of the Institution of Civil Engineers.*

The Ellesmere Canal, about 103 miles in length, was Telford's first great work, and led him to direct his attention almost solely to civil engineering. This canal crosses the Dee at an elevation of 70 feet, by an aqueduct bridge of 10 arches, each 40 feet span, the bed of the canal being of cast-iron plates instead of puddled clay and masonry. The Pont-y-Cysylte aqueduct bridge is still more remarkable, and consists simply of a trough of cast-iron plates flanged together, and supported on masonry piers 120 feet above low water. The Caledonian Canal is another of Telford's principal works, commenced in 1802 and opened in 1822. Its entire length (between the German and Atlantic Oceans) is 250 miles, of which 230 miles, friths and lakes, were already navigable; the canal itself is about 20 miles, and cost a million pounds sterling. We have not space to describe the other canals which Telford wholly or partially constructed. He executed many important drainage works, especially of Bedford Level. On the Continent he superintended the construction of the Gotha Canal in Sweden, for which he received a Swedish order of knighthood.

The works executed by Telford under the Commissioners of Highland Roads and Bridges are of the greatest importance; they intersect the whole of Scotland with 1000 miles of new road, and 1200 bridges, in a mountainous and stormy region; Telford also improved several harbors, and erected many Highland churches and manses.

Telford's most important harbor-work is the St. Katherine's Docks, London, which were constructed with unexampled rapidity. He also built many bridges of considerable size and improved construction; but the most perfect specimen of his skill as an engineer is the great road from London to Holyhead, and the works connected with it. The Menai Suspension Bridge is a noble example of his boldness in designing, and practical skill in executing, a work of novel and difficult character. It crosses the Menai Strait, where it connects Caernarvonshire with the Isle of Anglesea. The opposite shores being bold and rocky, allowed the roadway of the bridge to be 100 feet above high-water mark. The main chains, 16 in number, are supported on two stone pyramids above the roadway, the ends of the chains being secured in a mass of masonry built over stone arches between each of the pyramids, or piers, and the adjoining shores. The first stone was laid by W. A. Provis, resident engineer, August 15, 1819. In 1824 the works were so far advanced that the only remaining difficulty was, " How are the main chains to be put up ?" for no precise details had up to that time been determined upon; which was so far an advantage, that the engineer had the benefit of full consideration and experience, and many mistakes were obviated that must have happened had the details been all settled beforehand. In the beginning of May the cast-iron segments and saddles were carried up to the pyramids, but it was not till April 26, 1825, that the first chain was carried across. It was scarcely fixed, when one of the men got astride it, and then walked over 30 or 40 yards of the middle of the chain, only nine inches wide, its height being 125 feet above the water! After the second chain had been put up, it was found necessary to replace some of the bars which had been damaged; and owing to this, it was practically ascertained that if one or more links of a chain should at any time be injured they could be taken out and replaced.

During the progress of the work every piece of iron was carefully tested; and, to prevent any injury of the metal by oxydation, each piece, after its strength had been proved, was cleaned, heated, and, while hot, immersed in linseed oil; after remaining in the oil a few minutes, that the pores might be filled, the bar was taken out, and returned to the heating stove, in which the oil was dried by a moderate heat: the oil was thus converted into a thin coat of hard varnish, affording a complete protection from the atmosphere.

The massive iron castings which are imbedded in the rock to form an abutment for the chains are placed upon layers of coarse flannel saturated with white-lead and oil, which, with a few timber wedges, enables them to bear steadily against the rock. On the tops of the suspension towers are massive cast-iron saddles to receive the chains; and between these and the cast-iron beds which sustain them are inserted rollers, which allow the saddles to move under their immense load when the chains expand or contract. The operation of raising the portions of the chains between the suspension towers occasioned much anxiety, but was accomplished without great difficulty by joining several bars from the top of each tower by a hanging scaffold, and elevating the intervening portion of each chain from a raft 400 feet long and 6 feet wide by means of a capstan; and to check the vibration occasioned by high winds, the chains are tied together by transverse braces. The several chains being thus suspended, the roadway of oak planking, with felt and tar beneath, was bolted to the underneath;* and on Jan. 30, 1826, the mails drove over it for the first time. In February following repeated gales did much damage to the iron-work.

The main dimensions of the bridge are: extreme length of chains, about 1715 feet; height of roadway from water-line, 100 feet; height of each suspending pier from road, 53 feet; length from pier to pier, 553 feet; width of two carriage-ways and footpath in centre, 16 feet. The 16 chains consist each of 5 bars 10 feet long; width, 3 feet by 1 inch, with 6 connecting links at each joint, which weighs about 50 lbs.; bars in cross-section of chain, 80. Total weight of the iron-work, 1,373,281 lbs. The chains will bear without any risk 1245·5 tons, more than the strain produced by the weight of the bridge itself; or 732½ tons besides its own weight.

The thread-like appearance of the suspending rods, easily shaken by the wind or by the hand, the vast size and lightness of the whole, give the idea of a fairy's power having stretched a series of chains from the woods on the one side to the barren rocks on the other; and its fairy lightness is heightened by contrast with the gigantic massiveness of the Britannia Bridge at about a mile distant.

Telford left his autobiography, with an elaborate account of his labors of more than half a century, and other valuable contributions to engineering literature. He taught himself Latin, French, Italian, and German. He died in 1834, at the age of seventy-seven, and was buried near the middle of the nave of Westminster Abbey. He was the first president of the Institution of Civil Engineers, to whom he bequeathed his scientific books,

* It is related that, just previous to the fixing of the last bar, Mr. Telford withdrew to his private office at the works, and there knelt in fervent prayer to the Giver of all good for the successful completion of this great work.

prints, drawings, etc., and £2000 to provide annual premiums to be given by the Council. In their house is a fine portrait of Telford.

As we reflect upon the noble works which Telford left for posterity, we feel that the Eskdaile shepherd-boy has duly earned every honor he has received.

His services have been appreciated by the public, but by the public alone. He received the honor of knighthood from the King of Sweden, but no mark of distinction from the King of England—no memorial from a country whose scientific eminence he illustrated, and whose commercial power he enlarged. By subscription of a few of his friends and admirers, however, a marble statue of the great engineer has been placed in the Islip Chapel at Westminster Abbey. It is from the chisel of Baily, R.A., who received for it but £1000, a third of the sum usually charged for such a work. The Dean demanded £300 for permission to place the statue in the Abbey, but subsequently lowered it to £200, which demand was acquiesced in. But Telford's " various works are conspicuous ornaments to the country, and speak for themselves as the most durable monument of a well-earned fame. In number, magnitude, and usefulness, they are too intimately connected with the prosperity of the British people to be overlooked or forgotten in future times, and the name of Telford must remain permanently associated with that remarkable progress of public improvement which has distinguished the age in which he lived."*

* *Council of the Institution of Civil Engineers.* Two or three days before Mr. Telford's death, he caused to be completed, under his direction, the corrected MS. of the detailed account of the principal undertakings which he had planned or lived to see executed. This work, edited by Mr. John Rickman, one of Mr. Telford's executors, was published in 1838.

JOHN RENNIE: DOCKS AND BRIDGES.

FEW of the great masters in this mechanical age have executed such stately works for posterity as John Rennie, the designer of three of the noblest bridges in the world, in addition to numerous other monuments of engineering skill.

John Rennie was the son of a respectable Scottish farmer, and was born on June 7, 1761, at Phantassie, in the county of East Lothian. He was the youngest of nine children, and received the first rudiments of education at the school of his native parish; and to a trifling circumstance connected with his daily journeys thither his friends ascribe his acquisition of a taste for mechanics, which fixed the course of the future man. The school was situated on the opposite side of a brook, the usual mode of crossing which was by stepping-stones; but when the freshes were out, it was necessary to employ a boat, which was kept at the workshop of Mr. Andrew Meikle, a millwright, well known in Scotland for his improvements in the threshing machine. This led him to Mr. Meikle's workshop, where he learned his first lessons in mechanics; and, ere he had completed his eleventh year, he had constructed a wind-mill, a pile-engine, and a steam-engine. He subsequently received instruction in elementary mathematics at Dunbar, where, on the promotion of the master, he for a short time conducted the school. He did not pursue his studies far in pure mathematics, but applied himself chiefly to elementary mechanics, drawing machinery, and architecture. He also attended the courses of lectures on mechanical philosophy and chemistry which were given at Edinburgh by Drs. Robison and Black. Prepared thus with what books and professors could teach, he entered the practical world. Meanwhile, he had been employed by Mr. Meikle as a workman, under whose superintendence he assisted in the erection of some mills in the neighborhood; and he is said to have rebuilt on his own account a mill near Dundee. It is probable that soon after this work was finished, or about 1780, Rennie left Scotland for London, on his way visiting the great manufacturing towns in the north of England, and inspecting their principal works.

Soon after he was established in the metropolis Mr. Rennie was employed in the construction of two steam-engines, and the machinery connected with them, at the Albion Flour Mills, Blackfriars Bridge. These engines were of the kind called double, which Mr. Watt had just then patented; each of them was of fifty-horse power, and the two could turn twenty mill-stones. All the wheel-work was made of cast-iron instead of wood, which had before been used in such machinery. Mr. Rennie's skill was strikingly manifested in the methods which he adopted to render the movements steady; and by this great work he at once established his character as a machinist.

Mr. Rennie continued to the last to be employed in the construction of steam-engines, or of the different kinds of machinery to which, as a first mover, steam is applied; and in its execution he may be said to have been the first who made that skillful distribution of the pressures, and gave those just proportions to the several parts, which have rendered the work of Englishmen superior to that of any other people. He was likewise extensively engaged in designing or superintending various important public works. Between 1799 and 1803 he constructed the stone bridge at Kelso, below the junction of the Tweed and Teviot. This handsome structure consists of five elliptical arches, carrying a level roadway; and over each pier are two small columns which support the entablature. Mr. Rennie also built stone bridges at Musselburgh and other places in Scotland; but his masterpiece of this class is the Waterloo Bridge over the Thames, which has no parallel in Europe. This bridge was begun in 1811, and finished in six years: it is built of granite, "in a style of solidity and magnificence hitherto unknown. There elliptical arches, with inverted arches between them, to counteract the lateral pressure, were carried to a greater extent than in former bridges; and isolated coffer-dams upon a great scale, in a tidal river, with steam-engines for pumping out the water, were, it is believed, for the first time employed in this country; and the level roadway, which adds so much to the beauty as well as the convenience of the structure, was there adopted." Canova dignified this as "the noblest bridge in the world," adding that "it alone was worth coming from Rome to London to see."

Baron Dupin, in classic eulogy, styled it "a colossal monument worthy of Sesostris and the Cæsars." Wells, in his *History of the Bedford Level*, observes of this bridge, "that a fabric of this immensity, presenting a straight horizontal line stretching over nine large arches, should not have altered more than a few inches (not five in any one part) from that straight line, is an instance of strength and firmness elsewhere unknown, and almost incredible." The bridge itself cost about £400,000, which, by the approaches, was increased to a million of money.

Rennie, besides the elegant iron bridge over the Witham in Lincolnshire, designed and constructed the Southwark Bridge over the Thames. It consists of three cast-iron arches, the centre 240·feet span, and the two side arches 210 feet each; the ribs forming a series of hollow masses, or voussoirs, similar to those of stone; a principle new in the construction of cast-iron bridges, and very successful. The segmental pieces and braces are kept together by dovetailed sockets and long cast-iron wedges, so that bolts are unnecessary.

Rennie's chief work connected with inland navigation is the Kennet and Avon Canal, fifty-seven miles in length, and requiring all the skill of the engineer to conduct it through a very rugged country. He also gave a plan for draining the fens at Witham, which was executed in 1812.

The London Docks, and the East and West India Docks at Blackwall, were executed from Rennie's plans. He likewise formed the new Docks at Hull (where also he constructed the first dredging-machine used in this country); also the Prince's Dock at Liverpool, and those of Dublin, Greenock, and Leith, the latter with a stupendous sea-wall. Mr. Rennie also built the pier at Holyhead. To these works must be added the insular pier, or Breakwater, nearly a mile in extent, protecting Plymouth Sound from the tremendous force of the full roll of the Atlantic in southerly or southwestern gales. It is constructed on true hydrodynamical principles, and in its formation $3\frac{1}{2}$ million tons of stone have been deposited.

Rennie's last work was his design for the present London Bridge, unrivaled in the world "in the perfection of proportions and the true greatness of simplicity." He died in 1821; but the charge of its construction was confided to his son, now Sir John Rennie, who, in 1831, finished the magnificent work.

The principal undertakings in which Rennie was engaged are estimated from his reports to have cost forty millions sterling. His works were costly; but it has been well said that "they were made for posterity; they were never of slight construction, nor would he ever engage in any undertaking were a sufficiency of funds was not forthcoming to meet his views." His industry was untiring: he was rarely occupied in business less than twelve hours a day. Like Jesse Ramsden, he was strikingly clear in communicating information to others; yet rarely had either of them recourse for illustration to any other instrument than a two-foot rule, which each always carried in his pocket. Rennie owed his good fortune to no lucky accident or successful artifice, but to talent, industry, prudence, perseverance, boldness of conception, soundness of judgment, and habits of untiring application. His remains rest in the crypt of St. Paul's Cathedral, beside the grave of Robert Mylne, the architect of Blackfriars Bridge.

"THE FIRST PRACTICAL STEAM-BOAT."

THE story of the Steam-boat is one of the most interesting chapters in the records of human invention. The accounts of vessels propelled by machinery lead us through a retrospect of many centuries and the oldest countries. The paddle-wheel is stated to have been used by the ancient Egyptians, but not upon admissible authority. The wheel of a chariot in an Egyptian painting has often been mistaken for a paddle-wheel; a precisely similar mistake has been made in describing one of the sculptured slabs from Nineveh; and Sir H. Rawlinson and Mr. Layard assured Mr. Macgregor that in their Assyrian researches they have not discovered any indication of the use of machinery in propelling vessels.

We find some indistinct records of vessels propelled by wheels. An old work on China contains a sketch of a vessel moved by four paddle-wheels, perhaps in the seventh century; but the earliest distinct notice of this means of propulsion appears to be by Robertus Valterius, A.D. 1472, who gives several wood-cuts representing paddle-wheels. The account of Blasco de Garay's experiment in 1543 is now generally discredited (see page 275); but boats propelled by paddle-wheels are mentioned by many early writers, such as Julius Scaliger in 1558; Bourne in 1578; and Roger Bacon in 1597. Among the earliest projectors we find David Ramsey, one of the pages to King James I., who, with another, in 1618, obtained a patent for " divers newe apt formes or kinds of engines for ploughing without horse or oxen; as also to raise water," and " to make boats for carriages runnin upon the water as swift in calmes, and more safe in storms, than boats full sayled in great windes;" and in 1630, " to raise water from lowe pits by fire" (the steam-engine); " to make boats, ships, and barges to goe against the wind and tyde." Passing over a few similar inventions, we come to the Marquis of Worcester's patent (in 1661) of the application of a current to turn paddle-wheels on a vessel, which was propelled by winding up a rope. Edward Bushnell, in 1678, described a mode of rowing ships by connecting the oars on both sides with the heaving of a capstan.

In 1681, Papin, the improver of steam-engines for pumping, proposed to the Royal Society "a new-invented boat, to be rowed by oars moved with heat," which was recommended by Leibnitz. It is clear also that Papin conceived steam might be employed to propel ships by paddles; for, as early as 1690, in a paper published in the *Acta Eruditorum*, Papin says: "Without doubt, oars fixed to an axis could be most conveniently made to revolve by our tubes. It would only be necessary to furnish the piston-rod with teeth, which might act on a toothed wheel, properly fitted to it, and which, being fitted on the axis to which the oars are attached, would communicate a rotary motion to it." During Papin's residence in England, he witnessed an interesting experiment made on the Thames, in which a boat, constructed from a design of the Prince Palatine Robert, was fitted with revolving oars, or paddles, attached to the two ends of a long axle, going across the boat, and which received their motion from a trundle, working a wheel turned by horses. The velocity with which this horse-boat was propelled was so great that it left the king's barge, manned with sixteen rowers, far astern in the race of trial.

In 1682, a horse tow-vessel was used at Chatham: it had a wheel on each side, connected by an axle across the boat, the paddles being made to revolve by horses moving a wheel turned by a trundle fixed on the axle. In 1692, Anthony Duvivian patented "a very easy and not costly machine for making a ship go against wind and tide." In 1696, Thomas Savery patented his invention for moving a paddle-wheel on each side of the ship by men turning round the capstan. By some writers it is stated that Savery proposed to drive a paddle-wheeled vessel by his steam-engine, already described at page 279; whereas he merely believed that it might be very useful to ships, but dare not meddle with that matter. "It appears," says Mr. Bennet Woodcroft, "to be a proof of Savery's sound mechanical views that he knew his engine, although doubtless the most effective of its kind at that period, to be incapable of propelling a boat advantageously."

In 1724 John Dickens patented his contrivance by floats for moving ships; and in 1729, Dr. John Allen his engines for navigating ships in a calm, by forcing water through the stern of the ship, at a convenient distance under the surface of the water, as well as by firing gunpowder in vacuo, and applying its whole force to move the engines.

In 1736 Jonathan Hulls patented his machine. He placed a paddle-wheel on beams projecting over the stern, and it was turned by an atmospheric engine acting, in conjunction with a counterpoise weight, upon a system of ropes and grooved wheels. His mode of obtaining a rotary motion was new and ingenious, and would enable a steam-boat to be moved through water; but it was not practically useful. The cranks, as described by Hulls, receive rotary motion from the axis on which they are placed, and do not, as often stated, impart that motion to it; had he discovered this application of the crank, "there can be little doubt," says Mr. Woodcroft, "that the steam-engine would then have been applied not only to propel boats, but to various other useful purposes."

A prize being offered by the Academy of Sciences for the best essay on the manner of impelling vessels without wind, it was obtained in 1752 by Daniel Bernouilli, who proposed inclined planes moved circularly like the sails of a wind-mill, two at each side of the vessel, and two more behind, to be moved by men aboard, by steam-engines, or on rivers by horses placed to the barges. In 1760, J. H. Genevois, a clergyman of Berne, published his "Great Principle," to concentrate power by a series of springs to work oars for propelling vessels. He also proposed an atmospheric steam-engine to bend the springs, and the expansive force of gunpowder for the same purpose. He states that, since his arrival in England, he had learned that thirty years before a Scotchman had proposed to make a ship sail with gunpowder, but that thirty barrels of gunpowder had scarce forwarded the ship ten miles.

On January 5, 1769, JAMES WATT patented his improvements on the steam-engine, one of which, namely, the "fourth," was for causing the steam to act above the piston as well as below it. This was the first step by which the steam-engine was successfully used to propel a vessel; and this great "improvement," says Mr. Woodcroft, "was applied to the first practically-propelled steamboat, and is still used in the present system of steam navigation."

In 1774, the Comte d'Auxiron and M. Perrier are stated to have used a paddle-wheel steam-boat on the Seine, but with poor success. Desblanes, in 1782, sent the model to the Conservatoire (still there) of a vessel in which an endless chain of floats is turned by a horizontal steam-engine.

In 1779, Matthew Wasborough, an engineer of Bristol, added to Watt's improvement of the double-acting cylinder-engine by converting a rectilinear into a continuous circular motion; but it did not act well, and was superseded by the invention of James Pickard in 1780, which is no other than the present connecting-rod and crank, and a fly-wheel, being the second and last great improvement in the steam-engine, which enabled it to be of service in propelling vessels.

In the following year, 1781, James Watt patented his "sun and planet motion," or method of applying the vibrating or reciprocating motion of steam-engines to procure a continued rotative or circular motion round an axis or centre. In the same year the Marquis de Jouffroy constructed a steam-boat at Lyons, 140 feet in length, with which he is said to have experimented successfully on the Seine; but Mr. Macgregor states that no description of the machinery of this vessel is given before that published in 1816 by the Marquis de Jouffroy, who gives a sketch of the steam-boat, a copy of which is in our Great-Seal Patent-Office Library.

In 1785 Joseph Bramah patented a mode of propelling vessels by an improved rotary engine, by means either of a paddle-wheel, or what may be called a "Screw Propeller," or a wheel with inclined fans or wings, like the fly of a smoke-jack, or the vertical sails of a wind-mill, fixed on or beyond the stern, about where the rudder is usually placed; "its movement being occasioned by

R

means of a horizontal spindle or axletree, conveyed to the engine through or above the stern-end of the ship." "This," says Mr. Woodcroft, "was, without doubt, the best mode of steam-propelling that had been then suggested; for here the steam would so act as directly to produce a circular motion on the propeller-shaft. There is, however, no account of Bramah having tried this mode."

On June 5, 1785, William Symington, an engineer of Wanlock-head lead-mines, patented a mode of obtaining rotary motion from a steam-engine by chains, ratchet wheels, and catches; but it was inferior to the crank of Pickard, or the sun and planet wheel of Watt. Experiments conducted about the same time at Dalswinton, in Scotland, resulted, in 1787, in the successful use of a steam-engine by Miller, Taylor (tutor to Miller's sons), and Symington, to propel a vessel by paddle-wheels, which worked one before the other in the centre of the boat.

The first experiment was performed on the lake at Dalswinton in October, 1788, when the engine, mounted in a frame, was placed upon the deck of a double pleasure-boat. "We then proceeded" (says Taylor) "to action; and a more complete, successful, and beautiful experiment was never made by any man, at any time, either in art or science. The vessel moved delightfully, and, notwithstanding the smallness of the cylinders (4 inches dia.), at the rate of five miles an hour. After amusing themselves a few days, the engine was removed, and carried into the house, where it remained as a piece of ornamental furniture for a number of years."

The boat was twenty-five feet long and seven broad, and was propelled by two paddle-wheels, placed one forward and the other aft of the engine, in the space between the two hulls of the double boat. The engine, now become a curiosity, was most laboriously sought for by Mr. Bennet Woodcroft.

On the death of Mr. Miller in 1815, the engine came into the possession of his eldest son, Mr. Patrick Miller; and in 1828 it was sent by him, packed in a large deal case, to Messrs. Coutts and Co., bankers, 59 Strand, London. In this establishment the engine was kept until February 17, 1837, on which day it was removed to the store warehouse of Messrs. Tilbury and Co., 49 High Street, Marylebone. Here it remained till the 31st of January, 1846, and then it was forwarded to Mr. Kenneth Mackenzie, of 63 Queen Street, Edinburgh. Beyond this it could not be traced for a period of several years.

Mr. Bennet Woodcroft did not, however, relax in his search, and at length ascertained that the engine was sold by Mr. Mackenzie's direction to Mr. Kirkwood, a plumber of Edinburgh, who removed it from the framing, and threw it into a corner for the purpose of melting: this intention, however, was not carried into effect, doubtless owing to the death of Mr. Kirkwood. It was subsequently found in the possession of Messrs. William Kirkwood and Sons, from whom it was purchased, and dispatched to the Great-Seal Patent Office on the 19th of April, 1853. Subsequently it was transmitted to Messrs. Penn, engineers, of Greenwich, who gratuitously reinstated it in a frame, and put it again in working order, as an object of great public interest. The engine was returned to Mr. Woodcroft as good as new, January 4th, 1855, and on the 29th of January, 1857, it was removed from the Great-Seal Patent Office to the Patent Museum at South Kensington.

Symington's engine comprises several features of remarkable interest to engineers. The upper part of each cylinder is enlarged, so as to prevent the overflow of the water used for keeping the piston steam-tight, upon the plan used by Newcomen. The lower part of each cylinder is Watt's condenser and air-pump, not separated from the cylinder, as patented by Watt, but attached to it. The valves are opened and closed by an improved arrangement of Beighton.

For Mr. Taylor's efforts to introduce Steam Navigation, his widow received a pension from government of £50 per annum, granted by the then Lord Liverpool; and in 1837, each of his four daughters received a gift of £50 through Lord Melbourne. This is, however, but a miserable reward for the valuable services rendered. Mr. Miller sought no pecuniary reward, and fortunately he needed none: he had built eight vessels to improve naval architecture, but was refused a license to make experiments with one, it not being according to statute!

We have been led by the above curious story of the search for Symington's steam-engine somewhat out of the order of time. To return: in 1787, Mr. Miller described to the Royal Society experiments made by him in the Frith of Forth, in a double vessel 60 feet long, put in motion by his water-wheel, wrought by a capstan with five bars and a toothed wheel working in a trundle fixed on the axis of the water-wheel. The steam-boat was three-masted, and made sundry tacks in the Frith, with four men, at the rate of four miles an hour.*

* In 1787 Mr. Miller published a pamphlet (now scarce) on the subject of propelling boats by paddle-wheels turned by men, with drawings by Alexander Nasmyth. In 1825 Mr. Miller's son also published a pamphlet, in which he claims for his father the invention of Steam Navigation, and states that he (the father) had expended in experiments the sum of nearly £30,000. The pamphlet of Mr. Miller, sen., is reprinted in Mr. Woodcroft's work on Steam Navigation.

Meanwhile the subject was hotly pursued in the United States, where, in 1788, John Fitch and James Ramsey patented improvements in a steam-boat which went eighty miles in one day, worked by paddles perpendicularly. Fitch, however, was subsequently reduced to poverty by his project, and terminated his life by plunging into the Alleghany. Ramsey, being refused a patent in America, came to England, and here patented several improvements; but, just as he had completed his steamboat, he died; it was, however, floated in the Thames in 1793, against wind and tide, at four knots an hour. It appears that these two inventors had long conceived the project of propelling vessels by steam-power before they experimented; for, in 1784, Ramsey mentioned to General Washington the project of Steam Navigation, and Fitch showed the general a model of his proposed boat.

In the year 1801, Thomas Lord Dundas, of Kerse, who was acquainted with Mr. Miller's labors, and who was an extensive owner of property in the Forth and Clyde Canal, employed Mr. Symington to make experiments on steam-boats, to be substituted for the horses then employed to draw the vessels on the canal. These experiments in two years cost £7000; and the result was the production of the first practical Steam-boat, named the *Charlotte Dundas*, in honor of his lordship's daughter, the lamented Lady Milton. "This vessel," says Mr. Woodcroft, "might, from the simplicity of its machinery, have been at work to this day, with such ordinary repairs as are now occasionally required to all steam-boats." In the steamer there was an engine with the steam acting on each side of the piston (Watt's patented invention), working a connecting-rod and crank (Pickard's patent), and the union of the crank to the axis of Miller's improved paddle-wheel (Symington's patent). Thus had Symington the undoubted merit of having combined together for the first time those improvements which constitute the *present system of Steam Navigation.*

Although the experiments with this boat were highly successful, the proprietors of the Forth and Clyde Canal declined to adopt it, from an opinion that the waves it created would damage the banks. Lord Dundas, however, entertained a more favorable opinion on the subject, and recommended to the Duke of Bridgewater the adoption of Symington's steam-boat. His grace at first doubted the utility of the invention; but, after having seen a model, and received explanations from Mr. Symington, he gave him an order to build eight boats similar to the *Charlotte Dundas*, to ply on his canal. Symington returned to Scotland in high hope, but was doomed to disappointment; for, on the same day that the committee on the Forth and Clyde Canal refused to allow his boats to be employed, he received the intelligence of the death of the Duke of Bridgewater.

Unable longer to struggle against his misfortunes, and his resources being exhausted, Symington laid up his boat in a creek of the canal, where it remained a number of years exposed to public view. He next abandoned his own old engine, and obtained a patent for applying a Double-action Reciprocating Engine to a boat, and for placing his crank upon the axis of the paddle-wheel, which was a very important discovery and improvement. From the establishment of this combination of machinery to a boat, no improvement on his system has been effected either in this or any other country.

In the following year, 1789, Miller and his fellow-experimenters constructed an engine of about twelve-horse power (or twelve times the power of the first) at the Carron Works. This was mounted in the large double boat which had formerly run against the Custom-house boat at Leith. Except in size, this machine resembled the former model. This boat was tried on the Forth and Clyde Canal, performed very successfully, and attained a speed of nearly seven miles an hour; but the hull being much too slight for permanent use as a steam-boat, or for taking out to sea, it was, soon after the trial, dismantled.

Satisfactory as was the result of these experiments, they did not immediately lead to the introduction of Steam Navigation, and several unsuccessful schemes were tried in this country and North America before this was effected. One of these, Ramsey's on the Thames, has been already mentioned. About this time Dr. Cartwright contrived a steam-barge, and explained it to Fulton, as some say, in 1793, when he was studying painting under West; but others date it a few years later, when he was introduced to Dr. Cartwright during his journey to Paris in 1796. However this might be, it is evident that Fulton's attention was directed to the subject about this time. Colden, his biographer, states that he made drawings of an apparatus for Steam Navigation in 1793; he submitted them to Lord Stanhope, who, in 1795, made experiments in a steam-boat propelled by duck-feet paddles, with which, however, he could not obtain a greater speed than three miles an hour.

About the year 1825 Mr. Symington memorialized the Lords of the Treasury, in consequence of which the sum of £100 was awarded from his majesty's privy purse; and a year or two afterward, a farther sum of £50. He had cherished the hope that an annual allowance might be procured, but in this he was disappointed. He received a small sum from the London steam-boat proprietors, through the influence of Mr. James Walker; and in the decline of life, several kind relatives and friends contributed to Symington's support: among the number was Lord Dundas. Such was the fate of the inventor of "the first practical steam-boat."

Although Symington's experiments did not lead to the immediate adoption of steam-vessels for commercial purposes, they probably tended in no unimportant degree to their subsequent profitable establishment in America and Great Britain, for among the numerous individuals who inspected Symington's vessel with interest were Fulton and Bell. After Fitch and Ramsey, Chancellor Livingston attempted to build a steamer on the Hudson, and in 1797 he applied to the Legislature of New York for exclusive privileges to navigate boats by a steam-engine. Though his project excited much ridicule, the privilege was granted, on condition that he should within twelve months produce a steam-vessel which should attain a mean rate of four miles an hour. This he failed to accomplish, though assisted by an Englishman named Nesbit, and by Brunel (afterward Sir Mark Isambard); consequently his grant or patent became void. Shortly afterward, being at Paris,* as minister from the United

* While at Paris, Fulton submitted to Napoleon I. his plan of Steam Navigation, when, it was long said, Napoleon coldly received

States, Livingston conversed with Fulton on the subject of steam-boats, and they subsequently conjointly completed a boat of considerable size.

Meanwhile, in 1804, John Stevens, of Hoboken, near New York, tried a small boat 22 feet long, which attained, for short distances, seven or eight miles an hour.

Mr. Sime, M.A., has thus vividly narrated Fulton's important share in the success of Steam Navigation:

It was reserved for an American citizen to execute, and an American river to witness, an enterprise the honor of which properly belongs to Scotland. Robert Fulton visited Europe toward the close of the last century, and made several attempts, both in Britain and France, to propel vessels by steam. Watt and Boulton supplied him with machinery; and many of his ideas were borrowed from Miller of Dalswinton, and Symington, whose steam-boat he inspected when in Scotland.* From the first Fulton regarded the steamer as a means of developing the vast resources of the Western States of the Union, where 50,000 miles of river navigation, through a rich and fertile country, invited capital, enterprise, and population. Fourteen years elapsed before success crowned his labors; many difficulties and disappointments were encountered; and once, when a vessel which he had built was ready for an experimental trip on the Seine at Paris, the boat broke in two, and the machinery carried the fragments to the

the projector. Marshal Marmont, in his Memoirs, says that Bonaparte, who, from his education in the artillery, had a natural prejudice against novelties, treated Fulton as a quack, and would not listen to him. M. Louis Figuier also writes that Bonaparte refused to place the matter in the hands of the Academy. The following letter from Napoleon, dated from the Camp at Boulogne, 21st of July, 1804, and addressed to M. de Champagny, Minister of the Interior, proves the contrary:

"I have just read the project of Citizen Fulton, an engineer, which you sent me much too late, for it seems capable of *changing the face of the world.* At all events, I desire that you will immediately place the examination of it in the hands of a committee, composed of members of the Institute, for it is to them that the scientific men of Europe will naturally look for a decision on the question. A great physical truth stands revealed before my eyes. It will be for these gentlemen to see it, and endeavor to avail themselves of it. As soon as the report is made, it will be sent to you, and you will forward it to me. Let the decision be given in a week, if possible, for I am impatient to hear it.	NAPOLEON.

"*Camp of Boulogne, 21st July,* 1804."

* Before he returned to America Fulton visited England, and there induced Symington to afford him much information, and even to perform a voyage on his account, during which Fulton noted in a memorandum-book the particulars of the construction and effect of the machine, which Symington unhesitatingly afforded him.

bottom of the river.* In 1807 he launched his first steam-vessel on the Hudson, and inaugurated a new era in river and ocean navigation. The prejudices which rendered the multitude both of the wise and the ignorant skeptical before Fulton's ideas had been fully realized, and which drew them to the water-side to scoff at an expected failure, were destroyed in a few minutes by the steady motion of the vessel. Her first trip was made on the Hudson, between New York and Albany, a distance of 150 miles. When we look back on that voyage, fraught with unspeakable benefits to mankind, how amusing is almost every thing connected with it! The velocity of the steamer was only about five miles an hour; yet so rapid did this rate seem to those on board, that the ships they passed, moving with themselves, appeared as if at anchor. The pine-wood used as fuel sent forth a column of ignited vapor many feet above the flue; and so appalled were the crews of the ships on the Hudson as they saw this fiery monster moving toward them in the darkness against both wind and tide, that some abandoned their ships, and others thought their last hour was come. Between 1807 and 1812, the year in which the *Comet*, the first British steamer, began to ply on the Clyde, steam-boats were introduced on almost all the larger rivers of the United States.—*Edinburgh Essays*, 1856.

During the war between Great Britain and the United States, in 1814, Fulton proposed to defend the harbor of New York from attack by means of steam-frigates. That which he actually built, although it was not required, was pierced for thirty guns, and resembled the double-boats, or *twins*, constructed by Miller of Dalswinton. She was also fitted with machinery calculated to discharge an immense quantity of hot water through the port-holes of an enemy's ship, by which the ammunition would be rendered useless, and the crew scalded to death. Cutlasses without number were said to be moved by machinery; pikes, darted forth and withdrawn every quarter of a minute, would sweep the decks of our men-of-war; in short, the iron fingers of a modern Scylla would kill the sailors at their post. Little did either nation imagine that, before the lapse of forty years, Great Britain would depend on this very application of steam to maintain that supremacy at sea of which many supposed it had deprived her.

Fulton also formed two projects for submarine navigation: one, a carcass or box filled with combustibles, which was to be propelled under water, and made to explode beneath the bottom of a vessel; the other, a submarine boat, to be used for a similar destructive purpose; but for practical use both were failures. He appears, however, to have clung to the scheme with great perse-

---

* To this discouraging accident Mr. Scott Russell attributes one of the excellences of the American steam-boats—the strong and light framing, by which, though slender, they are enabled to bear the weight and strain of their large and powerful engines. To remedy the evil, Fulton almost reconstructed his vessel, when her shattered hull was raised.

verance, and not long before his death exhibited its power by blowing up an old vessel in the neighborhood of New York. Fulton's chest, which he named a Torpedo, or Nautilus, was, in his own words, "to blow a whole ship's company into the air;" it was nothing more than a chest containing gunpowder, which, by means of clock machinery, might be ignited at a given time under water, and, being placed under a ship's bottom, destroy her by the explosion. This application of gunpowder had before been made by Bushnell: it has been humorously described as " something like the scheme of children to catch swallows by applying salt to their tails." Fulton offered his invention to Bonaparte when First Consul, and he was sent to Brest under the promise of destroying the English blockading squadron; but he did nothing. He then offered his scheme to the British ministry, and, by way of experiment, blew to pieces in two days an old Danish brig in Walmer Roads; but his grand invention was the Catamaran expedition, as the trial of his machines against the Boulogne flotilla was called.

Fulton died in 1815; and so highly were his services appreciated in the United States, that, besides other testimonials of respect, the members of both houses of the Legislature wore mourning on the occasion of his death.

The practical application of Steam Navigation in Scotland did not take place until a few years after Fulton's success in America, when Henry Bell, of Helensburg, on the Clyde, a house-carpenter, had built the *Comet*, 40 feet keel, 25 tons burden, and three-horse power: her boiler is in the possession of Mr. Scott Russell. Mr. Bennet Woodcroft, in comparing these boats with their predecessors, emphatically says: Symington's boat, the *Charlotte Dundas*, was altogether superior in its mechanical arrangements to either Fulton's *Clermont* or Bell's *Comet*, as may be readily seen by inspection of the drawings. Next year, 1813, appeared the second steamer on the Clyde—the *Elizabeth;* and in 1814 was built the *Industry*, by Mr. Fyfe, of Fairlie: she is of wood, and her first engine was put on board by Duncan M'Arthur, of Glasgow. This vessel is now a luggage-steamer of the Clyde Shipping Company, and is stated to be *the oldest steamer afloat.*

R 2

In Ireland, a person named Dawson states that he had built a steam-boat of 50 tons burden, worked by a high-pressure steam-engine, as early as 1811, which he also named the *Comet*, after the great phenomenon of that year. In 1813 Dawson established a steam-boat on the Thames, to ply between Gravesend and London, "which was the first that did so for public accommodation; although Mr. Lawrence, of Bristol, who introduced a steam-boat on the Severn, soon after the successful operations on the Clyde, had her carried to London (through the canals) to ply on the Thames; but, from the opposition of the watermen to the innovation, he was in the end obliged to take her to her first station." Mr. Cruden, in his *History of Gravesend*, however, states the first Gravesend steam-boat to have been the *Margery*, built upon the Clyde in 1813 by the builders of the *Comet:* she started for Gravesend in 1815. In the previous year, 1814, a steamer began to ply between London and Richmond.

George Dodd, whose history is a melancholy instance of the poverty which often attends the most ingenious inventors, was, it appears, the first to undertake a considerable voyage by sea in a steam-vessel, built on the Clyde in 1813. She was 74 or 75 tons burden; 14 or 16 horse-power, with paddle-wheels 9 feet in diameter. Dodd brought her round to the Thames, by steam and sails, through rough weather, especially in the Irish Sea.

The first ocean steam-voyage of great length was made by the *Savannah*, of 350 tons, which arrived at Liverpool July 15, 1819, having made the voyage from New York in 26 days. She then went to St. Petersburg, touching at Copenhagen, and subsequently recrossed the Atlantic. Steam was, however, employed only during a part of these voyages.

The first steam voyage to India was made in the *Enterprise*, which sailed from Falmouth Aug. 16, 1825: for this feat the captain of the vessel received £10,000.

In 1838, the *Sirius*, of London, and the *Great Western* effected their first voyage to New York, almost simultaneously, from Bristol; and from the same port the *Great Britain*, propelled by a screw, made her first voyage out in July, 1845, thus establishing the usefulness of each mode of propulsion for the navigation of the ocean. Captain Ericsson appears to have accomplished for the screw propeller in America and England what Fulton did for the paddle-wheel in the former, and Bell in the latter country, namely, its practical introduction. To Francis Pettit Smith, the patentee of the Archimedean Screw Propeller, for the bringing into general use this system of propulsion, a magnificent plate testimonial and subscriptions, in the whole amounting to £2678, were presented at a festival in the summer of 1858, Mr. Robert Stephenson, M.P., presiding. The screw is specially adapted for war-steamers: it leaves a clear broadside for the guns, does not prevent the use of sails, and allows the machinery to be placed six or eight feet below the water-line, thus leaving the upper decks free for working the guns. The screw was first tried in the *Archimedes;* and in 1839 the first war-ship, the *Rattler*, was fitted with it.

The substitution of iron for wood in the building of steam-vessels insures their superior lightness and buoyancy, and has led to water-tight compartments and a multitude of other important changes.

SIR ISAMBARD M. BRUNEL: BLOCK MACHINERY AND THE THAMES TUNNEL.

THE name of Brunel has now for two generations, from the commencement of this century to the present time, been identified with the progress and the application of mechanical and engineering science.

The elder Brunel, ISAMBARD MARK, who displayed such diversity of genius for the minute and the vast, was born near Rouen in 1769, and from his earliest boyhood showed mechanical tastes. When sent to the seminary of St. Nicaise at Rouen, he preferred the study of the exact sciences, mathematics, mechanics, and navigation, to the classics, and loved to pass his holidays in a joiner's shop. At the age of twelve years he was proficient in turning, and in the construction of models of ships, machines, and musical instruments; he also made an octant, guided by the one belonging to his tutor and by a treatise on navigation; and at the age of fifteen he took such interest in astronomy as to observe the stars, greatly to the astonishment of the villagers. In 1786 he enlisted as a sailor, from which date up to 1793 he made several voyages to the West Indies, in which he used instruments of his own construction: he also made a piano-forte while the ship once lay at Guadaloupe.

Brunel's first engineering work was a survey for the canal which now connects Lake Champlain with the River Hudson at Albany. He afterward acted as an architect, and built one of the theatres at New York. He was employed on the forts erected for the defense of that city, and in the establishment of an arsenal and foundery; he also devised ingenious contrivances for boring cannon and moving large masses of metal with facility. He next visited England, where his first work was an autographic machine for copying maps, drawings, and written documents.

Brunel's next work was his invention and construction of the assemblage of machines in Portsmouth Dock-yard for the formation of Blocks employed in raising burdens, and particularly in the important service of moving the rigging of ships. There are sixteen different machines, all driven by the same steam-engine: seven cut or shape logs of elm or ash into the shells of blocks, while nine fashion stems of lignum-vitæ into pulleys or sheaves, and form the iron pin, which being inserted, the block is complete. Four men with this machine turn out as many blocks as fourscore did formerly, and at less cost; and the supply has never failed, even though 1500 blocks are required in the rigging of a ship of the line. By adjustments, blocks can be manufactured of one hundred different sizes: thirty men can make one hundred per hour; and the machinery, by Maudslay, in twenty-five years required no repairs. It cost £46,000; and the saving per annum, in time of war, has been £25,000. A second set of machinery was executed for the dock-yard at Chatham. This assemblage of machines contains so many ingenious processes for gaining the proposed ends with the utmost accuracy, and, at the same time, with the least possible labor, as to justify the opinion that it constitutes one of the noblest triumphs of mechanical skill. There is a set of magnificent models of this invention in the possession of the Navy Board: the machines work in succession, so as to begin and finish off a two-sheaved block, four inches in length, in the most perfect manner. A detailed account of the entire machinery is given in the *Penny Cyclopædia*, Supplement 1.

Mr. Brunel next built in Chatham Dock-yard the Steam Saw-mill, in which he introduced Circular Saws, subsequently improved for cutting veneers. He also invented a machine for making seamless shoes; for nail-making; for twisting, measuring, and forming sewing-cotton into hanks; for ruling paper; a contrivance for cutting and shuffling cards without the aid of fingers, produced in reply to a playful request of Lady Spencer; a hydraulic packing-press; new methods and combinations for suspension bridges; and a process for building wide and flat arches without centrings. He was employed in the construction of the first Ramsgate steamer; he was the first to suggest the advantage of steam-tugs to the Admiralty; and for ten years he carried on experiments in constructing a machine for using carbonic acid gas as a motive power.

A popular writer of forty years since has left this graphic picture of his visit to Brunel's workshops at Battersea: "In a small building on the left, I was attracted by the solemn action of a steam-engine of sixteen-horse or eighty-men power, and was ushered into a room where it turned, by means of bands, four wheels fringed with fine saws, two of eighteen feet in diameter, and two of nine feet. These circular saws were used for the purpose of separating veneers, and a more perfect operation was never performed. I beheld planks of mahogany and rosewood sawed into veneers the sixteenth of an inch thick, with a precision and grandeur of action which really was sublime. The same power at once turned these tremendous saws and drew their work from them. A large sheet of veneer, nine or ten feet long by two feet broad, was thus separated in about ten minutes, so even and so uniform that it appeared more like a perfect work of Nature than one of human art. The force of those saws may be conceived, when it is known that the large ones revolve sixty-five times in a minute; hence $18 \times 3\cdot14 = 56\cdot5 \times 65$ gives 3672 feet, or two thirds of a mile, in a minute; whereas, if a sawyer's tool gives thirty strokes of three feet in a minute, it is but ninety feet, or only the fortieth part of the steady force of Mr. Brunel's saws.

"In another building I was shown his manufactory of shoes, which, like the other, is full of ingenuity, and, in regard to subdivision of labor, brings this fabric on a level with the oft-admired manufactory of pins. Every step in it is effected by the most elegant and precise machinery; while, as each operation is performed by one hand, so each shoe passes through twenty-five hands, who complete from the hide, as supplied by the currier, a hundred pairs of strong and well-finished shoes per day. All the details are performed by the ingenious application of the mechanic powers; and all the parts are characterized by precision, uniformity, and accuracy. As each man performs but one step in the process, which implies no knowledge of what is done by those who go before or follow him, so the persons employed are not shoemakers, but wounded soldiers, who are able to learn their respective duties in a few hours. The contract at which these shoes are delivered to government is 6s. 6d. per pair, being at least 2s. less than what was paid previously for an unequal and cobbled article."— SIR RICHARD PHILLIPS's *Morning's Walk from London to Kew.*

Brunel is most popularly known by his great work of engineering construction—the Thames Tunnel, consisting of a brick-arched double roadway under the river, between Wapping and Rotherhithe.

In 1799 an attempt was made to construct an archway under the Thames, from Gravesend to Tilbury, by Ralph Dodd, engineer; and in 1804 the "Thames Archway Company" commenced a similar work from Rotherhithe to Limehouse, under the direction of Vasey and Trevethick, two Cornish miners: the horizontal excavation had reached 1040 feet, when the ground broke in under the pressure of high tides, and the work was abandoned, fifty-four engineers declaring it to be impracticable to make a tunnel under the Thames of any useful size for commercial progression.

In 1814, when the Allied Sovereigns visited London, Brunel submitted to the Emperor of Russia a plan for a Tunnel under the Neva, by which the terrors of the breaking up of the ice of that river in the spring would have been obviated. The scheme which he was not permitted to carry out at St. Petersburg he was destined to execute in London.

It was planned in 1823. Among the earliest subscribers to the scheme were the late Duke of Wellington and Dr. Wollaston; and in 1824 the "Thames Tunnel Company" was formed to execute the work. A brick-work cylinder, fifty feet in diameter, forty-two feet high, and three feet thick, was first commenced by Mr. Brunel, at 150 feet from the Rotherhithe side of the river; and on March 2, 1825, a stone with a brass inscription-plate was laid in the brick-work. Upon this cylinder, computed to weigh 1000 tons, was set a powerful steam-engine, by which the earth was raised, and the water was drained from within it; the shaft was then sunk into the ground *en masse*, and completed to the depth of 65 feet, and at the depth of 63 feet the horizontal roadway was commenced, with an excavation larger than the interior of the old House of Commons. The plan of operation had been suggested to Brunel in 1814 by the bore of the sea-worm *Teredo navalis* in the keel of a ship; showing how, when the perforation was made by the worm, the sides were secured, and rendered impervious to-water by the insect lining the passage with a calcareous secretion. With the auger-formed head of the worm in view, Brunel employed a cast-iron "Shield," containing thirty-six frames or cells, in each of which was a miner, who cut down the earth; and a bricklayer simultaneously built up from the back of the cell the brick arch, which was pressed forward by strong screws. Thus were completed, from Jan. 1, 1826, to April 27, 1827, 540 feet of the Tunnel. On May 18th the river burst into the works;

but the opening was soon filled up with bags of clay, the water pumped out of the Tunnel, and the work resumed. At the length of 600 feet the river again broke in, and six men were drowned.

The Tunnel was again emptied; but the work was discontinued for want of funds for seven years. Scores of plans were now proposed for its completion, and above £5000 were raised by public subscription. By aid of a loan sanctioned by Parliament (mainly through the influence of the Duke of Wellington), the work was resumed, and a new shield constructed, March, 1836, in which year were completed 117 feet; in 1837, only 29 feet; in 1838, 80 feet; in 1839, 194 feet; in 1840 (two months), 76 feet; and by November, 1841, the remaining 60 feet, reaching to the shaft which had been sunk at Wapping. On March 24 Brunel was knighted by the queen; on August 12 he passed through the Tunnel from shore to shore; and March 25, 1843, it was opened as a public thoroughfare. It is lighted with gas, and is open to passengers day and night, at one penny toll.

The Tunnel has cost about £454,000; to complete the carriage-descents would require £180,000: total, £634,000. The dangers of the work were many; sometimes portions of the shield broke with the noise of a cannon-shot; then alarming cries told of some irruption of earth or water; but the excavators were much more inconvenienced by fire than water, gas explosions frequently wrapping the place in a sheet of flame, strangely mingling with the water, and rendering the workmen insensible. Yet, with all these perils, but seven lives were lost in constructing the Thames Tunnel, whereas nearly forty men were killed during the building of new London Bridge. In 1833, Mr. Brunel submitted to William IV., at St. James's Palace, "An Exposition of the Facts and Circumstances relating to the Tunnel." Brunel has also left a minute record of his great work: it is well described and illustrated in Weale's *Quarterly Papers on Engineering.* A fine medal was struck at the completion of the work: *obv.* head of Brunel; *rev.* interior and longitudinal section of the Tunnel.

The width of the Tunnel is 35 feet; height, 20 feet; each archway and footpath, clear width, about 14 feet;

thickness of earth between the crown of the arch and the bed of the river, about 15 feet. At full tide the floor of the tunnel is 75 feet below the surface of the water.

Sir Isambard Brunel died at his house in Duke Street, Westminster, in 1849, aged 81. He left an only son, whose life and labors will be found recorded in a future sketch.

We engrave a section of the Tunnel Shield.

1. The Polling boards in front of the Shield.

2. The Jack-screws.

3. The Top Staves, securing the upper part of the excavation until the substitution of the brick-work: the sides of each division of the Shield were similarly defended.

4. Screws, to raise or depress the top staves.

5. The Legs, being Jack-screws, fixed by ball-joints to the Shoes, upon which the whole division stood.

6. The Shoes.

7 and 8. The Sockets, where the top and bottom horizontal screws were fixed to force the divisions forward as the work advanced.

Section of the Thames-Tunnel Shield.

GEORGE STEPHENSON, THE RAILWAY
ENGINEER.

In this practical age of physical comfort, it is scarcely possible to overestimate the value and importance of the railway, and the services of its far-seeing originator, GEORGE STEPHENSON. "It is not too much to say that the inventor (to all practical purposes) of the locomotive steam-engine,[*] and the founder of the railway system of the entire world, has done as much to promote human comfort and advantage as any single man that ever breathed; and, more particularly, we believe that there is hardly a man, woman, or child in Britain who is not reaping personal profit from the labors of this great and sterling Englishman; from the results of his wonderful ingenuity to devise, and his unparalleled perseverance in urging on his gigantic invention, at a time when great engineers, eminent lawyers, and leading members of Parliament were not ashamed to denounce him as an idiot, and to advise his consignment to Bedlam."[†] In every word of this honest tribute we heartily concur.

At a few miles west of Newcastle, in the colliery village of Wylam, on the north bank of the Tyne, amid slag and cinders, there still stands a red-tiled ordinary cottage, of two stories, divided into four dwellings. In one of these rooms, which has unplastered walls, bare rafters, and floor of clay, George Stephenson was born, on the

---

[*] It is believed to have been first remarked by George Stephenson, that the original source of the power of heat engines is the sun, whose beams furnish the energy that enables vegetables to decompose carbonic acid, and so to form a store of carbon, and of it combustible compounds, afterward used as fuel. The combination of that fuel with oxygen in furnaces produces the state of heat, which, being communicated to some fluid, such as water, causes it to exert an augmented pressure, and occupy an increased volume ; and these changes are made available for the driving of mechanism.—PROF. RANKINE'S *Manual of the Steam-engine*, 1859.

[†] *Saturday Review*, No. 87.

9th of June, 1781. At a few yards from the door is the line of rails which runs from the colliery toward New-castle, and has been put in place of the old *tram*-way,* along which the coal-wagons were formerly drawn by horses. Across the river the scenery is very beautiful, and the colliery appears in the distance; and at the back

Birthplace of George Stephenson, at Wylam.

of the house, the rich land, partly clothed with wood, rises steeply toward Haddon-on-the-Wall. George's pa-rents, Robert and Mabel Stephenson, were "honest folk, but sair haudden doun in the world." His father was fireman of the pumping-engine of the colliery, and his mother was "a rale canny body." There were six chil-dren, of whom George was the second, and the family was maintained on the fireman's wages of twelve shil-lings a week; food was so dear that neither of the chil-dren was sent to school, instead of which George was taken by his father bird-nesting, or told stories of Robin-son Crusoe, Sinbad the Sailor, etc. George's interest in birds'-nests never left him till his dying day, nor were other sights of his childhood less identified with the serious business of his life. In the rails of the wooden tram-road before his cottage, on which he saw the coal-

* Called *tram*-roads from having been first laid down by *Outram*, from whose name, omitting the first syllable, the word is said to have been derived.

wagons dragged by horses from the pit to the landing-quay, half the destiny of an age was latent, to be evolved by the very boy who, after his own probation was over, had to keep his younger brothers and sisters out of the way of the horses. He himself was, however, so little as to hide himself when the owner of the colliery came round, lest he should be thought unfit to earn his wages.

When little "Geordie Stephie" was eight years old, his father removed to Dewley Burn, about four miles distant; and George, to his great joy, obtained the place of cowboy, at 2d. a day. He spent much of his leisure in erecting Liliputian mills in the little streams that run into the Dewley Bog, and in making *clay engines*, along with a certain Thomas Tholoway; the boys found the clay in the adjoining bog, and the hemlock which grew about supplied them with imaginary steam-pipes; and the villagers to this day point out, "just aboon the east end," where the future engineer made his first models.

In due course, George had his wages doubled for hoeing turnips. He was next employed as "picker" or sorter of the coals. It was a proud day when he was advanced to be driver of the gin-horse at 8d.; "and there are those who still remember him in that capacity, as a 'grit bare-legged laddie,' whom they describe as full of tricks and fun." George was promoted to the post of assistant fireman when only fourteen years of age, at 1s. a day. At the colliery at Throckley Bridge he was advanced to 12s. a week, and at seventeen he became an engineman or plugman, while his father continued to stoke the fire; and on receiving his first week's wages, he said exulting-ly to a companion, "I am now a made man for life." At this time he was a big, raw-boned man, fond of displaying his strength and activity at the village feasts, but remarkable for his temperance, sobriety, industry, and good temper. He soon studied and mastered the working of his engine, which become a sort of pet with him. He delighted to find some one who could read to him by the engine-fire out of any book or stray newspaper; and having heard that the Egyptians hatched birds' eggs by artificial heat, he endeavored to do the same in his engine-house. He learned also that the wonderful engines of Watt and Boulton were to be found described in books, which induced him to attend a night-school at 3*d*. a week, to learn his letters and practice "pot-hooks," so that at eighteen he had learned to read, and at nineteen he was proud to be able to write his own name. He next went to the night-school of a Scotch domine, a skilled arithmetician, and there learned "figuring" much faster than his schoolfellows: he worked out his sums by the engine fire, and solved the arithmetical questions set him upon his slate by his master, so that he soon became well advanced in arithmetic. In 1801 he became brakesman at the colliery; and he began to increase his income by mending the workmen's shoes. He went on with his writing lessons; and by the next year, 1802, when he married a respectable young woman, Fanny Henderson, he signed his name in a good legible round-hand.

He now took up his abode in a humble cottage at Willington Quay, near Newcastle. He occupied his leisure in constructing little machines, and attempting to discover the perpetual motion. He soon advanced from mending shoes to making them; and an accident having obliged him to repair his own clock, he became the general clock cleaner and mender for the neighborhood, thus improving his own mechanical skill while adding to his income. At Willington, he made the first self-acting incline used in that district, by which the descending laden wagons on the tram-road were made to draw up the empty wagons. Here, on the 16th of November, 1803, was born his only son, Robert, who became second only to his father as a railway engineer. George Stephenson now became something more than a mere workman, by studying the principles of mechanism and the laws by which his engine worked. By steady conduct and saving habits, he procured the coveted means of educating his son, who, in after years, when he had risen to the highest scientific eminence, declared, with touching gratitude, "however extensive his own connection with railways, all he had known, and all he had himself done, was due to the parent whose memory he cherished and revered."

In 1804, George Stephenson removed to Killingworth Colliery, seven miles north of Newcastle; while there, his poor wife died. He spent the next year at a colliery near Montrose, in Scotland; and on his return, he found his aged father had been accidentally scalded and blinded by a discharge of steam, let in upon him while repairing an engine. He at once devoted all his savings to relieve the old man's distress, and place him in comparative comfort. So disheartened was Stephenson about this period, that he thought of emigrating to Canada. But his prospects brightened, through his perseverance in the colliery work, and by mending clocks and shoes, and even cutting out the clothes of the workmen. He also signalized himself by curing a wheezy engine, at which all the engineers of the neighborhood had failed: he got £10 for this job; from this day his services as an engineer came into request; and a vacancy occurring, he was appointed the engine-wright to the colliery, with a hundred pounds a year. He now began to turn his thoughts to the locomotive steam-engine.

Railways, consisting of wooden beams, tram, or wagon ways, were introduced as early as 1602 in the collieries in the north of England, to reduce the labor of drawing coals from the pits to the place of shipment. Lord-keeper North, in 1676, describes such *rails of timber* from the colliery to the river, exactly straight and parallel, with the rollers of bulky carts made to fit the rails. This "oaken way" first consisted of pieces of wood simply imbedded in the ordinary road. A century and a half elapsed before the rails were laid upon cross-pieces, or sleepers, to which they were fastened by pegs. In 1716, thin plates of malleable iron were nailed upon *portions* of the wooden rails. Next followed cast-iron rails. A wooden railway was used at the Coalbrookdale Ironworks about 1767, when, the price of iron becoming very low, it was determined, in order to keep the furnaces at work, to cast bars, which might be laid down upon the wooden rails to save their wear, but which it was proposed to take up, and sell as pigs of iron, in case of a sudden rise. This is confirmed by an entry in the Company's books of between five and six tons of cast-iron rails, but "only as an experiment, on the suggestion of one of the partners." A few years after, cast-iron rails, with an upright flange, were first used at the colliery of the Duke of Norfolk, near Sheffield, in 1776. Here we must leave the Railway for the Locomotive.

Various kinds of propelling power had been proposed for use on these *plate-ways*, as they were still called. Sails had their advocate. The application of the steam-engine to locomotion on land was, according to Watt, suggested by Robison in 1759. In 1784, Watt patented a locomotive engine, which, however, he never executed; and about the same time, Murdoch, assistant to Watt, made a very efficient model. In 1802, Trevethick and Vivian patented a locomotive engine, which, in 1804 or 1805, traveled at about five miles an hour, with a net load of ten tons. The use of fixed engines, to drag trains on railways by ropes, was introduced by Cook in 1808. Some years after, Mr. Blackett constructed an engine for the Wylam Colliery; but as it would only travel one mile an hour, it was soon laid aside.

Several other " traveling engines" were made by other engineers, with partial success; but it was left to Stephenson to render the locomotive practically useful. He pressed the matter on the lessees of the Killingworth colliery, and he made for them a locomotive, which was first tried on their railway July 25, 1814: it was very clumsy and ugly, but it drew thirty tons at four miles an hour. Some improvements were made in this engine, and next year Stephenson built a locomotive which contained the germ of all that has since been effected; " there being no material difference between the cumbrous machines that screamed and jolted along the coal tram-road in 1815, and the elegant and noiseless locomotive which now takes out the express train, gliding smoothly and swiftly as a bird through the air."

The engines which Stephenson constructed in 1815 worked away at Killingworth, but attracted little notice: their author always maintained that some day *such engines and railways would be well known* all over Britain, but he was regarded as an innocent enthusiast.

Meanwhile, a striking suggestion of uniting railway communication into a *system*, as connecting lines are now called, was made by an unprofessional writer, the first author on the subject to notice which was Mr. Smiles, in his admirable *Life of George Stephenson*.

This suggestion occurs in Sir Richard Phillips's *Morning's Walk from London to Kew*, and was written in 1813. On reaching the

Surrey Iron Railway, at Wandsworth, where a train of carriages was drawn by one horse, Sir Richard says, "I thought of the millions which have been spent at Malta, four or five of which might have been the means of extending *double lines of iron railway* from London to Edinburgh, Glasgow, Holyhead, Milford, Falmouth, Yarmouth, Dover, and Portsmouth. A reward of a single thousand would have supplied coaches, and other vehicles, of various degrees of speed, with the best tackle for readily turning out; and we might, ere this, have witnessed our mail-coaches running at the rate of ten miles an hour, drawn by a single horse, or *impelled fifteen miles an hour by Blenkinsop's steam-engine.*" The writer of these sagacious remarks lived until 1840, so that he had witnessed a triumph greater than his long-cherished hope.

In the interval, *i. e.*, in 1825, Sir Richard Phillips published the first *Treatise on Railways*, by Nicholas Wood, of Killingworth, wherein he deprecates any attempt at a greater speed than fourteen miles an hour upon railways. Yet this short-sightedness was exceeded by a writer in the *Quarterly Review:*

What (said the reviewer) can be more palpably ridiculous than the prospect held out of locomotives traveling twice as fast as stage-coaches! We should as soon expect the people of Woolwich to suffer themselves to be fired off upon one of Congreve's ricochet rockets, as trust themselves to the mercy of such a machine going at such a rate. We will back old Father Thames against the Woolwich Railway for any sum. We trust that Parliament will, in all railways it may sanction, limit the speed to eight or nine miles an hour, which we entirely agree with Mr. Sylvester is as great as can be ventured on with safety.

In 1819, Stephenson turned, for the owners of Hetton Colliery, their tram-road into a railway; and, taking advantage of the hilly country, formed self-acting inclines, the locomotive working on the level part: this line was opened in 1822.

In 1819, also, Mr. Edward Pease, supported by a number of Quaker friends, obtained, after much opposition, an Act of Parliament for the construction of a colliery railway from Stockton to Darlington. In 1821, George Stephenson applied to Mr. Pease to lay out the line. The wealthy Quaker was prepossessed in favor of Stephenson, "there was such an honest, sensible look about him, and he seemed so modest and unpretending." Mr. Pease had contemplated the use of horse-power upon his railway; but Stephenson assured him that the engine which had worked for years at Killingworth was worth fifty horses. He went and saw the engine, and George Stephenson was appointed engineer to the Stockton and Darlington Railway, with a salary of £300 a year; and he removed to Darlington with his family (he had married a second time in 1819) in the year 1823. He laid out every foot of the line; and he built, in a factory at Newcastle, three engines for use upon it, with £1000 given him by public subscription for his invention of a safety-lamp for use in coal-pits. The railway was opened September, 1825, when the first train, 38 carriages, with 600 passengers, was drawn by a single engine at from four to twelve miles an hour, the first passenger carriage being an old stage coach placed upon a wooden frame. The engines did their daily work admirably; and the little factory at Newcastle, founded mainly to bring together more skillful workmen than the country blacksmiths who had made the first locomotives, gradually grew into a gigantic establishment, which for many years supplied engines, drivers, and superintendents for all the railways of Europe.

The No. 1 engine made by Stephenson for the above railway, and which was the first machine ever run on a Parliamentary line, has been preserved, and was, in 1859, erected upon a pedestal at Darlington, as a public memorial of the commencement of the railway system; and it is a far more interesting object than the groups of monumental flattery which we are accustomed to see in public places.

The grand railway experiment of a line between Liverpool and Manchester was now commenced. It met with great opposition, especially from the authorities of the Bridgewater Canal. Nevertheless, a company was formed, and all the shares in it were immediately taken up. A line of railway was surveyed and mapped out, in spite of the furious resistance of land-owners; personal violence was threatened to the engineers employed, and the most absurd stories were circulated as to the dangerous nuisances to be apprehended from the passing engines. The best friends of the locomotive engine lamented that Stephenson should venture to predict that railway-trains would some day run at twelve and sixteen miles an hour; and members of the Parliamentary Committee whispered

S

doubts of the engineer's sanity. The Bill was thrown out by a majority of one; but early in the next session, 1826, an act was passed authorizing the construction of the railway, and Mr. Stephenson was appointed engineer, with a salary of £1000 a year. He set to work at once, and in June, 1826, began to make the road across Chat Moss, the great morass of four miles. Week after week, thousands of cubic yards were ingulfed, without the least apparent progress. At length the directors proposed to abandon it; but Stephenson persevered, and the four miles through Chat Moss now form the soundest part of the line. The expense was about £28,000, whereas an engineer had declared before Parliament that the cost must be at least £270,000. Stephenson organized all the works himself, there being then neither contractors nor navvies: he sent for his son Robert, who had been some years in America, for his aid and counsel in the great work.

The Railway had almost been completed before the motive-power to be employed on it was decided on. Stephenson stood alone in urging the directors to employ the locomotive; but other engineers who were consulted, without exception, recommended stationary engines, which should draw the trains by the help of ropes. Stephenson expostulated and entreated, and at length worried the directors into giving the locomotive a fair trial. A simple remark, made by him about this time, shows with what vivid reality the future passage-railway was present to his mind: "I said to my friends that there was no limit to the speed of such an engine, *provided the works could be made to stand.*" He had already, by his invention of the tubular boiler (in conjunction with his son), raised the speed of the engine from seven to thirty miles an hour.

A large heating surface is indispensable to generate the steam required; but the space allowed for the whole engine on the carriage being limited, Stephenson's ingenuity was exercised in providing the former without unduly increasing the latter. The flame and heated air leave the fire-box at a very high temperature, and much heat would be wasted if they were allowed to escape immediately into the atmosphere; but Stephenson had already supplemented the ordinary operation of the furnace by this heated air. As high-pressure engines are used, the steam escapes from the cylinder, after having done its work, at a high temperature; and being made to pass into the smoke-box, and then up the chimney, it acts as a powerful blast upon the fire. Instead of blowing the fire, it blows the chimney; and more air will, of course, enter the fire if the chimney be cleared more quickly. This, then, was Stephenson's great improvement, and it enabled him to give effect to another. Putting the chimney at one end of the boiler, and the fire-box at the other, he connected the two by a number of metal tubes passing from the back of the furnace to the smoke-box. Hot air escaping through these tubes heats the water by which they are surrounded, and enables engines to travel at the rate of twenty, sixty, or even seventy miles an hour.—JAMES SIME, M.A.; *Edinburgh Essays*, 1856.

Stephenson prepared an engine (the *Rocket*), constructed on this principle, to contend for the prize of £500 which the Railway Directors offered for the best engine, to be produced on a certain day, to draw a weight of twenty tons at ten miles an hour. The trial took place on October 6, 1829, at Rainhill. There were four engines, but Stephenson's *Rocket* won the prize: it drew thirteen tons at a maximum speed of twenty-nine miles an hour, and thus decided for ever the use of locomotive engines on railways.

The *Rocket* prize Locomotive Engine.

The opening of the Liverpool and Manchester Railway took place on the 15th of September, 1830. The Duke of Wellington, then prime minister, Sir Robert Peel, and other distinguished persons, were present; but the sad death of Mr. Huskisson, who fell beneath the train in motion, threw a gloom upon the day. Little passenger-traffic had been looked for; but, from the opening, the railway carried about 1200 passengers daily, and in five years afterward it carried half a million yearly. Stephenson's predicted ten miles rose to thirty miles an hour, and the net profit of the company exceeded £80,000 a year. The *Rocket* often attained a speed of sixty miles an hour; it weighed four and a quarter tons: the locomotive of the present day ranges from five to fifty tons weight, and its load from fifty to five hundred tons.

What has become of the *Rocket* engine? The French preserve with the greatest care the locomotive constructed by Cugnot, which is to this day to be seen in the Conservatoire des Arts at Paris. The *Rocket* has scarcely been so honored.

"Changing hands, says Professor G. Wilson, "more than once, and at length discarded, like an old horse as soon as it is unfit for work, it was finally purchased by the inventor's son, and is now preserved in the engine-works at Newcastle-on-Tyne. It can not always continue under filial guardianship; yet, when we consider that a century hence hundreds of curious pilgrims will gladly travel from distant lands to study the famous *Rocket* engine, if it be in existence to be studied, we can not but hope that at least it will not be willfully destroyed. We may have a thousand better engines, but we can never have the *Rocket* again. As the first of its race, the most infantile and the most venerable of engines, it has merits which no later engine can possibly possess."

From 1830 railways began to overspread England. In conjunction with his son, Stephenson was appointed the engineer of the London and Birmingham, the Grand Junction, the Midland and the North Midland, and other important lines. In 1840 he settled at Tapton House, near Chesterfield. His pupils became eminent engineers, among whom were Locke and Gooch, Swanwick and Birkenshaw. To his honor be it said, that Stephenson held aloof from all the schemes of the railway mania of 1845–6, and he strongly condemned the reckless spirit in which Parliament authorized lines which could not possibly remunerate the shareholders. In 1845 he visited Spain to survey a proposed line of railway, having previously laid out the government system of railways in

Belgium, for which he received a knighthood from King Leopold. He also constructed lines in Holland, France, Germany, and Italy. Stephenson's declining years were spent at Tapton, where he became an enthusiastic horticulturist, and began working the Claycross Collieries. He took great interest in the Mechanics' Institute in his neighborhood; and he was the founder and president of the Institution of Mechanical Engineers of Birmingham. His early fondness for all kinds of animals revived. He had many attached pets among his dogs, horses, and birds; and he was fond of rambling about the neighboring country, bird-nesting or nutting. Unfortunately, he spent too much time in the unwholesome air of his forcing-houses at Tapton, and he contracted an intermittent fever, which carried him off after a few days' illness, on the 18th of August, 1848, in the sixty-seventh year of his age. The shops of Chesterfied were closed, and all business was suspended, on the day of his funeral. A plain monument in Chesterfield Church marks his resting-place.

In 1844 a fine statue of Stephenson was erected in St. George's Hall, Liverpool, and in 1854 there was set up in the great hall of the terminus of the Northwestern Railway, London, Baily's colossal marble statue of Stephenson, purchased by the subscriptions of 3150 working-men and 178 private friends.

The genius and worth of George Stephenson are to be commemorated by a characteristic group of sculpture, to be erected at Newcastle-on-Tyne. It is to consist of a colossal statue of Stephenson upon an embellished pedestal. The model was completed by Mr. Lough, the sculptor, in the autumn of 1859. The height of the figure is seven feet eight inches, but the actual casting model will measure ten feet high. The figure is upright, and attired in modern costume, with a plaid crossing the chest from the left shoulder; the right hand, holding a pair of callipers, rests on the breast, and the left on a locomotive engine of very early form. The likeness is good, and the head is profoundly thoughtful. The pedestal intended for the support of this statue presents at its four angles types of the labor necessary to engineering works; these are accordingly a navvy, a blacksmith, a pitman, and an engineer.

There are countries where such a man would have been ennobled, and covered with ribbons and orders; here he died as he had lived, plain George Stephenson. But he has a most noble memorial in the great system of iron roads which converge to Britain's great cities, and are ramified away to her quietest country nooks.

ROBERT STEPHENSON AND RAILWAY WORKS.

THIS distinguished son of a distinguished father, George Stephenson, was born at Willington Quay, on the Tyne, about six miles below Newcastle, on November 16, 1803. Here, in his humble home, he was familiarized from his earliest years with the steady industry of his parents; for, when his father was not busy in shoemaking, or cutting out shoe-lasts, or cleaning clocks, or making clothes for the pitmen, he was occupied with some drawing or model, with which he sought to improve himself. Robert's mother very soon died; and his father, whose heart was bound up in the boy, had to take the sole charge of him. George Stephenson felt deeply his own want of education, and, in order that his son might not suffer from the same cause, sent him first to a school at Long Benton, and afterward to the school of a Mr. Bruce, in Newcastle, one of the best seminaries of the district. There young Robert remained for three years; and his father not only encouraged him to study for himself, but also made him, in a measure, the instrument of his own better education, by getting the lad to read for him at the library in Newcastle, and bring home the results of his weekly acquirements, as well as frequently a scientific book, which father and son studied together. They jointly produced a sun-dial, which was placed in the wall over the door of their cottage at Killingworth, and of which the father was always proud. On leaving school at the age of fifteen, Robert Stephenson was apprenticed to Mr. Nicholas Wood, at Killingworth, to learn the business of the colliery, where he served for three years, and became familiar with all the departments of underground work. His father was engaged at the same colliery, and the evenings of both were usually devoted to their mutual improvement. Mr. Smiles describes the animated discussions which in this way

GEORGE STEPHENSON.   ROBERT STEPHENSON.

SIR I. M. BRUNEL.   I. K. BRUNEL.

took place in their humble cottage, these discussions frequently turning on the then comparatively unknown powers of the locomotive engine daily at work on the wagon-way. The son was even more enthusiastic than the father on the subject. It was probably out of these discussions that there arose in George Stephenson's mind the desire to give Robert a still better education. He sent him in the year 1820 to the Edinburgh University, where Dr. Hope was lecturing on chemistry, Sir John Leslie on natural philosophy, and Professor Jameson on natural history. Though young Stephenson remained in Edinburgh but six months, it is supposed that he did as much work in that time as most students do in a three years' course. It cost his father some £80; but the money was not grudged when the son returned to Killingworth, in the summer of 1821, bringing with him the prize for mathematics, which he had gained at the University.

In 1822 Robert Stephenson was apprenticed to his father, who had by this time established his locomotive manufactory at Newcastle; but his health giving way after a couple of years' exertion, he accepted a commission to examine the gold and silver mines of South America. The change of air and scene contributed to the restoration of his health; and, after having founded the Silver Mining Company of Columbia, he returned to England in December, 1827, in time to assist his father in the arrangements of the Liverpool and Manchester Railway by placing himself at the head of the factory at Newcastle. About this time, indeed, he seems to have almost exclusively devoted his attention to the study of the locomotive engine, the working of which he explained, jointly with Mr. Locke, in a report replying to that of Messrs. Walker and Rastrick, who advocated stationary engines. How well he succeeded in carrying out the idea of his father was afterward seen, when he obtained the prize of £500 offered by the directors of the Liverpool and Manchester Railway for the best locomotive. He himself gave the entire credit of the invention to his father and Mr. Booth, although it is believed that the *Rocket*, which was the designation of the prize-winning engine, was entered in the name of Robert Stephenson.

Even this locomotive, however, was far from perfect, and was not destined to be the future model. The young engineer saw where the machine was defective, and designed the *Planet*, which, with its multitubular boiler, with cylinders in the smoke-box, with its cranked axletree, and with its external frame-work, forms, in spite of some modifications, the type of the locomotive engines employed up to the present day. About the same time he designed for the United States an engine specially adapted to the curves of American railways, and named it the *Bogie*, after a kind of low wagon used on the quay at Newcastle. To Robert Stephenson we are accordingly indebted for the type of the locomotive engines used in both hemispheres.

The next great work upon which Mr. Stephenson was engaged was the survey and construction of the London and Birmingham Railway, which he undertook in 1833, having already been employed in the execution of a branch from the Liverpool and Manchester Railway, and in the construction of the Leicester and Swannington line. The London and Birmingham line was completed in four years, and on the 15th of September, 1838, was opened. The difficulties of this vast undertaking were very formidable. In forming the Kilsby Tunnel, it was ascertained that about 200 yards from the south end there existed, overlaid by a bed of clay 40 feet thick, a quicksand. The contractor for the works is said to have died of fright in consequence of this discovery; and the danger was so imminent, that the tunnel would have been abandoned altogether but for the landholders in the vicinity of the line. Under these circumstances, Robert Stephenson accepted the responsibility of proceeding, and in the end conquered every difficulty. He worked with amazing energy, walking the whole distance between London and Birmingham more than twenty times in the course of his superintendence. Meanwhile, he had not ceased to devote his attention to the manufactory in Newcastle, convinced that good locomotives are the first step to rapid transit. His evidence before Parliamentary Committees was grasped at; and it may be said that, in one way or another, he became engaged on all the railways in England, while, in conjunction with his father,

he directed the execution of more than a third of the
various lines in the country.   Father and son were consulted as to the Belgian system of railways, and obtained
from King Leopold the Cross of the Order of Leopold in
1844.   For similar services performed in Norway, which
he visited in 1846, Robert Stephenson received the Grand
Cross of St. Olof.   So, also, he assisted either in actually
making or in laying out the systems of lines in Switzerland, in Germany, in Denmark, in Tuscany, in Canada, in
Egypt, and in India.   As the champion of locomotive in
opposition to stationary engines, he resisted to the utmost
the atmospheric railway system, which was backed with
the authority of Brunel, but it is now nearly forgotten.
In like manner, he had to fight with Mr. Brunel the battle of the gauges, the narrow against the broad gauge,
and he was successful also here.

It is, however, in the Bridges which Robert Stephenson erected for railway purposes that his genius as·an
engineer is most strikingly displayed, and by these he
will be best remembered.   Of his bridges, we refer to
the high-level one at Newcastle, constructed of wood and
iron; to the Victoria Bridge at Berwick, built of stone
and brick; to the bridge in wrought and cast iron across
the Nile; to the Conway and the Britannia Bridges óver
the Menai Straits; and to the Victoria Bridge over the
St. Lawrence.   The High-level Bridge, in which the
suspension and ordinary principles of a viaduct have been
combined in one structure, serves a two-fold object—a
bridge to accommodate Newcastle and Gateshead at the
same time that it carries the railway-lines above.

The idea of the Tubular Bridge was an utter novelty, and as carried
out was a grand achievement.   When, in 1844, Mr. Robert Stephenson undertook to construct a railway between Chester and Holyhead,
it was necessary to cross the Menai Straits from the main land to
Holyhead at such a height as to allow great ships to pass beneath it.
The Commissioners of the Admiralty would not consent to cast-iron
arches, and the principle of a suspension bridge was inadmissible.   Mr.
Stephenson then proposed to span the strait by a tunnel of wrought
iron, stretching from side to side, and allowing a passage for trains
through its interior.   The question then arose, Should the tubular
bridge be supported by chains, or left to itself? what should be the
form of the tube—elliptical, circular, or rectangular? where was the
most strength required? where least? and how could the greatest
strength be secured with the least expenditure of materials?   These

points were determined by careful experiments by Mr. Stephenson, assisted by Mr. Fairbairn, the eminent engineer; and the result was, it was seriously proposed to build an iron box, 460 feet long, 30 feet high, and 14 feet broad, on the banks of the Menai Straits; to float this mass of 1450 tons at high water to openings in piers prepared for its reception; to lift it upward of 100 feet, and build solid masonry underneath for its support; to rest it at its utmost height on cast-iron rollers, which would allow it to expand and contract as the sun rose and set, or as summer advanced and waned; then to make it a tunnel for the passage of railway trains weighing, perhaps, a hundred tons. Experiments made Mr. Fairbairn confident that there was no danger of the bridge giving way under its own weight; and numerous experiments upon a large scale proved the truth of his opinion. Chains were as unnecessary to support this bridge as intermediate piers, even if the latter could have been built. Its strength is derived from a different source from either. The roof consists of two platforms, 1 foot 9 inches apart, and 14 feet broad; this space is divided into eight equal parts by partitions running from end to end of the bridge, and the cells thus formed keep the tube from giving way to compression in the top, where the material is most liable to be injured.

Two of these stupendous bridges were constructed for the Chester and Holyhead line. The first was built on the banks of the Conway River in 1848, and now spans that stream not far from the suspension bridge erected by Telford on the Holyhead road about twenty years earlier. Two tubes of 400 feet span were required, one for each line of rails. A train of wagons, weighing altogether 301 tons, was placed in the middle of one of them, and the deflection in the centre amounted to 11 inches. The rollers on which the bridge rests allow the tubes to expand or contract with the ever-varying temperature of the day or season.

The Britannia Bridge over the Menai Straits (at about a mile distant from Telford's suspension bridge) was finished a year after, and is justly regarded as the greatest triumph of engineering skill that this or any other country has ever witnessed. A splendid tower rises to the height of 250 feet from a rock in the middle of the Straits; and four tubes, each 472 feet in length, stretch from it to smaller towers on the banks. Other four tubes, of 260 feet each, carry the railway to the high grounds on the east and west sides of the Straits. This magnificent bridge was the culminating point of railway enterprise and engineering, and half a century may elapse before necessity produces its rival.—JAMES SIME, M.A.; *Edinburgh Essays*, 1856.

The construction of this bridge was a vast labor. Any midway support was limited to a small area of the central rock; scaffolding below was impracticable, and the navigation was under no circumstances to be interfered with. To meet these requirements, the tubes were constructed upon the beach, and floated upon rapid tides; and, although weighing nearly 2000 tons each, were ultimately lifted by vast hydraulic presses into their place, to bear Mr. Stephenson's name with honor to posterity. Each of the tubes has been compared to a row of chimneyless houses, and, allowing it to have sky-lights in the roof, it would resemble the Burlington Arcade in Piccadilly; and the labor of placing each tube upon the piers has been likened to that of raising Burlington Arcade to the summit of the spire of St. James's Church, if surrounded with water. One of the tubes, if placed on its end in St. Paul's Church-yard, would reach 107 feet higher than the cross of the Cathedral. The masonry is Cyclopean. Mr. Stephenson tells us that no less than a million and a half of cubic feet, of which the piers and abutments are composed, were constructed within three years; and three cubic feet were accomplished per minute from the commencement.

Over the entrances to the tubes are massive lintels, consisting of single stones twenty feet long; and the approaches are marked by colossal lions couchant on pedestals, designed by Mr. John Thomas, and each composed of eleven pieces of limestone: they are each twenty-five feet long, twelve feet high, and weigh about thirty tons; and one of these lions was brought from a workshop at the base of the abutment, raised 100 feet, and put together complete on the pedestal in a single day.

The Britannia Tower is 221 feet 3 inches high; it contains 151,158 cubic feet of Anglesea limestone, 127,001 cubic feet of Runcorn sandstone, and 68,411 cubic feet of brick-work, in all weighing 24,700 tons. Including the bed-plates, it contains also 479 tons of cast-iron, and the weight from the two tubes is 4000 tons. The total weight at the foundations is thus 29,600 tons, or 16 tons per superficial foot of sectional area; whereas the weight required to crush the lower courses would be about 500 tons per superficial foot.

The security which Mr. Stephenson deemed it necessary to insure for the public in this wonderful structure may be illustrated by the following very extraordinary fact. It had been mathematically demonstrated, as well as practically proved by Mr. Fairbairn, that the strain which would be inflicted on the iron-work of the longest of Mr. Stephenson's aerial galleries, by a monster railway-train sufficient to cover it from end to end, would amount to six tons per square inch, which is exactly equal to the constant stress upon the chains of Telford's Menai Bridge when it has nothing to support but its own apparently slender weight.

The two tubular bridges constructed by Mr. Stephenson on the Egyptian railway are, one over the Damietta branch of the Nile, and the other over the large canal near Beaket-al-Sabá; they have this peculiarity, that the trains run, not, as at the Menai Straits, within the tube, but at the outside, upon the top.

Although the Britannia Bridge represented the most scientific distribution of material which could be devised at the date of its construction, it has since been improved upon by the same engineer in the Victoria Bridge now in the course of construction across the river St. Lawrence, near Montreal. The Victoria Bridge is, without exception, for gigantic proportions and vast length and strength, the greatest work of the kind in ancient or modern times. The entire bridge, with its approaches, is only sixty yards short of two miles; it is five times longer than the Britannia Bridge, and has twenty-four spans of 242 feet each, and one great central span—itself an immense bridge, of 330 feet. The road is carried within iron tubes, sixty feet above the level of the St. Lawrence, which runs beneath at a speed of about ten miles an hour, and in winter brings down the ice of 2000 miles of lakes and upper rivers. The weight of iron in the tubes will be upward of 10,000 tons, supported on massive stone piers. This gigantic work is upon the Grand Trunk Railway of Canada, which will be upward of 1100 miles in length.

Mr. Stephenson's labors were not confined to the construction and survey of railways. He made elaborate reports on the London and Liverpool system of Waterworks; he considerably aided with his counsel and experience his friend Sir Joseph Paxton in his design for the Great Exhibition Building in Hyde Park; and he was a member of the Royal Commission. In 1847 Mr. Stephenson was returned to Parliament for Whitby, in the Conservative interest, which he continued to represent until his death. His opinion upon scientific subjects was often sought by the House: this he gave impartially and with the modesty of true genius, and his information was exact. He took great interest in all scientific investigations. He was a Fellow of the Royal Society, and of other scientific institutions.

In 1856, when Mr. Piazzi Smyth was sent out, with

very limited means, on an astronomical expedition to Teneriffe, Mr. Stephenson, with a liberality and zeal for research worthy of the name he bore, placed at Mr. Smyth's disposal, for as long a time as the object he had in view might require, his yacht *Titania*, a finely-mould-vessel of the new school, of 140 tons burden, and manned with a picked crew of sixteen able seamen. As our observer went out and returned in this vessel, Mr. Stephenson must have abandoned its use for the whole summer and autumn; or, rather, as we have no doubt, he felt glad to find that an opportunity had occurred for enabling him to employ it so well.*

In the same spirit, in 1855, he paid off a large debt which the Newcastle Literary and Philosophical Society had incurred; his motive being, to use his own phrase, gratitude for the benefits which he himself had received from it in early life, and a hope that other young men might find it equally useful. And in 1858 he had taken down the cottage in which he was born at Willington, and erected upon its site a group of schools for girls, boys, and infants, a mechanics' institute, etc., at the cost of two thousand pounds.

As a member of the Institution of Civil Engineers, Mr. Stephenson's services were of the highest value; never had the council a more efficient *confrère*. As President in 1855–56, he presented an address, in which he applied himself with striking ability to the great question of British Railways. The data of this address are very important.

"Parliamentary legislation for railways," said the President, "is full of incongruities and absurdities. The Acts of Parliament which railways have been forced to obtain cost the country £14,000,000 sterling, the exclusive funds of Parliament, and of the system it enforced. The legislation of Parliament has made railways pay £70,000,000 of money to landowners for land and property, yet almost every estate traversed by a railway has greatly improved in value."

Referring to the benefits derived from the Institution, Mr. Stephenson observed that "it is the arena wherein have been exhibited that intelligence and familiar knowledge of abstract and practical science characterizing the papers and discussions. In consequence of the constant intercourse within its walls, professional rivalry and competition are now conducted with feelings of mutual forbearance and concilia-

* *National Review*, No. 18.

tion, and the efforts of the members are all directed in the path of enterprise, and toward the fair reward of successful skill. The business of the civil engineer, from a craft, has become a profession; and, by union and professional uprightness, a great field is opened to energy and knowledge."

In conclusion, Mr. Stephenson urged the duty devolving on civil engineers of improving and perfecting the vast railway system, with which his name, in consequence of his father's works, had been largely associated.

In the autumn of 1859, two months before he had reached his fifty-sixth year, Mr. Stephenson was struck down by death, in the maturity of his intellectual powers. His health is stated to have been impaired by the fatigues of his great work, the Britannia Bridge. He complained of failing strength just before his last journey to Norway. In Norway he became very unwell: his liver was so much affected that he hurried home; and when he arrived at Lowestoft, he was so weak that he had to be carried from his yacht to the railway, and thence to his residence in London, where his malady increased so rapidly as to leave from the first but faint hopes of his recovery. He had not strength enough to resist the disease, and he gradually sank, until at length he expired on October 12. He was interred in Westminster Abbey, on October 21, in the nave, next to Telford, the celebrated engineer of his day. Men of kindred genius and engaged in kindred enterprises, they lie at last side by side. Stephenson was wont to say that, had Telford been buried in some quiet country church-yard, he should have wished his remains to be interred along with him there; but, since he lay in Westminster Abbey, that was an idle wish.

Mr. Stephenson's remains were followed to the grave by his immediate relatives and friends, but the presence also of nearly two thousand persons at the interment gave the ceremony more of the character of a public than a private funeral. Among the spectators was a working-man from the South-eastern Railway, who many years ago drove the first locomotive engine, called "the Harvey Combe," that ran from London to Birmingham, Robert Stephenson standing at his elbow all the way. Westminster Abbey, as a place of sepulture, is commonly thought to have been reserved for sovereigns, warriors, and statesmen; but it must be remembered that here also rest many of our poets, and men of art and letters. The profession of an engineer almost belongs to our age; and Robert Stephenson, though neither warrior nor statesman, was not the less, if indeed not the more, a public benefactor in his many gigantic works. He was as good as he was great, and the man was even more to be admired than the engineer. His benevolence was unbounded, and every year he expended thousands in doing good unseen. His chief care in this way was for the children of old friends who had been kind to him in early life, sending them to the best schools, and providing for them with characteristic generosity. His own pupils regarded him with a sort of worship; and the number of men belonging to the Stephenson school who have taken very high rank in their peculiar walk shows how successful he was in his system of training, and how strong was the force of his example. Mr. Stephenson bequeathed by his will a large sum to various public institutions, located chiefly in Newcastle-upon-Tyne, in the vicinity of which he was born, and with which his life was so closely identified.

To conclude. Neither the originator of the Railway System, nor his son and coadjutor, were, in their day, honored with any national distinction in their own country, but their memory will live for ages in the hearts of a grateful people.

ISAMBARD KINGDOM BRUNEL: RAILWAY WORKS AND IRON SHIP-BUILDING.

ISAMBARD KINGDOM BRUNEL, the only son of Sir Isambard Mark Brunel (of whom see sketch, p. 396–401), was born at Portsmouth in 1806. He was educated at the College of Henri Quatre at Caen. As Normandy was the birthplace of both his parents, his mother being a Miss Kingdom, of Rouen, this choice of a school is easily explained. He was, as it were, born an engineer, about the time his father had completed the Block Machinery at Portsmouth. Those who recollect him as a boy remember full well how rapidly, almost intuitively, indeed, he entered into and identified himself with all his father's plans and pursuits. He was very early distinguished for his powers of mental calculation, and for his rapidity and accuracy as a draughtsman. His power in this respect was not confined to professional or mechanical drawings only: he displayed an artist-like feeling for and a love of art, which in later days never deserted him.

The bent of his mind when young was clearly seen by his father and by all who knew him. His education was therefore directed to qualify him for that profession in which he afterward distinguished himself. When he was about fourteen he was sent to Paris, where he was placed under the care of M. Masson, previous to entering the college of Henri Quatre at Caen, where he remained two years. He then returned to England, and commenced his professional career as his father's assistant in the Thames-Tunnel works. There are many of his fellow-laborers now living who well remember the energy and ability he displayed in that great scientific struggle against physical difficulties and obstacles of no ordinary magnitude, and it may be said that at this time the anxiety and fatigue he underwent, and an accident he met with, laid the foundation of future weakness and illness. In one of the irruptions the rush of the water carried him up the shaft.*

Upon this and a similarly trying occasion, he showed that zeal for his profession which characterized him to his dying day. Being an expert swimmer, he is known to have saved the lives of several of the workmen at the risk of his own. An eye-witness describes, while the tears ran down his cheeks, how the young man, still suffering from exposure and fatigue, paid a visit to the works during the men's dinner-hour. As soon as he appeared, they welcomed him with a hearty and respectful cheer. They crowded round him, stern, rugged men weeping like children, as they affectionately grasped his hand. While the wives of the men he had saved fell on their knees before him, imploring blessings upon him, others cut little pieces from his coat, which they long treasured as relics.†

Brunel displayed very early the resources not only of a trained and educated mind, but great, original, and inventive power. He possessed the advantage of being able to express or draw clearly and accurately whatever he had matured in his own mind. But not only that: he could work out with his own hands, if he pleased, the models of his own designs, whether in wood or iron. As a mere workman he would have excelled. Even at this early period Steam Navigation may be said to have occupied his mind, for he made the model of a boat, and worked it with locomotive contrivances of his own. Every thing he did, he did with all his might and strength, and he did it well: the same energy, thoughtfulness, and accuracy, the same thorough conception and mastery of whatever he undertook, distinguished him in all minor things.

Upon the stoppage of the Thames-Tunnel works by the irruption of the river, Mr. Brunel became employed on his own account upon various works. Docks at Sunderland and Bristol were constructed by him; and when it was proposed to throw a suspension bridge across the Avon at Clifton, his design and plan was approved by

---

* See page 400. Mr. Brunel's improved use of the Diving-Bell, after one of the Thames-Tunnel irruptions, is noticed at pages 60–61.

† *Atlas* journal, 1859.

Mr. Telford. This work was never completed: he thus became known, however, in Bristol; and when a railway was in contemplation between London and Bristol, and a company formed, Brunel was appointed their engineer. His earliest works were on the Bristol and Gloucestershire and the Merthyr and Cardiff tram-ways, in which works his mind was first turned to the construction of railways; and when he became engineer of the Great Western Railway Company, he recommended and introduced what is popularly called the Broad Gauge. Considering this line as an engineering work alone, it may challenge comparison with any other railway in the world for the speed and ease of traveling upon it, although the Narrow Gauge is more economical in working. Among the Great Western structures are the viaduct at Hanwell; the Maidenhead Bridge, which has the flattest arch of such large dimensions ever attempted in brick-work; the Box Tunnel, which, at the date of its construction, was the longest in the world; and the bridges and tunnels between Bath and Bristol, all more or less remarkable and original works. To these may be added the sea-wall of the South Devon Railway; and, above all, the tubular bridge over the Tamar, together with the similar bridge over the Wye at Chepstow.* On the South Devon Railway, Brunel adopted the plan which had been previously tried on the London and Croydon line, viz., of propelling the carriages by atmospheric pressure. The plan failed, but he entertained a strong opinion that this power would be found hereafter capable of adoption for locomotive purposes. It was in connection with the interests of the Great Western Railway that he first conceived the idea of building a steam-

* These bridges are imposing monuments of Mr. Brunel's boldness and skill. The principle upon which the bridges in question were planned has been much criticised, but the works as executed undoubtedly possess great strength and durability. The foundations of these bridges, under the customary modes adopted with such works, would have been extremely difficult of execution; but Mr. Brunel's ready appreciation of the merits of new discoveries enabled him to take full advantage of the pneumatic process, by a modification of which he established the foundations of the principal pier of the Saltash Bridge at a depth of water and soft mud at which no works of the kind had been previously founded.—*The Engineer*, No. 195.

ship to run between England and America. The *Great Western* was built accordingly. The power and tonnage of this vessel was about double that of the largest ship afloat at the time of her construction. Subsequently the *Great Britain* was designed and built under Mr. Brunel's superintendence. This ship, the result, as regards magnitude, of a few years' experience in iron ship-building, was not only more than double the tonnage of the *Great Western*, and by far the largest ship in existence, but she was more than twice as large as the *Great Northern*, the largest iron ship which at that time had been attempted. While others hesitated about extending the use of iron in the construction of ships, Mr. Brunel saw that it was the only material in which a very great increase of dimensions could safely be attempted. The very accident which befell the *Great Britain* upon the rocks in Dundrum Bay showed conclusively the skill he had then attained in the adaptation of iron to the purposes of ship-building. The means taken, under his immediate direction, to protect the vessel from the injury of winds and waves, attracted at the time much attention, and they proved successful, for the vessel was again floated, and is still afloat.

While noticing these great efforts to improve the art of ship-building at this date, it must not be forgotten that Mr. Brunel was the first man of eminence in his profession who perceived the capabilities of the screw as a propeller. From his experiments on a small scale in the *Archimedes*, he saw his way clearly to the introduction of that method of propulsion which he afterward adopted in the *Great Britain*. He next submitted it to the Admiralty, and succeeded in persuading the Board to give it a trial in her majesty's navy, under his direction. In the progress of this trial Brunel was much thwarted; but the *Rattler*, the ship which was at length placed at his disposal, and fitted with engines and screw by Messrs. Maudslay and Field, gave results which justified his expectations under somewhat adverse circumstances. She was the first screw ship which the British navy possessed, and her satisfactory performances led to numerous others being added.

The Bute Docks at Cardiff, and the North Dock at

Sunderland, were Brunel's work, as was also the elegant
Hungerford Suspension Bridge across the Thames.   To
his care, in 1850, was intrusted the Tuscan portion of the
Sardinian railway.   During the Russian war he was called upon to fit up the Renikoi hospitals on the Darda-
nelles : he laid on a special supply of water from the adjacent hills, and constructed short lines of railway, with
easy carriages, to facilitate the removal of the wounded
from the landing-place to the different wards.

We now approach Mr. Brunel's most stupendous work,
and which, without querulous remark, must be considered to have shortened his valuable life.   Prepared by experience and much personal devotion to the subject of
Steam Navigation by means of large ships, he, in the latter part of 1851, began to work out the idea he had long
entertained—that to make long voyages economically
and speedily by steam, required that the vessels should
be large enough to carry the coal for the entire voyage
outward, and, unless the facilities for obtaining coal were
very great at the outport, for the return voyage also;
and that vessels much larger than any then built could
be navigated with great advantages from the mere effects
of size.   Hence originated the *Great Eastern.*

The mere idea of a ship of a capacity six or eight times that of any
thing afloat had doubtless occurred to many an enthusiastic schemer,
for the sentiment of magnitude and immensity is innate in all in whose
character imagination is an element; but Mr. Brunel gave shape to
his idea by preparing plans and otherwise convincing himself of its
practicability.   As an example of naval construction, the *Great Eastern* is unquestionably the work of Mr. Scott Russell, every way as
much so as were the *Great Western* and *Great Britain* the works of
Mr. Patterson, of Bristol.   Yet Mr. Brunel's services were of hardly
less importance; and every one at all conversant with the organization of an establishment devoted to the construction of steam-vessels
is aware that the duties of the naval architect and builder, and those
of the engineer, are each clearly defined and in no way conflicting.
Certain it is that, whether the *Great Eastern* prove successful or otherwise, Mr. Brunel's name will be indelibly associated with her history
as long as that shall survive.—*The Engineer,* No. 195.

The success of this great work, in a practical point of
view, is admitted, as well as the strength and stability of
the construction of the vessel.   The difficulties attendant
on the launching of the ship in 1858 at one time seemed
insurmountable.   To a friend, who despondingly expressed his fears that the huge ship would never reach the

water, Brunel quietly replied, "Oh, she shall move—she must!" He never for a moment despaired of success. His health, however, had been undermined by these great exertions, and his death was hastened by the fatigue and mental strain caused by his efforts to superintend the completion of the great ship. We must not forbear to mention that for several years past Mr. Brunel had been suffering from ill health, brought on by over-exertion. Nevertheless, he allowed himself no relaxation from his professional labors; and it was during the period of bodily pain and weakness that his greatest difficulties were surmounted and some of his greatest works achieved.

By his death one more name has been added to the list of those who have been stricken down when their hopes were highest and victory within their grasp. By a coincidence, as it would appear, Mr. Brunel went on board the great ship for the last time on the first day when it could be said she was ready for sea. If not so in every detail, she was, as a whole, essentially completed, although still untried. On that day, the 5th of September, Mr. Brunel suffered an attack of paralysis, from which he never recovered. He sank until the evening of the 15th of September, when he passed away, still young, and upon the completion of the greatest work of his life.

Mr. Brunel died in his paternal house at Westminster. He was a fellow of the Royal Society, having been elected at the early age of twenty-six. In 1857 he was admitted by the University of Oxford to the honorary degree of Doctor of Civil Laws. He had filled the office of Vice-President of the Royal Society; he was Vice-President of the Institution of Civil Engineers, and of the Society of Arts; a Fellow of the Astronomical, Geographical, and Geological Societies, and a Chevalier of the Legion of Honor. But he had received no distinction from his own country—not even the knightly honor which his father bore. Yet he lived and died with a scientific reputation bounded only by the limits of the civilized world. He has left too many monuments of himself, raised both on land and sea, to permit of his being soon forgotten. It would be difficult to go far without finding something to recall the memory of Isambard Kingdom Brunel.

PHOTOGRAPHY AND THE STEREOSCOPE.

"With one touch virtuous
Th' arch-chemic sun, so far from us remote,
Produces."   MILTON's *Paradise Lost*, b. iii.

IF evidence were needed to show by what slow and gradual means the germs of great discoveries have been reared through a long lapse of years into full development, it might be found in the progress of Photography, since it has been the work of half a century of French, English, and German researches to suggest, apply, and finally develop the existence of the photographic element. The whole art, in all its varieties, rests upon the fact of the blackening effects of light upon certain substances, and chiefly upon silver, on which it acts with a decomposing power. The silver being dissolved in a strong acid, surfaces steeped in the solution become incrusted with minute particles of the metal, which in this state are darkened with increased rapidity. These facts were first ascertained and recorded, as regards silver combined with chlorine, in 1777, by Scheele, a native of Pomerania; and in 1801, in connection with nitrate of silver, by Ritter, of Jena. Here, therefore, were the raw materials for the unknown art. A very short time after Ritter's results, Dr. Wollaston made the same experiments, without having been informed what had been done on the Continent. This coincidence was, however, succeeded by the contemporary labors of three eminent experimenters. In conjunction with Mr. Thomas Wedgwood (the brother of Josiah), Sir Humphrey Davy, before June, 1802, succeeded, by means of a camera-obscura, in obtaining images upon paper, or white leather, prepared with nitrate of silver, by placing it behind a painting on glass exposed to the solar light, when the rays transmitted through the differently-painted surfaces produced distinct tints of brown and black, differing in intensity according to the shades of the picture; and, where the light was unaltered, the color became deepest. Thus was the first stain designedly traced upon the prepared substance. Mr. Wedgwood, by this method, took profiles or shadows of figures, and delineated the woody fibres of leaves, the delicate patterns of lace, and the beautiful wings of insects. But the charm, once set agoing, refused to stop; the slightest exposure to light continued the action, and the image was lost in the darkening of the whole paper. In short, there was wanting the next secret of *fixing* the images. The process seems, therefore, to have excited very little notice, and the experiment was left to be taken up by others, Sir Humphrey Davy prophetically observing, " Nothing but a method of preventing the unshaded parts of the delineation from being colored by the exposure to the day is wanted to render this process as useful as it is elegant."

The third worker then in the field was Dr. Thomas Young. In 1802, when Mr. Wedgwood was "making profiles by the agency of light," and Sir Humphrey Davy was " copying on prepared paper the images of small objects produced by means of the solar microscope," Dr. Young was taking photographs, upon paper dipped in a solution of nitrate of silver, of the colored rings observed by Newton; and his experiment clearly proved that the agent was not the luminous rays in the sun's light, but the invisible or chemical rays beyond the violet. This result is described in the Bakerian Lecture for 1803.

Meanwhile, in 1803, Dr. Wollaston proved the action of light upon gum guiacum, and in due time another experimenter entered the field, who availed himself of this class of materials. M. Nicephorus Niepce, a French gentleman of private fortune, who lived at Chalons-sur-Saone, and pursued chemistry for his pleasure—probably unacquainted with the labors of Davy and Wedgwood—like them, made use of the camera to cast his images; but the substance on which he received them was a polished plate of pewter, coated with a thin bituminous surface. He gained the important step of rendering his images permanent, which he was ten years in attaining, from 1814 to 1824. His pictures, on issuing from the camera, were invisible to the eye, and only disengaged by the application of a solvent, which removed those shaded parts unhardened by the action of the light. Nor did

T

they present the usual reversal of the position of light and shade known as a *negative* appearance, but, whether taken from nature or from an engraving, were identical in effect, or what is called *positive*. Nevertheless, Niepce's process was difficult, capricious, and tedious, and he never obtained an image from nature in less than from seven to twelve hours, so that the changes in lights and shadows necessarily rendered it imperfect. He therefore devoted his discovery mostly to copying engravings, and converted his plate, by means of an acid, into a surface for ordinary printing, as impressions still show.

Niepce seems to have obtained no definite results; but, foreseeing the value of his art, he went to England in 1827, and settled at Kew. He then drew up a short memorial, which he forwarded, with specimens, to the hands of George IV.; and at the close of the year Niepce submitted to the Royal Society a paper on his experiments, with several sketches on metal, of which communication the Society took no notice, it being their rule not to entertain a discovery which involved a secret. M. Niepce, therefore, returned to his own country, so chilled by the English indifference that, but for an accidental circumstance, he would not have proceeded farther. However, an optician having indiscreetly revealed to Niepce the secret that M. Daguerre,* the dioramic

---

* LOUIS JACQUES MAUDE DAGUERRE, who had long been known in England as one of the artists of the Dioramic Exhibition in the Regent's Park, died in Paris, January 10, 1851, in his sixty-second year. He was a member of the French Academy and of the Academy of St. Luke's; and many of his pictures are highly valued by his countrymen. It appears that the experiments which led to the discovery of the Daguerreotype were made by Daguerre while investigating the chemical changes produced by the solar radiations, in the hope of applying these phenomena to the production of peculiar effects in his dioramic paintings: such is the relationship of the Diorama and the Daguerreotype. The perfect illusion of the dioramic pictures has been thus explained: "When an object is viewed at so great a distance that the optic axes of both eyes are sensibly parallel when directed toward it, the perspective projections of it, seen by each eye separately, are similar; and the appearance to the two eyes is precisely the same as when the object is seen by one eye only. There is, in such case, no difference between the visual appearance of an object in relief and its perspective projection on a plane surface; hence pictorial representations of distant objects, when those circumstances which would prevent or disturb the illusion are carefully excluded, may be

artist, was pursuing researches analogous to his own at Paris, they entered into a copartnership in 1829. M. Niepce died in 1833, without having contributed any farther improvement to the now common stock; and M. Daguerre, taking into partnership Niepce's son, Isidore, discovered an essentially new process, which was named after its inventor, the *Daguerréotype*. By discarding the use of the bituminous varnish, and substituting a highly-polished plate of silver, he first availed himself of that great agent in photographic science, the action of iodine, by means of which he so increased the sensitiveness of his plate as to produce the image in fewer minutes than it had previously taken hours. At the same time, the invisible picture was brought to light by the fumes of mercury, after which a strong solution of common salt removed those portions of the surface which would otherwise have continued to darken, and would have rendered the impression permanent.

In 1839 the Daguerreotype came forth to the world. Men little thought how many years of patient research had been expended in arriving at this result. Daguerre and Niepce then applied to the French Chambers, stating that they possessed a secret which, if protected by patent, would be comparatively lost to society. A Commission was appointed by the French government, and the secret itself was intrusted to M. Arago, who succeeded at once in executing a beautiful specimen of the art. He then addressed the Chambers, urging the immense advantages which might have been derived, "for example, during the expedition to Egypt, by means of reproduction so exact and so rapid : to copy the millions and millions of hieroglyphics which entirely cover the great monuments at Thebes, Memphis, Carnac, etc., would require scores of years and legions of artists, whereas with the Daguerreotype a single man could suffice to bring this vast labor to a happy conclusion." M. Biot at the same time compared Daguerre's invention to the retina on the eye, the object being represented on one and the other surfaces with almost equal accuracy. The result

rendered such perfect resemblances of the objects they are intended to represent as to be mistaken for them. The Diorama is an instance of this."—PROFESSOR WHEATSTONE; *Philosophical Transactions*, 1838.

was, that a pension of 6000 francs (£250) was awarded to M. Daguerre, and 4000 francs (£166) to M. Niepce; and M. Arago declared that "France had adopted the discovery, and that from the first moment she had cherished a pride in liberally bestowing it *a gift to the whole world.*"* Nevertheless, the Daguerreotype was patented in England, which would have been thus restrained for eight years from the use of this important process; but the specification was afterward found defective, and the patent invalidated. All that has since been done for the Daguerreotype has not been any essential deviation from its process.

We now turn to England, where the undivided honor of having first successfully worked out the secret of Photography belongs to Mr. Fox Talbot, a private gentleman, who, in his delightful retreat at Lacock Abbey, in Wiltshire, pursued chemical researches for his own amusement. He took up the ground to which Davy and Wedgwood had made their way. Paper was the medium, which he made sensible to light by nitrate of silver, and then fixed the image by common salt. He first called his process *photogenic drawing;* then *calotype,* which his friends changed to *Talbotype,* in imitation of Daguerre's example. Mr. Fox Talbot is stated in the *Quarterly Review,* No. 202, to have sent his method to the Royal Society in the same month that Daguerre's discovery was made known (Jan., 1839); but Sir David Brewster dates Mr. Talbot's communication six months earlier.

As a new art, which gave employment to thousands, Mr. Talbot brought it to a high degree of perfection. He expended large sums of money in obtaining for the public the full benefit of his invention; and toward the termination of his patent he liberally surrendered to photographic amateurs and others all the rights which he possessed, with the one exception of taking portraits for sale, which he had conveyed to others, and which he was bound by law and in honor to secure to them. As Mr. Talbot had derived no pecuniary benefit from his patent, he had intended to apply for an extension of it to the Privy Council; but the art had been so universally practiced, that numerous parties interested in opposing the application combined

* Arago was so impressed with the vast importance of Photography in all its relations, that (Lord Brougham informs us) the last years of his life were chiefly occupied with whatever belonged to this subject.

with others to reduce the patent, and thus prevent the possibility of its renewal. Although we are confident that a jury of philosophers in any part of the world would have given a verdict in favor of Mr. Talbot's patent, taken as a whole, and so long unchallenged, yet we regret to say that an English judge and jury were found to deprive him of his right, and transfer it to the public. The patrons of science and of art stood aloof in the contest, and none of our scientific institutions, and no intelligent member of the government, came forward to claim from the state a national reward to Mr. Talbot. How different in France was the treatment of Niepce and Daguerre!

It is a curious fact, that Daguerre's patent for the sister art of the Daguerreotype was also invalidated by an English jury; "and," says Sir David Brewster, "it will never be forgotten in the history of art that the rights of property over the two noblest inventions of the age, which the patent laws were enacted to secure, were wrested from their owners by the unjust decision of an English jury, prompted by the selfish interests of individuals who had been fattening on the genius of the inventors."—*Encyclopædia Britannica*, 8th edit.

Next, in April, 1839, the Rev. J. B. Reade delineated objects of natural history by the agency of light, from their images taken by the solar microscope.

One of the earliest attempts in Paris was thus described: "A public experiment of the Daguerreotype was made by its inventor on Saturday last, in one of the halls of the hotel of the Quai d'Orsay. M. Daguerre described the mode of using his instrument to an assembly of about a hundred and twenty persons, and, in the course of an hour and a few minutes, produced a beautiful view of the river, the terrace, and the palace of the Tuileries." In 1839, however, the process at Paris occupied but from three to thirty minutes, and Daguerre was able to use the apparatus in the public streets without being noticed by the passengers. Still, the disappointment in the early plates was costly and mortifying, and reminded one of Uncle Toby's "here to-day and gone to-morrow." Many a plate for which ten guineas were paid disappeared in a corresponding number of days.

The first experiment made in England with the Daguerreotype was exhibited by M. St. Croix, on Friday, September 13, 1839, at No. 7 Piccadilly, nearly opposite the southern Circus of Regent Street, when the picture produced was a beautiful miniature representation of the houses, pathway, sky, etc., resembling an exquisite mezzotint. M. St. Croix subsequently removed to the Argyll Rooms, Regent Street, where his experimental results became a scientific exhibition. The discovery was patented by Mr. Miles Berry, who sold the first license to M. Claudet for £100 or £200 a year; and in twelve months after disposed of the patent to Mr. Beard, who, however, did not take a Daguerreotype *portrait* until after Dr. Draper had sent from New York a portrait to the editor of the *Philosophical Magazine*, with a paper on the subject.

The Talbotype process underwent various improvements by Herschel, Cundell, Bingham, Channing, Le Gray,

Martin, Müller, Stewart, Hunt, Fyfe, Furlong, Blanquart, Everard, Collen, Ryan, Woods, Horne, Saguer, Flacheron, and others; but the most important improvements were made by M. Victor Niepce and Mr. Scott Archer, the former substituting albumen, and the latter collodion, for paper. The albumen process can only be employed for statues and landscapes, and with it have been produced larger and more artistic pictures than by any other means. Mr. Archer generously threw his marvelous improvement open to the public. The birth and parentage of collodion are both among the recent wonders of the age. Gun-cotton is but a child in the annals of chemical science; and collodion, which is a solution of this compound in ether and alcohol, is its offspring: its first use was in surgery, its second in photography. Collodion may also be prepared from paper, flax, the pith of the elder, and many other vegetable substances. Not only does it provide a film of perfect transparency, tenuity, and intense adhesiveness; not only is it easy of manipulation, portable, and preservable, but it supplies that element of rapidity which, more than any thing else, has given the miraculous character to the art.

The *instantaneous process* of taking a picture on collodion in half a second has enabled the artist to delineate "a thoroughfare in London with its noon-day crowd." Farther than this the powers of Photography can never go: light is made to portray with a celerity only second to that with which it travels!

We have not space to do more than state that Mr. Norton's important application of bichromate of potash has led M. E. Becquerel to his photographic paper, with iodide of starch; Mr. Hunt to his chromatype; and the photographic property of this salt is also the foundation of M. Pretsch's photo-galvanography, and of some attempts at photo-lithography. Mr. Talbot, in 1841, patented a more sensitive photographic method; and subsequently an instantaneous process, photographic engravings, and the phoglyphic process.

Meanwhile, Sir John Herschel and Mr. Hunt found preparations of gold, platinum, mercury, iron, copper, tin, nickel, manganese, lead, potash, etc., more or less sensitive, and capable of producing pictures of beauty and distinct-ive character; and paper prepared with the juices of beautiful flowers was put in requisition.

Photography may be said to have depended for its perfection upon wonders only a little older than itself. Iodine, on which all popular photography rests, was not discovered until 1811; and bromine, the only other equally sensitive substance, not till 1826; and gun-cotton and chloroform only just preceded collodion. To these may be added the optical improvements purposely contrived or adapted for the service of the photograph, besides innumerable other mechanical aids. The value of photography, when kept perfectly distinct, as an auxiliary to the artist, is also unquestionably great, though only beginning to be duly and correctly appreciated.*

Although M. Biot, in 1840, considered it as an illusion to expect photographs to have *the color of the objects which they represent*, yet an important advance has been made to this result by M. Claudet and Sir John Herschel in copying the colors of nature. Mr. Hunt " produced colored images, not merely impressions of the rays of the spectrum, but copies in the camera of colored objects." But the most striking results have been obtained by M. Edmund Becquerel and M. Niepce St. Victor; the latter is said to have secured " all the colors of a picture by preparing a bath composed of the deuto-chloride of copper."

The most important application of Photography has certainly been to the Stereoscope, not only in reference to art, but to the great purposes of education, and to the illustration of works on every branch of knowledge. But perhaps one of the most curious applications of the art has been to Microscopic Portraits, by Mr. Dancer, of Manchester. Some of these are so small that ten thousand could be included in a square inch; and yet, when magnified, the pictures have all the smoothness and vigor of ordinary photographs.

Lord Brougham observes: "How vast an improvement of social life, and how valuable an addition to our power of executing the law, has been this optical discovery, by which we have made the sun our fellow-workman! It would have been deemed a romance had any one foretold, from observing the effect of light in discoloring certain

* See *Painting Popularly Explained*, p. 114–119.

substances, such a consummation as obtaining the most accurate portraits in a second; and the consequent power, not only of preserving the features of those most revered and beloved, but of preventing the escape of criminals, the commission of numberless frauds, and the defeat of the injured in seeking the recovery of their rights. In the sciences of astronomy, zoology, geology, meteorology, ethnology, electricity, and magnetism, Photography has been advantageously employed. The spots on the sun, the surface of the moon, the forms of the planets, and even groups of stars, have been delineated by their own light. M. de la Rue has obtained pictures of the moon analogous to binocular axes, which, when aided by the Stereoscope, exhibit her as a solid globe. The meteorologist registers photographically, in his absence, the indications of the barometer, thermometer, and hygrometer; the variations of the earth's magnetism are recorded every minute on chemically prepared paper: and the electricity of the atmosphere, brought down into the observatory, is made to exhibit on paper the number of its variations and the intensity of its action. The ethnologist has begun to collect accurate pictures of the different races of man. The zoologist has obtained forms of animal life which the painter had attempted in vain to preserve. The geologist has obtained delineations of phenomena which defied the highest efforts of the pencil. And the botanist has transferred to imperishable tablets those beautiful and complex forms of vegetable life which we seek in vain in the richest botanical collections."—SIR DAVID BREWSTER ; *Encyclopædia Britannica.*

Within a score of years from the first experiment exhibited by the Stereoscope, it has been advanced from a rude and imperfect apparatus to " one of the most popular and interesting instruments which science has presented to the arts.". It is employed for representing solid figures, by combining in one image two plane representations of the object as seen by each eye separately; or, in other words, two pictures of any object, taken from different points of view, are seen as a single picture of that object, having the actual appearance of relief or solidity. Hence the name, from two Greek words signifying *Solid I view.*

That we see with two eyes, yet that only a single representation of the object is presented to the mind, and that the picture of bodies seen by both eyes is formed by the union of dissimilar pictures formed by each, must have been very early observed, and the cause was speculated on by the earliest Greek philosophers. Euclid knew these palpable truths more than two thousand years ago, and showed by means of a sphere that each eye sees a dissimilar representation of an object. Five centuries later, Galen endeavored to explain the matter by stating that the dissimilar pictures are not seen at the same instant, but successively; and that these rapidly-succeeding pictures produce on the mind the impression which is conceived of the object. In looking at the diagram given by Galen, we recognize at once not only the principle, but the construction of the Stereoscope.*

As the vision of the object was obtained by the union of these dissimilar pictures, an instrument only was wanted to take such pictures, and another to combine them. "The Binocular Photographic Camera," says Sir David Brewster, "was the one, and the Stereoscope the other."

Baptista Porta repeats the proposition of Euclid on the vision of a sphere with one and both eyes, but, believing that we only see with one eye at a time, he denies the accuracy of Euclid's theorems; and while he admits the correctness of Galen's views, he endeavors to explain them upon other principles. The Greek physician, therefore (Galen), and the Neapolitan philosopher (Porta), who has employed a more distinct diagram, certainly knew and adopted the fundamental principle of the Stereoscope, and nothing more was required for producing pictures in full relief than a simple instrument for uniting the right and left hand dissimilar pictures.

We next find, in the treatise on Painting which Leonardo da Vinci left behind him in manuscript, a distinct reference to the dissimilarity of the pictures seen by each eye as the reason why "a painting, though conducted with the greatest art, and finished to the last perfection with regard to its contours, its lights, its shadows, and its colors, can never show a *relievo* equal to that of the natural objects, unless these be viewed at a distance, and with a single eye," which he proceeds to demonstrate. Aguilonius, the learned Jesuit, who published his *Optics* in 1613, next attempts to explain, but without success, why the two dissimilar pictures of a solid seen by each eye do not, when united, give a confused and imperfect view of it; but, down to our time, natural philosophers

---

* *The Stereoscope; its History, Theory, and Construction.* By Sir David Brewster. 1856.

T 2

have been almost universally content to adopt the opinion that we see with only one eye at a time.

Thus the matter rested until, in 1838, Mr. Wheatstone reopened the question of vision by one or by two eyes by arguing that the appearance of relief and solidity which we obtain in looking at objects in nature arises from there being a dissimilar picture of the object projected simultaneously on the retina of each eye, the optic axes of which are not parallel, whereas in viewing a pictorial representation two similar pictures are projected on the retinæ, and hence the resultant flatness; and Mr. Wheatstone sought to illustrate this theory by the ingenious instrument known as the Stereoscope. Its principle has been thus simplified by Mr. R. Hunt, F.R.S.:

When we look at any round object, first with one eye, and then with the other, we discover that with the right eye we see most of the right-hand side of the object, and with the left eye most of the left-hand side. These two images are combined, and we see an object which we know to be round.

This is illustrated by the *Stereoscope*, which consists of two mirrors placed each at an angle of 45 degrees, or of two semi-lenses turned with their curved sides toward each other. To view its phenomena, two pictures are obtained by the camera on photographic paper of any object in two positions, corresponding with the conditions of viewing it with the two eyes. By the mirrors or the lenses these dissimilar pictures are combined within the eye, and the vision of an actually solid object is produced from the pictures represented on a plane surface.

The Stereoscope excited considerable interest among scientific persons when first exhibited; the pictures prepared for it were almost exclusively dissimilar outlines of various geometrical solids; but it has been almost superseded by the Refracting Stereoscope, in which the simple principle of the Stereoscope is combined with, or rather aided by, photography. This principle might have been discovered a century ago, for the reasoning which led to it was independent of all the properties of light: but it would never have been illustrated, far less multiplied as it now is, without photography. A few diagrams, of sufficient identity and difference to prove the truth of the principle, might have been constructed by hand for the gratification of a few sages; but no artist, it is to be hoped, could have been found possessing the requisite ability and stupidity to execute two portraits, or two groups, or two interiors, or two landscapes, identical in the most elaborate detail, and yet differing in point of view by the inch between the two human eyes, by which the principle is brought to the level of any capacity. Here, therefore, the accuracy and insensibility of a machine could alone avail; and if in the order of things the cheap popular toy which the Stereoscope now represents was necessary for the use of man, *the photograph was first necessary for the service of the Stereoscope.**

Sir David Brewster, in a series of elaborate experiments to establish his theory of binocular vision, as distinguished from that of Professor Wheatstone, invented the *Lenticular Stereoscope,* which he has fully illustrated in his able volume on the Stereoscope. It consists of a pyramidal box, blackened on the inside, and having a lid for the admission of light when the pictures are opaque. The box is open below, in order to let the light pass through the pictures when they are transparent. The top of the box consists of two portions, in one of which is the right-eye tube, containing a semi-lens or quarter-lens, and in the other the left-eye tube, also containing a semi-lens or quarter-lens. The two dissimilar pictures (or slide) are placed in a groove in the bottom of the box, when, on looking through the eye-tubes, the pictures are seen united into one single picture; and the object or objects, if a proper amount of light is obtained, stand out with an almost magical appearance of relief and solidity. Thus has the employment of photography for the stereographs wonderfully extended the range of the instrument, and rendered it one of the most popular means of social amusement, and, rightly used, an extremely valuable means of instruction. We have said that each of the eye-pieces contains a semi-lens: it is by means of these semi-lenses, transferring the two dissimilar pictures or stereographs to a middle point, and their union thereon, that the stereoscopic effect is produced.

A *detective* application, similar to that of photographic portraits, has been devised for the Stereoscope. In 1859, it was ascertained by experiment that if two thoroughly

* *Quarterly Review,* No. 202.

identical copies of ordinary print be placed side by side in the Stereoscope, they will not offer any unusual appearance. But if there be the slightest, although inappreciable, difference—as, for instance, in the interval separating the same words—the difference will be made evident in the stereoscope by the elevation into relief (or the reverse) of the corresponding space above the adjoining parts. Professor Dove, of Berlin, proposes the above as an infallible means of distinguishing a forged bank-note from a genuine one, etc.

SIPPONIA ELASTICA.

FICUS ELASTICA.

CAOUTCHOUC AND ITS MANUFACTURES.

THE remarkable substance known as Caoutchouc is produced by many different plants, and its manifold applications within comparatively few years are certainly one of the marvels of our scientific age. "How curious, how wonderful," says an acute writer,* "is it to find a milky juice which exudes from certain trees on the banks of the Amazon, or from vines in the jungles of India, transformed by the ingenuity of man, on the banks of the Thames or the Irwell, into such a vast variety of useful and interesting objects! But it is still more curious and still more wonderful to reflect that this milky juice, with the many uses to which it is put, forms a necessary part of the progress of civilization, and tends to unite all the human race into one great and glorious family."

Caoutchouc was first introduced into Europe early in the last century; but its origin was unknown till the visit of the French Academicians to South America in 1715. They ascertained that it was the inspissated juice of a Brazilian tree, called by the natives Hhvé; and an account of the discovery was sent to the Academy by M. de la Condamine in 1736. One-and-thirty years later, 1767, a specimen was first brought to England, and was sent to Mr. Canton by Sir Joseph Banks as "two balls of the new Elastic Substance." In 1772, Dr. Priestley thus speaks of the new substance in his *Introduction to Perspective:* "I have seen a substance excellently adapted to the purpose of wiping from paper the marks of a black-lead pencil. It must therefore be of singular use to those who practice drawing. It is sold by Mr. Nairne, mathematical-instrument maker, opposite the Royal Exchange. He sells a cubical piece of about half an inch for three shillings, and he says it will last several years."

From this first application arose the name of *India-rubber*. The "new substance" engaged, as soon as it was

* Mr. Thomas Hodgskin.

known, the "attention of philosophers." They immersed
it in all kinds of solvents, tried its influence on sounds,
found in it a confirmation of the celebrated theory of
latent heat, ascertained its elements according to the then
knowledge of the elements; but they made nothing of
it. For more than 120 years they had it in their hands
and in their laboratories, thought it a wonderful substance, which might be converted to all kinds of uses,
but got no farther than to ascertain that by boiling it in
water its edges became soft, and that pieces of it then
pressed together could be united into one homogeneous
whole, which led to the formation of flexible tubes and a
few surgical instruments.

About the year 1820, however, Mr. Thomas Hancock,
afterward of the firm of Mackintosh and Co., being engaged in mechanical pursuits, began to take great interest in Caoutchouc. He wondered that such a curious
substance should have been put to little or no other use
than rubbing out pencil-marks; his wonder excited his
exertions; chemical knowledge he had none, and trying,
like the chemists, to find out a solvent, he failed. Then,
taking a more simple means, he cut Caoutchouc into narrow slips, inclosing them in a case of thin leather or cotton; and elastic springs for gloves, braces, etc.—that before were formed only of metal wire in a spiral form—
were made of this substance. This was the original new
application, in 1820, of Caoutchouc. Mr. Hancock followed up his success. He was always at work with his
rubber. He cut it into shreds; he rent it into pieces;
he invented machines for chewing it and pounding it
into a mass; he stewed it in digesters; he baked it; he
made it into solid blocks; he spread it into sheets almost
as thin as the finest textures of the animal frame: he
found one solvent for it, which had before been frequently tried, but only under the new mechanical form which
he gave it did oil of turpentine (camphine) answer the
purpose. Other persons found other solvents. From
1820 the new applications of this curious substance were
numerous and successive—in other countries, especially
in America, as well as here.

Mr. Hancock has been truly called the "father of this
important and wonderfully-increasing branch of the arts;"

but it had many nurses. In 1823 Mr. Macintosh applied the naphtha obtained from coal-tar to dissolve rubber, thus making a water-proof varnish; he invented and brought into use the garments and the cloth which bear his name.

The manufacture of Caoutchouc has three principal branches: 1. The condensation of the crude lumps or shreds of Caoutchouc, as imported, into compact homogeneous blocks, and the cutting of these blocks into cakes or shreds, for the stationer, surgeon, shoemaker, etc. 2. The filature of either the India-rubber bottles, or the artificial sheet Caoutchouc, into tapes and threads, which, being clothed with silk, cotton, linen, or woolen yarns, form the basis of elastic tissues of every kind. 3. The conversion of the refuse cuttings and coarser qualities of Caoutchouc into a viscid varnish, which, being applied between two surfaces of cloth, constitutes the well-known double fabrics, impervious to water and air.

It is curious to read that this application of Caoutchouc to water-proofing was known in South America upward of a century since. In a work entitled *La Monarchia Indiana*, printed at Madrid in 1723, we find described "very profitable trees in New Spain, from which there distill various liquors and resins." Among them is described a tree called *ulquahuill*, which the natives cut with a hatchet, to obtain the white, thick, and adhesive milk. This, when coagulated, they made into balls, called *ulli*, which rebounded very high when struck to the ground, and were used in various games. The author continues: "Our people (the Spaniards) make use of their *ulli* to varnish their *cloaks*, made of hempen cloth, *for wet weather;* which are good to resist water, but not against the sun, by whose heat and rays the *ulli* is dissolved." India-rubber is not known in Mexico at the present day by any other name than that of *ulli;* and the oiled-silk covering of hats very generally worn throughout the country by travelers is always called *ulli*. Shoes (worn in some countries as over-shoes) have also long been made of Caoutchouc in its native country. This is done by dipping the wooden lasts in the Caoutchouc milk, and then drying them over the smoke of a fire made with palm-nut. The coatings are repeated until the shoes are sufficiently thick, a greater number being given to the bottom or sole.

The grand improvement in the texture and qualities of the substance, by which its applicability to different purposes has been greatly enlarged, is called *vulcanizing*, and was not made till 1843, and seems then to have been brought about by something like an accident. In 1842, Mr. Hancock was shown small bits of Caoutchouc, which an American agent said would not *stiffen* by cold, and were not much affected by solvents, heat, or oil. To give Caoutchouc the property of remaining flexible under all circumstances and changes was most desirable. Mr. Hancock was again set wondering, or was stimulated by the assertion; the small bits of Caoutchouc so changed smelt of sulphur. He made all kinds of experiments in the direction thus indicated, and at length ascertained that the desired alteration was effected in the Caoutchouc by exposing it to the action of sulphur at a high temperature. "Had I known," he says, after he had ascertained the fact, "the simple mode by which this result could be produced, I might have made the discovery at once."

Caoutchouc, thus acted on by sulphur, retains its perfect elasticity in all temperatures, and, vulcanized under pressure, can be made in all forms hard and durable. It can be turned in a lathe, and cut into screws. It has been made into flutes, which sound easily and sweetly, and are so polished as to resemble ebony. Of it are made walking-sticks and picture-frames, and delicate mountings of all descriptions. A collection of beautifully made articles of this class can be seen in the "Vulcanite Court," at the Crystal Palace, Sydenham. It is converted into whips, hard, like wood, at the handle, and flexible, like the finest kind of leather, at the thong. It has some most remarkable properties. A ball will pass through it; and the hole closes so completely that persons who have tried the experiment would not believe the fact till it was demonstrated by the ball striking objects beyond the rubber. A piece two inches thick and a foot square was laid on an anvil under Mr. Nasmyth's steam-hammer; a six-inch round shot was placed on the rubber; the hammer was then made to fall on the shot with tremendous force, which was broken to pieces, while the rubber on which it was laid remained as elastic and uninjured as when it was placed on the anvil; nay, more extraordinary still, the shot had come into contact with the anvil, and was flattened slightly, but the rubber had retained, or immediately resumed, its original form and condition.

When Mr. Hancock showed the first piece of his "solid rubber" to an old gentleman, it was returned with the prescient remark, "the child is yet unborn who will see the end of that."[*] Ever since, the trade and the manufacture have been progressive here and in every other part of the civilized world. Within the memory of this generation—in less than forty years—an entirely new art has grown up from India-rubber bottles, and it is forever increasing. It is by no means the only art which has come into existence in the time, and attained an astonishing perfection. Moreover, all these new arts —the manufacture of rubber, photography, railways, telegraphs, etc.—are already common to all the civilized world.

The great consumption of Caoutchouc has naturally led to its being sought in other regions than that in which it was first found. It was at first principally imported from Para; but considerable quantities have since been brought from Java, Penang, Singapore, and Assam. In the latter country it has been obtained from trees in vast forests 100 feet high and 74 feet in girth.

[*] *Personal Narrative of the Origin and Progress of the Caoutchouc or India-Rubber Manufacture, etc.* By Thomas Hancock.

Note.—There is dispute as to the discovery of the processes by which these difficulties were surmounted. On one side, Mr. Charles Goodyear is said to have labored for five years in the research of the secret, and at last to have discovered it by accidentally placing some pieces of rubber against a hot stove, and noticing that they charred instead of melting. This discovery, combined with Mr. Hayward's previous adaptation of sulphur as a dryer of rubber, is said by Mr. Goodyear's friends to have led him to invent the vulcanizing process. Mr. Goodyear claims to have made his discovery in 1839; but the first authentic evidence of the fact is the patent obtained in 1844. Of course, we do not pretend to adjudicate a dispute which has exercised so many lawyers' wits. It is curious, however, that both discoverers should have hit upon the same name for their discovery— vulcanization.—*Am. Ed.*

GUTTA PERCHA AND ITS MANUFACTURES.

THIS wonderful substance appears to have been brought for the first time into England in the days of Tradescant, "King's Gardener" to Charles I.; and it is believed to have been shown in Tradescant's Museum, at South Lambeth, as a curious product, under the name of Mazerwood, of which bowls and goblets were formerly made. Subsequently it was often brought from China and other parts of the East, in the form of elastic whips, sticks, etc. The specimens of two centuries since probably lay in Tradescant's Museum neglected, and the knowledge of its importance and value in the arts seems to have been reserved for the age of the Electric Telegraph, since the use of this substance for inclosing its metallic wires entitles it to a share in the success of the Submarine Telegraph, by means of which the great cities of the world are now brought within a few minutes of each other.

The reappearance of Gutta Percha in our times resembles a rediscovery. It is obtained from the Isonandra Gutta plant, of the order *Sapotaceæ*, and was found by Mr. Thomas Lobb while on a botanical mission in Singapore and the Malay peninsula, where forests of the Percha trees grow to an enormous size, this discovery being made more than three centuries after the country had been frequented by Europeans. Early in 1843, Dr. William Montgomerie, in a letter to the Bengal Medical Board, commends Gutta Percha as likely to prove useful for some surgical purposes; and in the same year he transmitted to the Society of Arts in London a specimen of the Gutta Percha, at one of their evening meetings: the Society then simply acknowledged the receipt of the gift, but subsequently presented to Dr. Montgomerie their gold medal. It was ascertained from Sir James Brook, the Resident at Sarawak, that the tree is indigenous to that place, and is known to the natives by the name of Niato; and the doctor's curiosity was first aroused by noticing the handle of a chopper in the hands of a Malay woodman made of this novel material, which he found could be moulded into any form by immersing it into boiling water until it was thoroughly heated, when it became plastic as clay, and regained when cold its original hardness and rigidity. In its native country it is commonly used for whips, and it was by the introduction of a horsewhip made of this substance that its existence was made known in Europe. Specimens shown in the Great Exhibition of 1851 proved that the Malays knew also how to appropriate Gutta Percha to the manufacture of vases, and that European industry had little more to do than to imitate their processes. The first articles manufactured of it in England, in 1844, were a lathe-band, a short length of pipe, and a bottle-case, which had been made by hand, the concrete substance being rendered sufficiently plastic by immersion in hot water; casts from medals were also early taken with it. Mr. Francis Whishaw thus early discovered the valuable property which Gutta Percha possesses for the conveyance of sound, and accordingly made of it the Telakouphanon, or Speaking Trumpet, through which, by simply whispering, the voice could be audibly conducted for a distance of three quarters of a mile, and a conversation by this means kept up. Another of its early applications was as the soles of shoes. In its pure state, Gutta Percha is indestructible by water, and is an excellent non-conductor of electricity; hence it was used in making a tube for the conveyance of the wires of the submarine telegraph, and was first so employed across the Hudson River, New York.

Gutta Percha is, like Caoutchouc, a carburet of hydrogen, and isomeric with that substance; and while it possesses a great number of the properties which characterize Caoutchouc, it also exhibits certain special properties which admit of its being applied to particular uses to which Caoutchouc is not adapted.

In 1845, only 20,000 lbs. of Gutta Percha were imported into England; now the consumption has increased to millions of pounds annually. Its manufacture into an endless variety of articles demands new processes, new machines, and new tools, in which the steam-engine plays the most important part. The rough blocks of gum are first cut into slices by a vertical wheel, faced with knives or blades, and revolving 200 times a minute; the slices are then cleaned from stones and other impurities, and boiled in waste steam from the engine. The mass is next put into an iron box, or teaser, in which an iron cylinder with teeth rapidly revolves, and tears it into shreds, throwing it into vats of cold water. There the Gutta Percha floats at the top, and the impurities sink to the bottom. It is then transferred to tanks of boiling water, and thence removed into boxes, and kneaded like dough; and next rolled between heated iron cylinders into sheets, which are then cooled by passing between steel rollers. The sheets are cut by a knife-edged machine into bands or strips. For making tubes and pipes, the soft mass of kneaded Gutta Percha is passed through heated iron cylinders, and is drawn by the drawing-mill into cylindrical cords, and tubes of various diameters. This, however, is but a glimpse of the complicated machinery and processes by which Gutta Percha is fashioned into a legion of articles. Among the applications are breast-coating for water-wheels, galvanic batteries, shuttle-beds for looms, packing for steam-engines and pumps, cricket-balls, noiseless curtain-rings, whips and sticks, policemen's staves, plugs or solid masses used in buildings, buffers for railway-carriages, gunpowder canisters, sheet-covering for damp walls, lining for ladies' bonnets, jar-covers, bobbins for spinning machines, book-covers, moulds for stereotype and electrotype, coffin-linings, and stopping for hollow teeth. These are but a small number of the myriads of uses to which we have extended the application of the vegetable product which was used by the Malays ages since for a few common purposes.

It may be interesting to add that both Gutta Percha and Caoutchouc plants may be seen growing in the Royal Gardens of Kew, and cases of articles made of the two substances are shown there in the Museum of Economic Botany.

In estimating the various aids and appliances to the success of the Submarine Telegraph, it is scarcely possible to overrate the properties of Gutta Percha. It would seem as though one were sent to perfect the other; for the coating of the telegraph-wire with Gutta Percha, thereby insuring its entire insulation, is a most important provision.

The employment of Gutta Percha in electrical experiments was first noticed by Faraday in 1848, who stated its use to depend upon the high insulating power which it possesses under ordinary conditions, and the manner in which it keeps this power in states of the atmosphere which make the surface of glass a good conductor. The telegraph-wire is not only coated with Gutta Percha, but is closed in tubing made of it. For this purpose the Gutta is dissolved in bisulphuret of carbon; the wire is passed over pulleys through the solution, and then through a tube lined with brushes, which remove any thing superfluous; and when the wire reaches the second pulley, the bisulphuret has evaporated, and left a thin coating of Gutta Percha. Where the wire is to be roughly handled, it is covered with cotton, and then passed through the solution; but the tubing is still more effective. Great feats of dispatch have been accomplished in this application. One day, in 1849, a coil of copper wire 12,200 feet long was coated at the Company's works in the City Road with sulphureted Gutta Percha, and shipped for the Russian government, within twenty-four hours of its arrival at the works.

THE ELECTRIC TELEGRAPH.

THE great secret of *instantaneous transmission* has long exercised the ingenuity of mankind in various romantic myths; and the discovery of certain properties of the loadstone gave a new direction to those fancies, the majority of which can scarcely be traced. Many of the ancient stories of ubiquity which we find related as facts are doubtless of this fabulous origin; and in the present instance, credulity being, as it were, backed by science, there was some method in the popular belief. To such a source may be traced in modern times the earliest anticipation of the Electric Telegraph, the marvel of the science of the present age; the discoveries in which, and their application to useful ends almost as soon as made, give this science a peculiar interest. The anticipation to which we have just·referred occurs in the *Prolusiones* of the learned Italian Jesuit Strada in 1617, who supposes the existence of "a species of loadstone which possesses such virtue that if two needles be touched with it, and then balanced on separate pivots, and the one be turned in a particular direction, the other will sympathetically move parallel to it." He then directs each of these needles to be poised and mounted parallel on a dial having the letters of the alphabet arranged round it. Accordingly, if one person has one of the dials and another the other, by a little prearrangement as to details, a correspondence can be maintained between them at any distance by simply pointing the needles to the letters of the required words. Strada, in his poetical reverie, dreamed·that some such sympathy might one day be found to exist in the magnet; but his conceit does not seem to have caught Bishop Wilkins, who, in his book on Cryptology, strangely fears lest his readers should mistake Strada's fancy for fact, it being altogether imaginary, having no foundation in any real experiment.

Addison, in the 241st number of the *Spectator*, 1712, describes Strada's "Chimerical correspondence;" and adds that, "if ever this invention should be revived or put in practice," he "would propose that upon the lover's dial-plate there should be written not only the four-and-twenty letters, but several entire words which have always a place in passionate epistles, as flames, darts, die, language, absence, Cupid, heart, eyes, being, drown, and the like. This would very much abridge the lover's pains in this way of writing a letter, as it would enable him to express the most. useful and significant words with a single touch of the needle."

When electricians had become acquainted with the new force by friction, then the only known method of generating electricity, they renewed their experiments. In 1729, one Stephen Gray, a pensioner of the Charter House, made electrical signals through a wire 765 feet long; yet, in those dull times, this success did not excite much attention. Next, Le Monnier's account of his feeling the electric shock through an acre of water at Paris by means of an iron chain, led Dr. Watson, and other Fellows of the Royal Society, in 1745, to make a series of experiments to ascertain how far electricity could be conveyed by means of conductors.

They caused the shock to pass across the Thames at Westminster Bridge, the circuit being completed by making use of the river for one part of the chain of communication. One end of the wire communicated with the coating of a charged phial, the other being held by the observer, who in his other hand held an iron rod, which he dipped into the river. On the opposite side of the river stood a gentleman, who likewise dipped an iron rod in the river with one hand, and in the other held a wire, the extremity of which might be brought into contact with the wire of the phial. Upon making the discharge, the shock was felt simultaneously by both the observers.—PRIESTLEY'S *History of Electricity.*

In 1747, the same persons made experiments near Shooter's Hill, when the wires formed a circuit of four miles, and conveyed the shock with equal facility; "a distance which, without trial," they observed, "was too great to be credited." These results established two great principles: 1, that the electric current is transmissible along nearly two miles and a half of iron wire; 2, that the electric current may be completed by burying the poles in the earth at the above distance. These experiments were performed at the expense of the Royal Society, and cost £10 5s. 6d. In the paper detailing them, printed in the 45th volume of the *Philosophical Transactions*, occurs the first mention of Dr. Franklin's

U

name, and of his theory of positive and negative electricity.

In the following year, 1748, Benjamin Franklin performed his celebrated experiments on the banks of the Schuylkill, near Philadelphia, which being interrupted by the hot weather, they were concluded by a picnic, when spirits were fired by an electric spark sent through a wire in the river, and a turkey was killed by the electric shock, and roasted by the electric jack before a fire kindled by the electrified bottle. In two years Franklin made his more celebrated experiment to determine the identity of Lightning and Electricity, as described at p. 336, 337.

In the year 1753 there appeared in the *Scots' Magazine* definite proposals for the construction of an electric telegraph requiring as many conducting wires as there are letters in the alphabet; it was also proposed to converse by chimes, by substituting bells for the balls. A similar system of telegraphing was next invented by Joseph Bozolus, a Jesuit, at Rome, and mentioned by the great Italian electrician Tiberius Cavallo, in his treatise on Electricity.

In 1787, Arthur Young, when traveling in France, saw a model working telegraph by M. Lomond: "You write two or three words on a paper," says Young; "he takes it with him into a room, and turns a machine inclosed in a cylindrical case, at the top of which is an electrometer, a small, fine pith-ball; a wire connects with a similar cylinder and electrometer in a distant apartment; and his wife, by remarking the corresponding motions of the ball, writes down the words they indicate, from which it appears that he has formed an alphabet of motions. As the length of the wire makes no difference in the effect, a correspondence might be carried on at any distance."

On January 31, 1793, Volta announced to the Royal Society his discovery of the development of electricity in metallic bodies. Galvani had given the name of Animal Electricity to the power which caused spontaneous convulsions in the limbs of frogs when the divided nerves were connected by a metallic wire. Volta, however, saw the true cause of the phenomena described by Galvani, which have passed under his name as Galvanism by an

error similar to that which gave the name of Amerigo
Vespucci, instead of Columbus, as the discoverer of the
New World. Observing that the effects were far greater
when the connecting medium consisted of two different
kinds of metal, Volta inferred that the principle of ex-
citation existed in the metals, and not in the nerves of
the animal; and he assumed that the exciting fluid was
ordinary electricity, produced by the contact of the two
metals. The convulsions of the frog consequently arose
from the electricity thus developed passing along its
nerves and muscles. Hence the term Voltaic Electricity.

The following year, according to *Voigt's Magazine*,
Reizen made use of the electric spark for the telegraph;
and in 1798, Dr. Salva, of Madrid, constructed a similar
telegraph, which the Prince of Peace exhibited to the
King of Spain with great success.

In 1802 it was discovered that the earth might be substituted for the return wire of a voltaic circuit.

In 1809, Soemmering exhibited to the Academy of
Sciences at Munich an electro-telegraphic apparatus, in
which the mode of signaling consisted in the development of gas-bubbles from the decomposition of water
placed in a series of glass tubes, each of which denoted a
letter of the alphabet. In 1813, Mr. Hill, of Alfreton, in
Hampshire, devised a *voltaic* electric telegraph, which he
exhibited to the Lords of the Admiralty, who spoke ap-
provingly of it, but declined to carry it into effect. And
in the following year Soemmering constructed a similar
telegraph, but with this inconvenience—that there were
as many wires as signs or letters of the alphabet.

The next invention is of much greater practical worth.
Upon the suggestion of Cavallo, Francis Ronalds constructed a perfect electric telegraph, employing frictional
electricity, although Volta's discoveries had been known
in England for sixteen years. This telegraph was exhibited at Hammersmith in 1816, the very year in which
Andrew Crosse, the electrician, said, " I prophesy that
by means of the electric agency we shall be enabled to
communicate our thoughts instantaneously with the
uttermost parts of the earth." Ronald's telegraph consisted of a single insulated wire, the indication being by
pith-balls in front of a dial: when the wire was charged

the balls were divergent, but collapsed when the wire was discharged; at the same time were employed two clocks with lettered disks for the signals. Ronald's success was complete; nevertheless, the government of the day refused to avail itself of his telegraph.

In 1819, Professor Oersted, of Copenhagan, who had for some years asserted the identity of chemical and electrical forces, announced his great discovery of the intimate relation existing between magnetism and electricity, in consequence of his having, while lecturing to his class, observed that a magnet, when placed near a wire conducting a voltaic current, was strangely deflected. And upon the Copley Medal being adjudicated to Oersted for his discovery, he demonstrated that " there is always a magnetic circulation round the electric conductor; and that the electric current, in accordance with a certain law, always exercises determined and similar impressions on the direction of the magnetic needle, even when it does not pass through the needle, but near it." Thus Oersted laid the foundations of the science of electro-magnetism, and led the way to its practical application to the Electric Telegraph, although, in the popular accounts of the invention, we hear much more of the adapters of his researches than of Oersted himself, to whom the main merit is due. " Nothing," says Professor Owen, " might seem less promising of profit than Oersted's painfully-pursued experiments with his little magnets, voltaic pile, and bits of copper wire, yet out of these has sprung the Electric Telegraph."

Dr. Hamel, of St. Petersburg, states that Baron Schilling was the first to apply Oersted's discovery to telegraphy by actually producing an electro-magnetic telegraph simpler in construction than that which Ampère had *imagined*.

Sturgeon next conceived the idea of involving soft iron with copper wire, and, by circulating voltaic electricity through these convolutions, of rendering it powerfully magnetic. The experiment proved the correctness of the thought, and electro-magnets of enormous power have been the result. These have enabled Faraday to discover and enunciate the laws of voltaic and magneto-electric induction. Light and magnetism are proved to be mysteriously related, and all bodies in nature have been shown to exist in one of two conditions—they are either *magnetic*, as iron is, or they are *dia-magnetic*, like bismuth and glass.

In 1835, Gauss and Weber established electro-telegraphic communication between the Observatory at Gottingen and the University. In Professor Airy's experiments with the Electric Telegraph, several years after, to determine the difference of longitude between Greenwich and Brussels, the time spent by the electric current in passing from one observatory to the other (270 miles) was found to be rather more than the ninth part of a second, this determination resting on 2616 observations. Such a speed would "girdle the globe" in ten seconds. During all this time the Voltaic Battery was gradually improved, and its powers vastly augmented, by Daniell and Grove.

In 1836, Professor Muncke, of Heidelberg, who had inspected Schilling's telegraphic apparatus, explained the same to William Fothergill Cooke, who in the following year returned to England, and subsequently, with Professor Wheatstone, labored simultaneously for the introduction of the Electro-magnetic Telegraph upon the English railways, the first patent for which was taken out in the joint names of these two gentlemen.

In 1844, Professor Wheatstone, with one of his telegraphs, formed a communication between King's College and the lofty shot-tower on the opposite bank of the Thames: the wire was laid along the parapets of the terrace of Somerset House and Waterloo Bridge, and thence to the top of the tower, about 150 feet high, where a telegraph was placed; the wire then descended, and a plate of zinc attached to its extremity was plunged into the mud of the river, while a similar plate attached to the extremity at the north side was immersed in the water. The circuit was thus completed by the entire breadth of the Thames, and the telegraph acted as well as if the the circuit were entirely metallic. Shortly after this experiment, Professor Wheatstone and Mr. Cooke laid down the first working Electric Telegraph on the Great Western Railway, from Paddington to Slough.

In 1845, by the Electric Telegraph, then laid from Paddington to the Slough Station, on the Great Western Railway, John Tawell was captured on suspicion of having murdered Sarah Hart at Salt Hill on Jan. 1. Tawell left Slough by the railway on that evening; and at the same instant, by Telegraph, his person was described, with instructions to the police to watch him on his arrival at Paddington. Thus, while the suspected man was on his way to London at a fast rate, the Telegraph, with still greater rapidity, sent along the wire which skirts the road the startling instructions for his capture; and in the metropolis he was followed, apprehended, and identified. This early employment of the Telegraph produced in the public mind an intense conviction of the vast utility of this novel application of man's philosophy to the protection of his race.

The first newspaper report by Electric Telegraph appeared in the *Morning Chronicle*, May 8, 1845, detailing a railway meeting held at Portsmouth on the preceding evening. On April 10, in the same year, a game of chess was played by Electric Telegraph between Captain Kennedy, at the Southwestern Railway terminus, and Mr. Staunton, at Gosport: the mode of playing was by numbering the squares of the chess-board and the men; and in conveying the moves, the electricity traveled backward and forward during the game upward of 10,000 miles.

On Nov. 13, 1851, the Submarine Electric Telegraph between Dover and Calais was first worked for the public; and the opening and closing prices of the Paris Bourse were transmitted to the Stock Exchange, London, during business hours.

In America, the Submarine Electric Telegraph was invented by Professor Morse, who, in 1822, while on his passage from Liverpool to New York, maintained the passage of electricity through wire to be instantaneous to any distance, and that it might be made the means of conveying and recording intelligence. For thirteen years he pursued his experiments, and in 1835 patented his "Recording Electric Telegraph," in the same year that Wheatstone in England, and Steinheil in Bavaria, invented a Magnetic Telegraph of entirely different construction. Morse uses the steel point for indenting the paper, and renders the instrument more powerful and certain by substituting electro-magnets for needles. Morse next attempted Submarine Telegraphing between Governor's Island and Castle Garden, New York; and in October, 1842, interchanged messages, and laid the first cable of copper wire, one twelfth of an inch in diameter, insulated by hemp coated with tar, pitch, and India-rubber. From this success Morse inferred that a telegraphic communication upon his plan might be established across the Atlantic. In 1844 he completed the first Electric Telegraph in the United States, and in 1856 his claim to the invention of the writing apparatus was accorded.

Before the Atlantic Telegraph was finally decided on here, 2000 miles of subterranean and submarine telegraph wires, ramifying through England and Ireland, under the Irish Sea, were connected, and through this distance of 2000 miles 250 distinct signals were recorded and printed in one minute. In 1857 the Atlantic Cable was completed, the length of iron and copper wire spun into it being 332,500 miles, or sufficient to engirdle the earth thirteen times: the cable weighed about a ton per mile, and was incased in Gutta Percha. A submarine cable, when in the water, is virtually a lengthened-out Leyden jar; it transmits signals while being charged and discharged, instead of merely allowing a single stream to flow evenly along it. The electro-magnetic current possesses treble the velocity of simple voltaic electricity; and with a single pair of zinc and silver plates (1-20th of a square inch large), charged by a single drop of liquid, distinct signals have been effected through 1000 miles of the cable, and each signal was registered in less than three seconds of time. The Perpetual Maintenance Battery, for working the cable at the bottom of the sea, consisted of large plates of platinated silver and amalgamated zinc, mounted in ten cells of Gutta Percha, each cell containing 2000 square inches of acting surface, worked at the cost of one shilling per hour. This voltaic current was the primary power used to call up a more speedy apparatus of " Double Induction Coils," while a fresh battery did the printing labor.* The attempts to lay this cable in August, 1857, failed through stretching it so tightly that it snapped and went to the bottom, at a depth of 12,000 feet, forty times the height of St. Paul's. The cause of this failure was frankly confessed. " The best workmen," said the engineers, " were worn out with fatigue; the second-best took their places, and put on the brakes unskillfully; the cable snapped; and that is the long and short of the matter."

This great work was resumed in August, 1858, and on the 5th the first signals were received through *two thou-*

* By Mr. Wildman Whitehouse, the eminent electrician.

sand and fifty miles of the Atlantic Cable, when the engineer-in-chief, Mr. Charles Bright, was knighted. And it is worthy of remark, that just 111 years previously, on the 5th of August, 1747, Dr. Watson astonished the scientific world by practically proving that the electric current could be transmitted through a *wire hardly two miles and a half long.*

The success, however, lasted but for a few days, for on September 1st the Cable ceased to work, and it has continued useless up to the present time.

A little north of the 50th parallel of latitude, at the bottom of the Atlantic Ocean, where the plateau is unbroken by any great depression, and on a soft bed of mud, constantly thickening, and composed almost entirely of carbonate of lime, there lies now some 1500 miles of disabled telegraphic cable, deposited in the summer of 1858, at a depth varying from 10,000 to 15,000 feet. The wire was sufficiently thick to resist any strain it was thought likely to have to bear. Whether, however, it may not, where partially injured, have become melted by the intense heat evolved during the passage of magnetic storms through the earth, and even of the strong magnetic currents employed in communicating the early messages, is a question that has not yet been answered; but, at any rate, it is in the highest degree probable that in the course of time the copper would have become reduced to the crystalline state, and the cohesion of the metal reduced so as to render it incapable of resisting even a very small strain. These and other difficulties may arise, and will have to be overcome. Meanwhile the great problem of telegraphy is solved.* The power that attracts the needle to the pole, and has for centuries guided the navigator across the surface of the water, is now rendered available in providing means of communication through its hitherto unfathomed depths, and the girdle is being put round the world which will, at no distant time, unite all civilized nations into one great brotherhood. —*Westminster Review*, October, 1859.

In soundings taken along the telegraph plateau, specimens of the animals and vegetables found at the bottom of the Atlantic have been brought up, of which the

---

* The following statement of the actual number of messages that passed across the Atlantic during the time when the condition of the wire was still doubtful, will show clearly how complete was the success, and how great the certainty that submarine lines will ultimately be laid. Exclusive of conversations among the clerks, 97 messages, consisting of 1002 words and 6476 letters, were sent from Valentia to Newfoundland, and duly comprehended; while 269 messages, of 2840 words and 13,743 letters, were received from Newfoundland in Ireland. This gives a total of 366 messages, consisting of 3942 words, made up of 20,219 letters, actually transmitted.— *Westminster Review*, No. 32, N. S.

accompanying engraving represents a highly-magnified group. It includes *Foraminifera*, beings which secrete many-chambered calcareous shells, each the habitation of a group of individuals so minute as to require the highest powers of the best microscope to perceive them. With these are intermixed *Diatomaceæ*, the simplest tribes of the simplest plants, whose remains form a sensible proportion of the silicious part of the ooze on which the telegraph cable rests.

Highly-magnified animals and plants brought up from the Atlantic Telegraph plateau.

The *Applications of Electricity to the Arts* are too numerous to be specified here; but a few of the more prominent instances must be noticed. The new arrangement of Franklin's discovery by Sir Snow Harris, in lightning conductors, has already been mentioned. The firing of gunpowder by electricity beneath the water, as an agent in blasting and exploding, has led to the safer and more economical recovery of sunken property, and the execution of vast engineering works. Hopes have been strongly excited that the electro-magnetic current may be so modified as to act as a moving-power for machinery, and in lieu of steam, wind, water, and animal power; locomotive carriages by land, and small vessels on rivers, have been impelled by electro-magnetism, but at too great a cost for adoption. As the moving power of clocks, electricity is employed with great success for indicating exactly the same time in any number of places distant from each other. Electro-metallurgy, or the working in metals by electrical agency, was first illustrated by Professor Jacobi, of St. Petersburg, and Mr. Spencer, of Liverpool, and the precipitation of the precious metals from the solution led to electro gilding and plating, in place of the usual process of gilding and plating. By this process watch-springs are electro-gilded, to prevent oxydation; and the great metal dome of St. Isaac's Cathedral at St. Petersburg, which weighs nearly 2000 tons, has been electro-gilded with 274 lbs. of ducat gold. To Mr. Spencer we also owe the application of electricity to the multiplying copies of works of art—in the electrotype, a valuable improvement also upon stereotype for printing surfaces. Plates are etched and multiplied by electricity. The uses of electricity as a curative agent, or as a means of physiological investigation, are very striking. The electro-luminous experiments have led to the introduction of the Electric Light for public purposes; but its costliness greatly restricts its popular service. This wonderful illuminating power has been adapted to light-houses; and in 1859 the upper South Foreland lighthouse, near Dover, was lighted by the electro-magnetic light, by Professor Holmes. The electricity is not evoked by a voltaic battery, but is the result of magneto-electric induction, the current being obtained by about 85 revolutions per minute. The light is visible for 27 miles, and can be seen from the tops of the light-houses on the coast of France.

GENERAL INDEX.

THE END.